Lecture Notes in Physics

383

J.D. Barrow A.B. Henriques
M.T.V.T. Lago M.S. Longair (Eds.)

The Physical Universe: The Interface Between Cosmology, Astrophysics and Particle Physics

Proceedings of the XII Autumn School of Physics
Held at Lisbon, Portugal, 1–5 October 1990

Springer-Verlag
Berlin Heidelberg GmbH

Editors

John D. Barrow
Astronomy Centre, University of Sussex
Falmer, Brighton BN1 9QH, United Kingdom

Alfredo B. Henriques
IST, Universidade Técnica de Lisboa, Dep. de Física
Av. Rovisco Pais, P-1096 Lisboa Codex, Portugal

Maria Teresa V. T. Lago
Centro de Astrofísica, Universidade do Porto
Rua do Campo Alegre 823, P-4100 Porto, Portugal

Malcom S. Longair
Cavendish Laboratory, Madingley Road
Cambridge CB3 OHE, United Kingdom

ISBN 978-3-662-13815-1 ISBN 978-3-540-47544-6 (eBook)
DOI 10.1007/978-3-540-47544-6

Originally published by Springer-Verlag Berlin Heidelberg New York in 1991
Softcover reprint of the hardcover 1st edition 1991

2153/3140-543210 – Printed on acid-free paper

PREFACE

The XII Autumn School of Physics "The Physical Universe: The Interface between Cosmology, Astrophysics and Particle Physics" was held in Lisbon, 1-5 October 1990.

As the title suggests, the school was dedicated to the interface between cosmology, astrophysics and particle physics, an exciting area of research at present developing rapidly, and one of its main purposes was to put young physicists and postgraduate students in contact with these new subjects. To achieve this, the school was organized around introductory courses, complemented by more specialized seminars, and was fortunate to count among its lecturers some of the most active scientists in the field, whom we would like to thank for their invaluable contribution.

We acknowledge and thank the financial support given to the organization by INIC, JNICT, GTAE and the British Council, as well as the help of the Banco Português do Atlântico, Nestlé Portugal, Região de Turismo do Oeste and Óbidos Town Hall. We also thank the Reitoria da Universidade de Lisboa, where the School took place, for permission to use their premises.

We would like to express our gratitude to Mrs. Maria de Fátima Maia de Loureiro for her dedication through all stages of the organization; special thanks are due to our colleagues Orfeu Bertolami, José Mourão, Paulo Moniz and Paulo Sá for their constant help.

Finally, we are very grateful to Springer-Verlag for publishing this volume.

Lisbon, April 1991

The Editors
J. D. Barrow
A. B. Henriques
M. T. Lago
M. S. Longair

CONTENTS

Gravitation and Hot Big-Bang Cosmology 1
J. D. Barrow

Euclideanized Einstein-Yang-Mills Equations, Wormholes and the Ground-State Wave Function of a Radiation Dominated Universe 21
O. Bertolami and J. M. Mourão

Experiments with Neutrinos 39
P. Bordalo

Topological Defects in the Early Universe 47
E. Copeland

The Space Telescope and the Problems of Cosmology 102
P. Crane

Self-Gravitating Magnetic Monopoles, Global Monopoles and Black Holes 110
G. W. Gibbons

Understanding Large-Scale Cosmic Structure 134
B. J. T. Jones and R. van de Weygaert

Metric Space as a Model of Spacetime: Classical Theory and Quantization 159
C. J. Isham, Yu.A. Kubyshin and P. Renteln

Galaxy Formation - An Update 174
M. S. Longair

The Hypothesis of the Expansion of the Universe and the Global Tests 197
M. Moles

The Early Universe Behaviour with Non-minimal Coupling 227
P. Moniz, P. Crawford and A. Barroso

Stability of Compactification in Einstein-Yang-Mills Theories 237
O. Bertolami, Yu.A. Kubyshin and J. M. Mourão

Multivariate Analysis and Pattern Recognition Methods: A Short Review and Some Current Directions 253
F. Murtagh

Cooling of Neutron Stars 265
C. J. Pethick

Experiments in High-Energy Physics: A Brief Introduction 275
M. Pimenta

Nucleosynthesis in Big-Bang Models 281
F. D. Santos

The Cosmological Constant, Third Quantization and All That 292
G. Lavrelashvili

Structure of the Inflationary Universe 305
M. I. Zelnikov

LIST OF PARTICIPANTS

Regina C. Arcuri -
Augusto A. Barroso - Universidade de Lisboa
John D. Barrow - University of Sussex
Luís Bento - Universidade de Lisboa
Francis Bernardeau - CEN, Saclay
Orfeu Bertolami - GTAE, Lisboa
Luís Miguel A. Bettencourt - IST, Lisboa
Duarte Nuno Borba - IST, Lisboa
Paula Bordalo - LIP and IST, Lisboa
José M. Braga - Universidade da Beira Interior, Covilhã
José M. Chamiço - IST, Lisboa
Pedro F. Conceição - IST, Lisboa
Edmund J. Copeland - University of Sussex
Philippe Crane - ESO, Garching
Paulo Crawford - Universidade de Lisboa
Jorge Dias de Deus - IST, Lisboa
João Alberto M. Dias - IST, Lisboa
João M. Fernandes - Universidade do Porto
Filipe C. Freire - IST, Lisboa and Imperial College, London
Clemente M. Garrido - Instituto de Astrofísica de Canarias
Gary W. Gibbons - University of Cambridge
Yuri A. Kubyshin - Moscow State University
Bernard J. T. Jones - University of Sussex
Jaime Julve - Instituto de Estrutura de la Materia, Madrid
Alfredo B. Henriques - IST and CFMC, Lisboa
Maria Teresa V. T. Lago - Universidade do Porto
Luís Miguel V. Lapão - IST, Lisboa
Fernando P. Lau - IST, Lisbon
George Lavrelashvili - Tbilisi Mathematical Institute
Malcon S. Longair - University of Cambridge
Paulo G. Macedo - Universidade do Porto
Joaquim S. Marcos - CFMC, Lisboa
Luís E. Mendes - IST, Lisboa
Rui Vilela Mendes - CFMC, Lisboa
José Pedro Mimoso - Universidade de Lisboa and University of Sussex
Mariano Moles - Instituto de Astrofísica de Andalucia, Granada

Paulo V. Moniz - Universidade de Lisboa
José C. Mourão - CFN and IST, Lisboa
Fionn Murtagh - ST - ECF, ESO, Garching
Filomena Nunes - IST, Lisboa
Margarida S. Nunes - Universidade do Porto
Nino Panagia - STSI, Baltimore
Christopher Pethick - NORDITA, Copenhagen
Mário Pimenta - LIP and IST, Lisboa
J. Pinto da Costa - Universidade do Porto
João M. Pulido - CFMC and IST, Lisboa
Paula S. Sá - GTAE, Lisboa
Carlos A. Santos - Universidade da Beira Interior, Covilhã
Filipe Duarte Santos - Universidade de Lisboa
Luís Miguel O. Silva - IST, Lisboa
Menahem Simhoby - Hebrew University, Jerusalem
Luís F. Teodoro - IST, Lisboa
Roland Triay - CNRS - Luminy, Marseille
Maxim I. Zelnikov - P. N. Lebedev Physical Institute, Moscov.

GRAVITATION AND HOT BIG-BANG COSMOLOGY

John D Barrow
Astronomy Centre
University of Sussex
Brighton BN1 9QH
UK

1. Introduction

This lecture deals with mathematical and formal aspects of Newtonian and general relativistic gravitation as well as some higher-order lagrangian theories of gravity. It introduces the Friedman metrics and places them in the context of all solutions to Einstein's equations. The emphasis is upon basic principles and ideas that do not lean too heavily upon specific models.

2. Newtonian Gravitation

2.1 *Newtonian versus General Relativistic Gravity*

Most of intuition regarding the behaviour of self-gravitating systems, like universes, derives from our experience with Newtonian gravitation and so it is important to appreciate the differences between Newtonian and general relativistic theories of gravity. These are summarized in the table below. Most of the contrasts are self-explanatory but some, like the ill-posedness of Newtonian cosmology, require a little more explanation and this will be given below. We stress that if a cosmology were based upon special relativity then it would be defined by a metric (Minkowski space-time, η) and a 4-manifold (R^4) whereas in general relativity the pair (η, R^4) is generalized to (g, M) where g is any Riemannian metric and M any 4-manifold (although additional properties might be required - eg orientability, differentiability etc).

General relativity	Newtonian Gravity
* 10 field equations	1 field equation
* 10 potentials	1 potential
* field equations non-linear (sum of 2 solutions not necessarily a solution)	Field equations linear (sum of solutions is a solution)
* intrinsically geometrical	absolute space (R^3) and time (R)
* permits infinite space	requires finite space
* all energy sources gravitate	only density gravitates
* cosmological problem well-posed	cosmological problem ill-posed
* consistent with special relativity	instantaneous signal propagation
* singularities of space-time	singularities in space-time
* horizons and black holes	no horizons or black holes
* gravitational waves	no gravitational waves

We remark that general relativity is non-linear in an additional sense to that listed above. It is self-interacting in that the carrier of the gravitational force (the graviton) also carries the gravitational "charge" and feels the gravitational force. Gauge theories of this sort are non-Abelian. For example, QCD has this property because gluons carry the colour charge, whereas U(1) electromagnetism does not because the photon does not carry electric charge. As a consequence there could not be a gauge theory of photons alone but there could be a gauge theory of gluons only, or a vacuum theory of general relativity. Another subtlety in the tabulated properties is the veto on black holes in Newtonian theory. The Newtonian "black hole" of Mitchell and Laplace (see Barrow & Tipler 1986a for original references and discussion) is often employed for explanatory purposes but it should be recognised, aside from the illegal extrapolation of the theory to deal with potentials of order c^2, that if a Newtonian object has a radius $R = 2GM/c^2$ then this merely prevents light reaching infinity. There is no veto upon photons crossing the radius $2GM/c^2$, whereas in the general relativistic Schwarzschild solution the barrier created by the event horizon at $2GM/c^2$ is absolute. Finally, the intrinsic coupling of the geometry of space-time to its matter content ensures that infinities in physical quantities create singularities in the fabric of space-time in general relativity and not merely infinities on some indestructible absolute space as is the case in Newtonian theory. Space and time can come to an end in general relativity; they cannot in Newtonian theory.

2.2 The Newtonian Potential and the Cosmological Constant

Most textbooks regard the cosmological constant as an embarrassing creation of general relativity that should be excluded by a demand that the Newtonian limit of general relativity reduce to Poisson's equation. However, it is instructive to see that the cosmological constant also arises inevitably in the right approach to Newtonian theory. Suppose, like Newton, we were concerned to be able to treat spherical masses as though they were point masses located at their centres; then we could ask, what is the most general form of the gravitational potential $\Phi(r)$ such that the external gravitational field of a spherical shell is the same as that of an equal point mass located at its centre? If we consider a shell of radius a, mass M, and surface density σ, and equate the potential due to the shell at some point $r > a$ from its centre to that of a point mass $M = 4\pi a^2\sigma$ at $r = 0$ then we have an integral equation for the potential $\Phi(r)$.

$$M\Phi(r) + 2\pi\sigma a\lambda(a) = \frac{2\pi\sigma a}{r}\int_{r-a}^{r+a} x\Phi(x)dx \qquad (2.1)$$

where λ is a constant which can depend upon the (fixed) value of a. Differentiating (2.1), the solution is given by $M = 4\pi\sigma a^2$, $\lambda = 2Ba^2$, B constant, and

$$\Phi = \frac{-GM}{r} + \frac{1}{6}\Lambda r^2 \qquad (2.2)$$

where Λ is an arbitrary constant that we have labelled and prefixed by a numerical factor in order to make it exactly correspond to the weak-field limit of general relativity with a cosmological constant Λ. The potential (2.2) derives from the generalization of Poisson's equation to

$$\nabla^2\Phi + \Lambda = 4\pi G\rho \qquad (2.3)$$

On some occasions (*eg* when interpreting the Newtonian limit of static, straight cosmic strings, (Vilenkin 1985)) it is useful to have a generalization of (2.3) that includes the contributions of the three principal pressures, p_i. In this case (2.3) becomes

$$\nabla^2\Phi + \Lambda = 4\pi G\left[\rho + \sum_{i=1}^{3}\frac{p_i}{c^2}\right] \qquad (2.4)$$

In the standard derivations of general relativity (Weinberg 1972), a linear combination of second-rank tensors involving g_{ab}, Rg_{ab} and R_{ab} is set proportional to the energy momentum tensor T_{ab} and the constant of proportionality ($8\pi G/c^4$) is derived from the Newtonian limit after associating g_{00} with Φ/c^2. The 'miracle' of general relativity is that a purely mathematical assembly of second-rank tensors should have anything to do with

Newtonian gravity in any limit. In fact, this limit could have failed to exist. The 'Newtonian limit' of Einstein's equations in 2 + 1 dimensional space-time does not lead to the analogue of (2.3) in two space-dimensions (Barrow, Burd & Lancaster 1985), but to $\nabla^2\Phi + \Lambda = 0$. However, in N + 1 dimensional space-time with N > 2 the correct Newtonian limit always arises.

2.3 Newtonian Cosmology - fluid formulation

Newtonian cosmology is often employed for ease of explanation in cosmological expositions and calculations. (Detailed treatments can be found in Heckmann & Schücking 1955, 1959, Raychaudhuri 1955, Zeldovich 1965, Ellis 1971, Shikin 1971, 1972 Narlikar & Kembhavi 1980, Barrow & Götz 1989.) For this reason it is important to appreciate when and why it can be believed.

First, let us formulate the cosmological problem as a problem in non-relativistic fluid mechanics in which there exists an external gravitational field $\Phi(\underset{\sim}{x})$. We consider a pressureless fluid with density ρ, velocity vector $\underset{\sim}{v} = \{v^\alpha\}$ and so the three equations defining the theory are:

$$\dot{\rho} + \underset{\sim}{\nabla}.(\rho\underset{\sim}{v}) = 0 \tag{2.4}$$

$$\dot{\underset{\sim}{v}} + (\underset{\sim}{v}.\underset{\sim}{\nabla})\underset{\sim}{v} = -\underset{\sim}{\nabla}\Phi \tag{2.5}$$

$$\nabla^2\Phi + \Lambda = 4\pi G\rho \tag{2.6}$$

The Poisson equation (2.6) is elliptic and Newtonian gravity acts instantaneously. As a result, boundary conditions must be prescribed for Φ at infinity in order to determine the solutions uniquely for given values of $\rho(\underset{\sim}{x})$ and $\underset{\sim}{v}(\underset{\sim}{x})$ at $t = 0$. Unfortunately, there is no mathematical or physical reason *a priori* to dictate which boundary conditions should be chosen. This is in essence the so-called 'gravitational paradox'. We can understand it more fully if we analyse (2.4)-(2.6) further in a homogeneous Euclidean space with Cartesian coordinates $\underset{\sim}{x} = \{x^\alpha\}$ and absolute time, t. This ensures that there exists a simply transitive group of motions for the Euclidean metric space $\delta_{\alpha\beta}dx^\alpha dx^\beta$, and hence that the velocity field obeys a generalized 'Hubble law'.

$$v^\alpha = h^{\alpha\beta}\, x^\beta \tag{2.7}$$

The matrix $h^\alpha{}_\beta$ can be decomposed into symmetric ($\theta_{\alpha\beta} = \theta_{\beta\alpha}$) and antisymmetric parts ($\omega_{\alpha\beta} = -\omega_{\beta\alpha}$):

$$h^\alpha{}_\beta = \theta^\alpha{}_\beta + \omega^\alpha{}_\beta \tag{2.8}$$

and then $\theta^\alpha{}_\beta$ can be split into its trace free (shear) and trace (volume dilation) parts

$$\theta^{\alpha}{}_{\beta} = \sigma^{\alpha}{}_{\beta} + \delta^{\alpha}{}_{\beta}\frac{\dot{a}}{a} \quad : \quad \sigma^{\alpha}{}_{\alpha} = 0 \tag{2.9}$$

where the volume is taken as a^3 for some characteristic length-scale factor $a(x^{\alpha},t)$. If $\omega_{\alpha\beta} \neq 0$ then the flow has non-zero vorticity; if $\sigma_{\alpha\beta} \neq 0$ then it has non-zero shear (ie distortion at constant volume). If $\sigma_{\alpha\beta} = \omega_{\alpha\beta} = 0$ then it will either expand or contract isotropically if $\dot{a} \neq 0$. Using these kinematic quantities we can rewrite the continuity, momentum equations (2.4)–(2.5), as

$$M \equiv \frac{4\pi}{3}\rho a^3 = \text{constant} \tag{2.10}$$

$$3\ddot{a} = -4\pi Ga(\rho + \Lambda - \sigma_{\alpha\beta}\sigma^{\alpha\beta} + \omega_{\alpha\beta}\omega^{\alpha\beta}) \tag{2.11}$$

In addition, the integrability condition for Φ,

$$\frac{\partial^2\Phi}{\partial x^{\alpha}\partial x^{\beta}} = \frac{\partial^2\Phi}{\partial x^{\beta}\partial x^{\alpha}}\,, \tag{2.12}$$

leads to a vorticity equation whose first integral prescribes the conservation of angular momentum

$$\dot{\omega}_{\alpha\beta} + \frac{2\dot{a}}{a}\omega_{\alpha\beta} + \omega_{\alpha\beta}\sigma^{\gamma}{}_{\beta} + \omega_{\gamma\beta}\sigma_{\alpha}{}^{\gamma} = 0 \tag{2.13}$$

from which we see that if our fluid is shear-free ($\sigma_{\alpha\beta} = 0$) then $\omega \propto a^{-2}$ and the angular momentum $I\omega \propto Ma^2\omega$ is constant (note that such a state is impossible in general relativity, Ellis 1971).

The gravitational potential is a sum of four pieces

$$\Phi = \frac{2\pi G}{3}\rho r^2 + \frac{\Lambda r^2}{6} + \Phi_1 + \Phi_2 \tag{2.14}$$

where $r^2 = \delta_{\alpha\beta}\, x^{\alpha}\, x^{\beta}$. The first term on the right of (2.14) is just the potential for a finite homogeneous sphere of constant density ρ and radius r. The second term, also isotropic, is the contribution of the cosmological constant (see (2.2)). The third term, Φ_1, satisfies Laplace's equation, $\nabla^2\Phi_1 = 0$, and is determined by $\sigma_{\alpha\beta}$ and $\omega_{\alpha\beta}$ as

$$\Phi_1 = -\tfrac{1}{2}\Big(\dot{\sigma}_{\alpha\beta} + \frac{2\dot{a}}{a}\sigma_{\alpha\beta} + \sigma_{\alpha\gamma}\sigma_{\beta}{}^{\gamma} - \frac{1}{3}\delta_{\alpha\beta}\sigma_{\gamma\epsilon}\sigma^{\gamma\epsilon} + \delta^{\kappa\lambda}\omega_{\alpha\kappa}\omega_{\lambda\beta} - \frac{1}{3}\,\delta_{\alpha\beta}\omega_{\kappa\lambda}\omega^{\kappa\lambda}\Big)x^{\alpha}x^{\beta} \tag{2.15}$$

It is not associated with presence of matter and resembles the (quadratic) potential within a homogeneous ellipsoid. The function Φ_2 is an arbitrary function of time and is a pure integration constant of the Poisson equation.

The equations (2.10)–(2.13) display the 'Newtonian paradox' explicitly. There are only five equations for ten unknowns and the five components of $\sigma_{\alpha\beta}$ remain completely undetermined by the differential equations. We see from (2.15) that this indeterminancy is associated with Φ_1, which accordingly contains five undetermined functions. If homogeneous and isotropic solutions ($\sigma_{\alpha\beta} = \omega_{\alpha\beta} = 0$) solutions are sought (see Milne & McCrea (1934) or Bondi (1960)) then the indeterminancy is removed and Newtonian cosmology is well-posed. However, when the universe is anisotropic ($\sigma_{\alpha\beta} \neq 0$ and, or $\omega_{\alpha\beta} \neq 0$) the cosmological problem is ill-posed. A good deal of the effort expended in studying Newtonian cosmology has focussed upon ways of closing this general ill-posedness; for example, by appending to the theory the shear propagation equations which arise in general relativity, or by adding some other condition which ensures that solutions are the weak-field limit of some general relativistic solution.

Because of the 'gravitational paradox' we have just described there exist Newtonian solutions with no general relativistic analogue (for example there are solutions of (2.10)–(2.13) with $\sigma_{\alpha\beta} = 0$ and $\omega_{\alpha\beta} \neq 0$). Of course, there are also many general relativistic solutions which have no Newtonian analogue. Thus we might schematize the solution spaces of general relativity and Newtonian gravity as sets of which have an intersection but there are Newtonian solutions with no relativistic analogue (which we would regard as unphysical) as well as relativistic solutions with no Newtonian analogue (for example those which possess a non-zero "electric' part of the Weyl curvature tensor associated with anisotropic spatial curvature).

2.4 The Newtonian N-body Problem

The indeterminacy of Newtonian cosmology is not a problem peculiar to the fluid description. It is possible to draw some interesting conclusions about it by studying the Newtonian N-body problem in general. No symmetry assmptions are made here. We assume simply that N is finite. Let m_k, $\underline{r}_k$ and $\underline{v}_k$ denote the mass, position vector and velocity vector of the k + h particle and let $r_{ij} = |\underline{r}_i - \underline{r}_j|$. The total kinetic energy T is defined by

$$T = \tfrac{1}{2}\Sigma m_k v_k{}^2 \tag{2.16}$$

where Σ denotes a sum over k from 1 to N. The total energy, h, is the sum of T and the potential energy, $-U$, so

$$T = U + h \tag{2.17}$$

We define the moment of inertia of the system by

$$I = \Sigma m_k r_k^2 \qquad (2.18)$$

and the angular momentum by

$$\underline{J} = \Sigma m_k(\underline{r}_k \times \underline{v}_k) \qquad (2.19)$$

The virial theorem of Lagrange and Jacobi is

$$\ddot{I} = 4T + 2\Sigma \underline{r}_k . \frac{\partial U}{\partial \underline{r}_k} \qquad (2.20)$$

If we define $M = \Sigma m_k$ to be the total mass then,from (2.18), we see that

$$IM = \Sigma^* m_j m_k r_{jk}^2 \qquad (2.21)$$

where Σ^* denotes the restricted double sum of j and k over $1 \leq j < k \leq N$

Using these definitions we can prove an interesting general singularity theorem. First, take the modulus of equation (2.19), so

$$J \leq \Sigma m_k |\underline{r} \times \underline{v}| \leq \Sigma m_k r_k v_k = (\Sigma m_k^{\frac{1}{2}} r_k)(\Sigma m_k^{\frac{1}{2}} v_k) \qquad (2.22)$$

now use the Cauchy–Schwarz inequality so

$$J \leq (\Sigma m_k r_k^2)(\Sigma m_k v_k^2) = 4IT \qquad (2.23)$$

This is Sundman's inequality (Pollard 1966).

We will say a <u>strong singularity</u> occurs if all $r_{jk} \to 0$. (This is the analogue of a strong curvature or crushing singularity in general relativity, see Tipler, Ellis & Clarke 1980). From (2.21) we see that it requires $I \to 0$. This cannot occur at $t = \infty$ because if so then since for Newtonian gravity (with $\Lambda = 0$)

$$U = -\Sigma^* \frac{Gm_j m_k}{r_{jk}} \qquad (2.24)$$

$r_{jk} \to 0$ implies $U \to \infty$ hence, by (2.20), that $\ddot{I} \to \infty$ and so $I \to \infty$ as $t \to \infty$ in contradiction to $I \to 0$. So the strong singularity must occur at some $t = t_* < \infty$. Now, for some interval $t_1 < t < t_*$ we must have $\dot{I} \leq 0$ so we can multiply (2.23) by $-\dot{I}/I \geq 0$ and integrate to obtain

$$\tfrac{1}{4}J^2\log(I^{-1}) \leq EI + k \tag{2.25}$$

where E and k are constants. Now, since $I \to 0$ as $t \to t_*$, the inequality (2.25) ensures that we must have $J^2 \to 0$, but since J is a constant of the motion we must have $J \equiv 0$. Thus an all-encompassing singularity requires zero angular momentum in Newtonian gravity. No such restriction on the rotation is required in general relativity (see Tipler *et al* 1980).

It is possible to ascertain various general results about the asymptotic behaviour of the N-body problem in Newtonian gravity, both with $\Lambda = 0$ and $\Lambda \neq 0$, (see Barrow & Götz 1989). These results all amount to a determination of the asymptotic behaviour of the moment of inertia I, defined by (2.18), as a function of time. The fact that no information is available about the detailed shape of the system, only its coarse-grained volume evolution, is the discrete manifestation of the 'gravitational paradox' and incompleteness of the problem we found in the fluid picture.

3. General Relativity

In this section we shall introduce natural units (Barrow 1983) such that the fundamental constants G, c, $\hbar$ and k_B have values

$$8\pi G = c = \hbar = k_B = 1 \tag{3.1}$$

There are a variety of ways to derive Einstein's equations. The most familiar textbook derivation (*eg* Weinberg 1972), following Einstein, is to equate the most general linear combination of second-rank tensors involving the metric g_{ab} and Ricci curvature tensor R_{ab}, to the energy momentum tensor T_{ab}. Use of the contracted Bianchi identities and the fact that T_{ab} is covariantly constant then gives (in N-dimensional space-time)

$$R_{ab} - \tfrac{1}{2}Rg_{ab} = T_{ab} \tag{3.2}$$

where we have suppressed the cosmological constant because it can be reinterpreted as an effective part of T_{ab} and we have used the units (3.1) after obtaining correspondence with the Newtonian limit to set the Einstein constant $8\pi G/c^4 = 1$.

It is important to note that (3.2) gives a <u>linear</u> relation between the curvature and T_{ab}. Because of this it is possible to contract (3.2) and obtain an equivalent form of the Einstein equations.

$$R_{ab} = T_{ab} - \frac{1}{(N-2)} g_{ab}T \tag{3.3}$$

This elementary step has profound consequences. All the singularity theorems (see

Hawking & Ellis 1973) make no use of the Einstein equations. They use the geodesic equations to establish that if $R_{ab}V^aV^b \geq 0$ for a timelike vector V^a then a singularity is inevitable (given some other technical conditions). In general relativity – and only general relativity – the linearity of (3.2) permits the re-expression (3.3) and hence the Einstein equations can be used to translate the sufficient condition geometrical $R_{ab}V^aV^b \geq 0$ into a statement of physics:

$$\left[T_{ab} - \frac{g_{ab}T}{N-2}\right] V^aV^b \geq 0 \qquad (3.4)$$

which is known as the *strong energy condition* in the N = 4 case. In more complicated theories of gravity, for example those derived from quadratic lagrangians, this nice retranslation of the timelike convergence condition $R_{ab}V^aV^b \geq 0$ into a condition on T_{ab} is not possible.

A second point to make about the equations (3.2), or equivalently (3.3), is that they are quasi-linear. All the familiar equations of physics have this property. It means that they have a generic form.

$$f(g_{ab}, \partial g_{ab})\partial^2 g_{ab} + F(g_{ab}, \partial g_{ab}) = 0 \qquad (3.5)$$

Where ∂ denotes derivatives with respect to any of the space-time coordinates and f and F are some continuous functions.

The essential feature of (3.5) is that the highest (*ie* second) derivatives of g_{ab} occur linearly. Quasi-linear equations have a variety of 'nice' mathematical properties and well-behaved initial value problems.

The second, and perhaps most fundamental, route to Einstein's equations is via an action principle. In the presence of a cosmological constant the gravitational lagrangian is taken as the linear function of the curvature

$$L_g = -2\Lambda + R \qquad (3.6)$$

The matter lagrangian, L_m, is related to the energy-momentum tensor by

$$T^{ab} = \frac{2}{\sqrt{-g}} \frac{\partial}{\partial g_{ab}} (\sqrt{-g}L_m) \qquad (3.7)$$

The standard Hilbert action employed to derive the Einstein equations, with $\Lambda \neq 0$, is

$$S_H = -\tfrac{1}{2}\int \sqrt{-g}\, d^Nx(Lg + L_m) \qquad (3.8)$$

However, it has only been shown surprisingly recently (York 1972, Madsen & Barrow 1989) that the variation of S_H leaves an unwanted and uncancelled surface contribution equal to $2\int K d\Sigma$, where K is the extrinsic scalar curvature of the boundary surface Σ. The Einstein equations are therefore obtained from the stationary variation, not of S_H, but of

$$\bar{S}_H = S_H - 2\int K d\Sigma \tag{3.9}$$

provided only that the variations of the metric, δg_{ab}, vanish on the boundary Σ. The surface element in (3.9) could be rewritten as $dx^{N-1}\sqrt{h}$ where h is the determinant of the spatial metric in (N-1) dimensions. The presence of the "boundary term" in the action (3.9) is an important consideration in attempts to quantize general relativity.

4. Higher-order lagrangians

The derivation of general relativity from an action principle is suggestive of a natural generalization of Einstein's theory to a (possibly quantum) regime of large space-time curvature. We can replace the linear Hilbert lagrangian (3.6) by a higher-order polynomial in one or more curvature scalar invariants. For example, to quadratic aorder we could consider the lagrangian (see Barrow & Ottewill 1983)

$$Lg = -2\Lambda + R + \alpha R^2 + \beta R_{ab}R^{ab} + \gamma R_{abcd}R^{abcd} \tag{4.1}$$

Where α, β and γ are arbitrary constants. However, in four space-time dimensions we have a variational identity

$$\frac{\delta}{\delta g_{ab}}\int d^4x \sqrt{-g}\,(R^{abcd}R_{abcd} - 4R_{ab}R^{ab} + R^2) = 0 \tag{4.2}$$

which means that we can set either α, β or γ equal to zero in (4.1) without loss of generality because its contribution to the variation differs from that of the other two quadratic pieces by only a pure divergence. In fact, if our space-time is also homogeneous and isotropic (for example, a Friedman or de Sitter universe) then we have also (in 4-dimensional space-time) that

$$\frac{\delta}{\delta g_{ab}}\int d^4x \sqrt{-g}\,(3R^{ab}R_{ab} - R^2) = 0 \tag{4.3}$$

and so (4.2) and (4.3) allow us to set β and γ zero without loss of generality.

If we allow the lagrangian L_g to be an arbitrary analytic function of R alone, then the vacuum field equations that result are

$$0 = f'(R)R_{ab} - \tfrac{1}{2}f(R)g_{ab} - \nabla_a\nabla_b f'(R) + g_{ab}\Box f'(R) \tag{4.4}$$

where ∇_a is the covariant derivative with respect to x^a and $\Box = g_{ab}\nabla^a\nabla^b$. It has been shown by Barrow & Cotsakis (1988 that under a conformal transformation of the metric g_{ab} the field equations (4.4) become those of general relativity with a scalar field matter source. The scalar field has a potential $V(\varphi)$ which is a function of f and f'. This result is of considerable interest for studies of inflation since it reveals which higher-order lagrangian corrections to general relativity produce inflationary behaviour without the presence of explicit scalar matter field sources.

The structure of the boundary term in these lagrangian generalizations of general relativity is extremely complicated. Recently, it has been shown by Madsen and Barrow (1989) that the boundary term can only be expressed in very particular cases. One of them is the case of general relativity that we have just examined (equ (3.9)) but if we have a higher-order lagrangian theory then the boundary term can only be evaluated for particular space-times with high symmetry. For example, if we take $L_g = f(R)$ again, then the boundary term can be calculated if the space-time is maximally symmetric. This result suggests that quantum gravity theories may only be well-posed in very particular cases (like general relativity) or when there is a high level of symmetry.

Solutions possessing maximal symmetry (like de Sitter) are characterized by the property that all the curvature invariants can be expressed in terms of the Ricci scalar:

$$R_{klmn} = \frac{1}{N(N-1)} R(g_{kl}g_{mn} - g_{kn}g_{ml}) \qquad (4.4)$$

$$R_{mn} = \frac{1}{N} Rg_{mn} \qquad (4.5)$$

If the gravitational lagrangian is an arbitrary function F(X,Y,Z) where $X = R$, $Y = R_{ab}R^{ab}$ and $Z = R_{abcd}R^{abcd}$ then the condition for the vacuum gravity theory derived from this lagrangian to possess a maximally symmetric (de Sitter) solution (or solutions) is that,

$$\frac{1}{2}F - \frac{1}{N}\frac{\partial F}{\partial X} - \frac{z}{N^2}R^2\frac{\partial F}{\partial Y} - \frac{4R^2}{N^2(N-1)}\frac{\partial F}{\partial Z} = 0 \qquad (4.6)$$

All the higher-order lagrangian theories we have been discussing are fourth-order theories as can be seen from the field equations (4.4) of the f(R) theory. It is of some interest to ask the form of the most general theory that gives rise to second-order field equations. This question has been answered by Lovelock (1971). The Lovelock lagrangian in N-dimensional space-time, is

$$L_* = \sum_{0 \le r < \frac{1}{2}N} \alpha_r L_{(r)} \qquad (4.7)$$

where the α_r are arbitrary constants with dimensions of $[\text{length}]^{-2r-2}$ and

$$L_{(r)} = \frac{1}{2^r}\,\delta^{i_1\ldots i_{2r}}_{j_1\ldots j_{2r}}\,R^{j_1j_2}{}_{i_1i_2}\cdots R^{j_{2r-1}j_{2r}}{}_{i_{2r-1}i_{2r}} \qquad (4.8)$$

Here $0 \leq i,\ j \leq N-1$ and $\delta^{i_1\ldots i_{2r}}_{j_1\ldots j_{2r}}$ is the generalized Kronecker symbol. Given the dimension N, we can see from (4.7) that L_* is always a finite sum. In particular $L_{(0)} = 1$, $L_{(1)} = R$, $L_{(2)} = R_{abcd}R^{abcd} - 4R_{ab}R^{ab} + R^2$. When $N = 4$ the term $L_{(2)}$ is a total divergence (cf (4.2)) and $L_{(r)}$ vanishes for $r > 2$. The form of $L_{(r)}$ for $R \geq 3$ in $N \geq 6$ dimensions is prohibitively complicated. The vacuum field equations arising from the Lovelock lagrangian are not quasi-linear, but they do have some special properties that make the initial value problem comparitively well-behaved; they have the form

$$0 = \sum_{0\leq r<\frac{1}{2}N} \alpha_r \overset{(r)}{G}{}^m{}_n \qquad (4.9)$$

where

$$\overset{(r)}{G}{}^m{}_n = -\frac{1}{2^{r+1}}\,\delta^{m\,i_1\ldots i_{2r}}_{n\,j_1\ldots j_{2r}}\,R^{j_1j_2}{}_{i_1i_2}\cdots R^{j_{2r-1}j_{2r}}{}_{i_{2r-1}i_{2r}} \qquad (4.10)$$

So for example,

$$\overset{(1)}{G}{}^m{}_n = R^m_n - \tfrac{1}{2}\,\delta^m_n\,R \qquad (4.11)$$

The Lovelock lagrangian has recently attracted considerable interest because of the form of the field theory limit of superstring theories. It appears that the first two terms in the resulting gravitational theory have the form of $L_{(1)}$ and $L_{(2)}$, however it does not appear that the full gravitational theory is given by the Lovelock lagrangian. Various investigations of the solutions of the Lovelock theory up to $L_{(2)}$ order have been made. Cosmological solutions have been found, as well as black hole solutions. One of the novelties of the latter appears to be the possibility of the black hole specific heat becoming positive so that the Hawking evaporation does not evaporate the black hole completely. It ends after a finite time leaving a stable massive relic (Myers & Simon 1988). Finally, we should add that the Lovelock lagrangian is one for which the boundary term in the associated gravitational action can be calculated.

5. The Initial Value Problem

We have introduced both general relativity and several classes of higher-order lagrangian theory with far more complex field equations. In order to appreciate the depth of these theories as well as the likely generality of any particular cosmological solution of them that we may be using it is useful to have a characterization of the general initial data that may be prescribed in each theory. Let us consider a metric in synchronous coordinates

$$ds^2 = dt^2 - g_{\alpha\beta}dx^\alpha dx^\beta; \quad 1 \leq \alpha, \ \beta \leq 3 \qquad (5.1)$$

then on any space-like hypersurface of constant t-time we require the values of the six independent components of $g_{\alpha\beta}$ and six independent components of $\dot{g}_{\alpha\beta}$. If we assume that the matter source is a perfect fluid with density ρ, pressure $P(\rho)$ and normalized 4-velocity U^a, then the Cauchy problem also requires that we specify three components of U^α and $p(\rho)$. Thus we require 16 spatial functions of 3 variables in 4-dimensional space-time. However, we may eliminate eight of these by using the four Einstein constant equations ($G^0{}_a$ equations) and the four coordinate covariances of the theory and this leaves the general solution specified by eight functions of three spatial variables (four in vacuum). In higher-order lagrangian theories the generic situation (excluding particular forms like the Lovelock one) yields fourth-order field equations and so the initial value problem requires the specification of $g_{\alpha\beta}$, $\dot{g}_{\alpha\beta}$, $\ddot{g}_{\alpha\beta}$ and $\dddot{g}_{\alpha\beta}$ on a t = constant slice. The remaining constraints remain the same and so a four-dimensional perfect fluid model is prescribed by 20 spatial functions of 3 variables. If there are D dimensions of space then the number of arbitrary functions of D variables that must be specified on a spacelike slice of constant time are given in the following table:

Gravitation Theory	Vacuum	Perfect Fluid
general relativity	(D-2)(D+1)	D^2-1
higher-order lagrangian	$2(D^2-1)$	(D+1)(2D-1)

For a further discussion, including exceptional degenerate cases like the pure R^2 lagrangian see Starobinsky & Schmidt (1987) and Barrow & Sirousse-Zia (1989).

6. The Friedman Metric

Most cosmological interest focusses upon the isotropic and homogeneous universes first investigated by Friedman and Robertson and Walker (see Weinberg 1972). They have the metric

$$ds^2 = dt^2 - a^2(t)\left[\frac{dr^2}{1-kr^2} + r^2(d\theta^2 + \sin^2\theta d\varphi^2)\right] \tag{6.1}$$

where the constant k can be set equal to 0 or ±1 without loss of generality (by using the coordinate transformations $r \to |k|^{\frac{1}{2}}r$ and $a(t) \to |k|^{-\frac{1}{2}}a(t)$ if $|k| \neq 1$). When $k = 0$ the space-like slices of constant t-time have Euclidean geometry. If the spatial topology is R^3 then the volume is infinite. However, with 3-torus topology (identify (x, y, z) with (x + A, y + B, z + C) for some constants A, B, C) the volume is finite. When $k = -1$ the coordinate transformation $r = \sinh\chi$ yields

$$ds^2 = dt^2 - a^2(t)\{d\chi^2 + sh^2\chi\ (d\theta^2 + \sin^2\theta d\varphi^2)\} \tag{6.2}$$

The t = constant slices are spaces of constant negative curvature. The spatial volume is infinite if the natural R^3 spatial topology is chosen but would be finite for the 3-torus topology. When $k = +1$ the transformation $r = \sin\chi$ produces the form

$$ds^2 = dt^2 - a^2(t)\ \{d\chi^2 + \sin^2\chi(d\theta^2 + \sin^2\theta d\varphi^2)\} \tag{6.3}$$

The geometry of the t = constant slices is that of a space of constant positive curvature. The natural topology is S^3 and the volume of space is equal to $2\pi^2a^3(t)$.

In all of these space-times the 3-space may be considered as a curved 3-dimensional surface relative to an 'artificial' flat 4-dimensional space. The scale factor, a(t), is the curvature radius of this surface in the fourth dimension. The time derivative $\dot{a}$ describes the expansion speed of the universe in this 'extra' dimension. This expansion does not take place "into" any external 3-space. The scale factor a(t) is not an observable.

Einstein's equations for the Friedman metric when the matter source is a perfect fluid with energy momentum tensor

$$T_{ab} = (\rho + p)U_aU_b - pg_{ab}: \quad U_aU^a = 1 \tag{6.4}$$

and $\Lambda = 0$, reduce to the pair ($8\pi G \equiv 1 \equiv c$)

$$\ddot{a} = -\frac{a}{6}(\rho + 3p) \tag{6.5}$$

and

$$\frac{\dot{a}^2}{a^2} = \frac{\rho}{3} - \frac{k}{a^2} \tag{6.6}$$

The vanishing covariant divergence of T_{ab} gives the fluid conservation equation

$$\dot{\rho} + \frac{3\dot{a}}{a}(\rho + p) = 0 \tag{6.7}$$

However, this is not independent of (6.5)–(6.6) and any two of (6.5)–(6.7) imply the third. We recognise (6.6) as the conservation of energy for a Newtonian cosmology, but there the constant k has no geometrical interpretation. It merely determines the sign of the total energy of the Newtonian system. Equation (6.5) is a generalization of Newton's second law of motion in which the role of the Newtonian mass $4\pi\rho a^3/3$ is played by the combination $4\pi(\rho + 3p)a^3/3$ because of the special relativistic requirement that all sources of energy (including pressure) have an associated mass (via $E = Mc^2$) and hence gravitate.

If the universe contains many non-interacting fluids, p_i, ρ_i, then ρ and p in (6.5)–(6.6) are replaced by $\Sigma_i\rho_i$ and $\Sigma_i p_i$ and each pair ρ_i, p_i satisfies an equation of the form (6.7).

The coordinates in the Friedman metric are called 'comoving coordinates'. Each point retains its (r, θ, φ) label as the universe expands or contracts with changing a(t). This is equivalent to the choice of 4-velocity $U_a = \delta_a{}^0$, (cf eqn. (6.4)).

In order to solve the system (6.5)–(6.7) we need to specify an equation of state linking p and ρ. If we do not do this then the quations have no content because <u>any</u> function a(t) will solve (6.5)–(6.7) for some p and $\rho > 0$. If we specify a perfect fluid

$$p = (\gamma - 1)\rho: \quad \gamma \text{ constant} \tag{6.8}$$

then $\rho \propto a^{-3\gamma}$ and there exists a collection of exact solutions involving elementary functions

<u>k = 0</u>: $\quad a(t) \propto t^{2/3\gamma}$

$$\rho = \frac{4}{3\gamma^2 t^2} \qquad \gamma \neq 0 \tag{6.9}$$

$$a(t) = \exp(H_0 t): \qquad \gamma = 0 \tag{6.10}$$

$$\rho = 3H_0{}^2 = \text{constant} \tag{6.10a}$$

<u>k ≠ 0</u>: $\quad a(t) \propto (t - kt^2)^{\frac{1}{2}}: \qquad \gamma = 4/3 \qquad (6.11)$

$$a(t) \propto t \qquad \gamma = 2/3 \tag{6.12}$$

The solutions (6.11)–(6.12) are the only values of $2 \geq \gamma > 0$ where solutions for all k can be found for a(t). When $\gamma = 1$ the solutions with $k \neq 0$ can be found in terms of elliptic functions or, more transparently, in parametric form

k = +1: $a(\eta) \propto (1 - \cos\eta)$ (6.13)

$t(\eta) \propto (\eta - \sin\eta)$ (6.14)

k = -1: $a(\eta) \propto (\cosh\eta - 1)$ (6.15)

$t(\eta) \propto (\sinh\eta - \eta)$ (6.16)

where the parameters η and t have been chosen so that $\eta = 0$ and $t = 0$ when $a = 0$. Equations (6.13)–(6.14) are the parametric equations of a cycloid in the $t - a$ plane.

It is also worth noting that the only vacuum ($\rho = p = 0$) solution of the Friedman equations with $a \neq$ constant occurs when $k = -1$

$$a(t) \propto t \tag{6.17}$$

This is called the Milne universe. All the $k = -1$ models with $\rho + 3p > 0$, $\rho > 0$ approach (6.17) as $t \to \infty$ (eg (6.15)–(6.16) as η or $t \to \infty$).

With the equation of state (6.8) and $\gamma > 2/3$, $\rho > 0$ the qualitative behaviour of the Friedman models is dictated by the sign of k. When $k \leq 0$ the models begin at an initial singularity and expand forever. When $k > 0$ they reach an expansion maximum and recollapse. It is conventional to call the $k < 0$, $k > 0$ and $k = 0$ Friedman models 'open', 'closed' and 'flat' respectively. However, one must beware of being brainwashed by this terminology. The $k = +1$ Friedman models do not recollapse if $0 \leq \gamma \leq 2/3$ (one can see from (6.5) that this ensures $\ddot{a} > 0$ and hence a(t) can have no expansion maximum when $\rho + 3p < 0$). Technically, a closed universe is one with compact spatial slices of constant time and in general it need not have either positive curvature or an expansion maximum (Barrow & Tipler 1985, Barrow, Galloway & Tipler 1986, Barrow 1988). It so happens that for the isotropic Friedman models the curvature is always of one sign (determined by ka^{-2} only) and if $\rho + 3p > 0$ the sign of k determines whether there is an expansion maximum or not. In anisotropic cosmologies this correspondence does not necessarily hold.

In general, when there is no equation of state we can show that if $\rho > 0$, $p + 3p > 0$ and $|p| \leq C|\rho|$, C positive constant (or alternatively, if $dp/d\rho$ continuous) then $k > 0$ Friedman models recollapse to a second singularity (Barrow, Galloway & Tipler 1986). The last condition on p and ρ is necessary to prevent a pressure singularity occurring before recollapse is achieved. Under these conditions there is always an initial singularity with $p = \infty$. When $p > 0$ and $p + 3p < o$ there will never be an expansion maximum for any sign of k, and there may or may not be an initial singularity.

6.1 Conformal Time

The most transparent representation of the Friedman universe containing perfect fluid (6.4), (6.8) is obtained by introducing the conformal time, τ, where $dt = a d\tau$, so (Barrow & Tipler 1986a, Barrow 1987),

$$\tau = \int \frac{dt}{a(t)}; \qquad {}' \equiv d/d\tau \tag{6.18}$$

and defining

$$y = a^{(3\gamma-2)/2} \tag{6.19}$$

Then the Friedman equations reduce to a single equation, the simple harmonic oscillator, for $y(\tau)$

$$y'' = -k\left[\frac{3\gamma-2}{2}\right]^2 y \tag{6.20}$$

For example, the k = +1 models with a(0) = 0 have y(0) = 0, hence

$$a(\tau) = a_{max}\left\{\sin\left[\left[\frac{3\gamma-2}{2}\right]\tau\right]\right\}^{\frac{2}{3\gamma-2}} \tag{6.21}$$

$$t(\tau) = a_{max}\int_0^\tau \left[\sin\left[\frac{3\gamma-2}{2}\right]x\right]^{\frac{2}{3\gamma-2}} dx \tag{6.22}$$

When $3\gamma - 2 > 0$, the k = +1 models all have the same total life-time in τ-time whereas they do not have this property in t-time. In τ-time they are distinguished only by their size at the expansion maximum. This general reduction to the simple harmonic oscillator is a useful property to exploit in quantum problems.

6.2 Observable Parameters

For astronomical purposes it is expedient to work with functions of a(t) which are observable. The scale factor a(t) is not an observable and its value may be set equal to unity at any time t by a redefinition of coordinates. We define the *Hubble parameter*, H,

$$H(t) = \frac{\dot{a}}{a} \tag{6.23}$$

and *deceleration parameter*, q

$$q(t) = -\frac{\ddot{a}a}{\dot{a}^2} \tag{6.24}$$

and the *redshift*, z,

$$1 + z = \frac{a_0}{a} \tag{6.25}$$

where the subscript '0' denotes the present day value of a quantity. If we evaluate the equations (6.5)–(6.6) at the present time, t_0, with $p = 0$ then they become the algebraic relations

$$H_0{}^2(2q_0 - 1) = ka_0{}^2 \tag{6.26}$$

$$2q_0 = \frac{8\pi G}{3H_0{}^2}\rho_0 \tag{6.27}$$

where we have not set $8\pi G = 1$ explicitly in (6.27). The *critical density*, ρ_c, is defined by

$$\rho_c = \frac{3H_0{}^2}{8\pi G} \tag{6.28}$$

and the *density parameter*, Ω_0, by

$$\Omega_0 = \frac{\rho_0}{\rho_c} \tag{6.29}$$

so

$$2q_0 = \Omega_0 \tag{6.30}$$

we see from (6.26) that the sign of k is determined by the sign of $(q_0 - \frac{1}{2})$ or, equivalently, by the sign of $(\Omega_0 - 1)$.

When $\Lambda \neq 0$ the Friedman equations (6.5)–(6.6) generalize to

$$\frac{\ddot{a}}{a} = -\frac{1}{6}(\rho + 3p) + \frac{\Lambda}{3} \tag{6.31}$$

$$\frac{\dot{a}^2}{a^2} = \frac{\rho}{3} - \frac{k}{a^2} + \frac{\Lambda}{3} \tag{6.32}$$

and so, for $p = 0$, the algebraic relations (6.26)–(6.27) generalize to

$$\frac{\Lambda}{3} = \frac{4\pi G}{3}\rho_0 - q_0 H_0{}^2 \qquad (6.33)$$

$$\frac{k}{a_0{}^2} = H_0{}^2(2q_0 - 1) + \Lambda = \frac{H_0{}^2}{2}(3\Omega_0 - 2q_0 - 2) \qquad (6.34)$$

so the sign of k is fixed by that of $(3\Omega_0 - 2q_0 - 2)$ and $\Omega_0 \neq 2q_0$ when $\Lambda \neq 0$.

Acknowledgements

I would like to thank Alfredo Henriques for his invitation to participate in the Lisbon School and for his efforts which were responsible for making our stay in Lisbon so enjoyable.

References

Barrow, J.D. (1983) Quart. Jl. Roy. astron. Soc. 24, 24.

Barrow, J.D. (1988) Nucl. Phys. B 296, 697.

Barrow, J.D. & Ottewill, A. (1983) J. Phys. A 16, 2757.

Barrow, J.D., Burd, A. & Lancaster, D. (1985) Claas. Q. Grav. 3, 551.

Barrow, J.D. & Tipler, F.J. (1985) Mon. Not Roy. astron. Soc. 216, 395.

Barrow, J.D. & Tipler, F.J. (1986a) *The Anthropic Cosmological Principle*, OUP: Oxford.

Barrow, J.D., Galloway, G. & Tipler, F.J. (1986) Mon. Not. Roy. astron. Soc. 223, 835.

Barrow, J.D. (1987) in *Gravitation in Astrophysics*, eds B.Carter & J.B. Hartle, Reidel: Dordrecht.

Barrow, J.D. & Cotsakis, S. (1988) Phys. Lett. B 214, 515.

Barrow, J.D. & Götz, G. (1989) Class. Q. Grav. 6, 1253.

Barrow, J.D. & Sirousse-Zia, H. (1989) Phys. Rev. 39, 2187.

Bondi, H. (1960) *Cosmology*, CUP: Cambridge.

Ellis, G.F.R. (1971) in *General Relativity and Cosmology*, ed R.Sachs, Academic; NY.

Hawking, S.W. & Ellis, G.F.R. (1973) *The large scale structure of space-time*, CUP: Cambridge.

Heckmann, O. & Schücking, E. (1955) Zeit. Astrophys. 38, 95.

Heckmann, O. & Schücking, E. (1959) in *Handbuch der Physik* 53, 489, Springer: Berlin.

Lovelock, D., (1971) J. Math Phys. 12, 498.

Madsen, M. & Barrow, J.D. (1989) Nucl. Phys. B 323, 242.

Milne, E. & McCrea, W.H. (1934) Quart. J. Math. 5, 73.

Myers, R.C. & Simon, J.Z. (1988) "Black Hole Thermodynamics in Lovelock Gravity", ITP preprint.

Narlikar, J.V. & Kembhavi, A.K. (1980) Fund. Cosmic Phys. 6, 1.

Starobinskii, A.A. & Schmidt, H.-J. (1987) Class. Q. Grav. 4, 695.

Pollard, H. (1966) *A Mathematical Introduction to Celestial Mechanics*, Prentice Hall: NJ.

Raychaudhuri, A.K. (1955) Phys. Rev. 98, 1123.

Shikin, I.S. (1971) Sov. Phys. JETP 32, 101.

Shikin, I.S. (1972) Sov. Phys. JETP 34, 236.

Tipler, F.J., Ellis, G.F.R. & Clarke, C.J.S. (1980) in *General Relativity, One Hundred Years After the Birth of Albert Einstein,* Vol 2, Plenum: NY.

Vilenkin, A. (1985) Physics Reports 121, 263.

Weinberg, S. (1972) *Gravitation and Cosmology*, Wiley: NY.

York, J. (1972)

Zeldovich, Y.B. (1965) Sov. Phys. JETP 21, 656.

EUCLIDEANIZED EINSTEIN-YANG-MILLS EQUATIONS, WORMHOLES AND THE GROUND-STATE WAVE FUNCTION OF A RADIATION DOMINATED UNIVERSE

O. Bertolami[1] and J.M. Mourão[2] [3]

1) Centro de Física da Matéria Condensada
Av. Prof. Gama Pinto 2, 1699 Lisboa Codex, PORTUGAL

2) Centro de Física Nuclear
Av. Prof. Gama Pinto 2, 1699 Lisboa Codex, PORTUGAL

3) Departamento de Física, Instituto Superior Técnico
Av. Rovisco Pais, 1000 Lisboa, PORTUGAL

Abstract: SO(4)-symmetric (anti-)selfdual solutions of euclideanized Yang-Mills equations are used for constructing the ground-state wave function of the Universe. We discuss also SO(4)-symmetric wormhole solutions of Einstein-Yang-Mills systems and some physical implications of our minisuperspace ground-state wave function of the Universe.

1. Introduction

Developments in high energy physics and cosmology have raised important issues concerning the very early Universe. However, progress in the understanding of the very first instants after the Big-Bang requires a consistent unification of quantum and gravitational phenomena. Among the several possibilities for constructing a quantum theory of gravity one finds Quantum Cosmology [1]. This approach is based on the canonical Hamiltonian formalism of quantization and allows one to extrapolate most of the framework of the cosmological standard model back to energies of the order of the Planck mass, $M_P = 1.2 \times 10^{19}$ GeV.

Within Quantum Cosmology the state of the Universe is determined by a wave function. The latter is a functional. of the three-metric $h_{ij}(\mathbf{r})$ and of matter field configurations, generically denoted by $\Phi(\mathbf{r})$, and satisfies the Wheeler-DeWitt equation [2,3]:

$$\left[G_{ijkl}\frac{\delta^2}{\delta h_{ij}\,\delta h_{kl}} - h^{1/2}[{}^3R(h) - 2\Lambda^{\cdot}] + k^{-2}\,\hat{T}_{00}\,(-\,i\,\frac{\delta}{\delta\Phi},\Phi\,)\right]\psi[h_{ij},\Phi] = 0 \quad (1.1)$$

where G_{ijkl} is the metric on the space of all three-geometries, which is given by

$$G_{ijkl} = \frac{1}{2}\,h^{-1/2}\,(\,h_{ik}\,h_{jl} + h_{il}\,h_{jk} - h_{ij}\,h_{kl}\,)\;, \qquad (1.2)$$

$h = \det[\,h_{ij}\,]$, ${}^3R$ is the scalar curvature of the spatial hypersurface, Λ stands for the cosmological constant, $k^2 = M_p{}^2/16\pi$ and $\hat{T}_{00}\,(-\,i\frac{\delta}{\delta\Phi},\Phi\,)$ is the time-time component of the energy-momentum tensor promoted into an operator.

A striking breakthrough in quantum cosmology was due to Hartle and Hawking [4]. The Wheeler-DeWitt equation can be seen as the gravitational analogue of the Schrödinger equation. Since the ground-state solution of the latter equation can, in quantum mechanics, be obtained by the Euclidean path integral over the configuration space, Hartle and Hawking suggested that the minimal excitation solution of the Wheeler-DeWitt equation could be found via the functional integral over four-metrics ${}^4g$ and matter fields on a four-manifold M weighted by the euclidean action :

$$\psi[h_{ij},\Phi] = \int_M \mathcal{D}\,{}^4g\;\mathcal{D}\,\Phi_0\,\exp(\,-\,S_E\,[\,{}^4g,\,\Phi_0]\,) \qquad (1.3)$$

such that the three-metric h_{ij} and Φ are restrictions of ${}^4g$ and Φ_0 on Σ, a boundary of M. Moreover, Hartle and Hawking proposed that the boundary conditions of (1.1) should be fixed in the most economical way, i.e. by considering that M in (1.3) is a compact manifold with no other boundaries than Σ.

Quantum Cosmology becomes specially useful in the study of the early Universe when one considers minisuperspace models. These are obtained by freezing all but a finite number of modes of the three-metric h_{ij} and then using the canonical quantization methods to treat the unfrozen modes - for a recent discussion on the validity of the minisuperspace quantization see ref. [5]. In the simplest models only a single gravitational mode a(t) corresponding to the radius of a Friedmann-Robertson-Walker Universe and an homogeneous scalar field $\phi(t)$ - t being the cosmic time - are considered. This framework allows a satisfactory description of for instance, the inflationary phase which seems to be present for the various scalar field potentials that have been analyzed in the litearature [6-12].

Hartle and Hawking have also advanced a concrete proposal for the ground-state wave function of the Universe [4]. They considered a Universe with a positive cosmological constant and a single free massless conformally coupled scalar field.

This choice implies that the field dynamics of the scalar field conformally invariant effectively decouples from gravity. Moreover, the scalar field action turns out to be the Euclidean action for the harmonic oscillator, yielding that the corresponding wave function is the Gaussian function associated with the ground-state of the harmonic oscillator.

In this contribuition we discuss an alternative proposal for the construction of the ground-state wave function of the Universe [13]. This proposal arises in the context of Einstein-Yang-Mills (EYM) systems where one finds (anti-)selfdual solutions which have vanishing energy-momentum tensor and features similar to the solutions discussed by Hartle and Hawking for the free massless conformally coupled scalar field. Furthermore, we believe that a setting in which the very early Universe - earlier than the inflationary period - is dominated by gravity and non-abelian gauge fields is more realistic from the physical view point than the one of ref. [4] where the scalar field plays together with gravity a prominent rôle.

Recently solutions of euclideanized EYM systems have been studied with interest. These are gravitational-instanton solutions, referred to also as wormhole-type solutions to emphasize that they correspond to a narrow tunnel connecting disjoint regions of space-time. Although, wormhole solutions are usually seen as solutions of the euclideanized classical field equations, it has been argued that wormholes can be obtained from the quantum mechanical Wheeler-DeWitt equation with appropriate boundary conditions [14]. Wormhole-type solutions of EYM systems with SU(2) gauge group have been obtained by various authors [15-17]. The generalization to arbitrary SO(n) and SU(n) gauge groups has been obtained in ref. [18] and in ref. [19]. In these references a key ingredient in the generalization of the EYM solutions is the use of the theory of symmetric fields on homogeneous spaces to obtain SO(4)-symmetric ansätze for the fields in a $\mathbb{R} \times S^3$ topology. This method, which is extensively used in the dimensional reduction of multidimensional models (see e.g. refs [20]) and in the theory of spontaneous compactification of extra dimensions (see [21] and references therein) allows the classification of all gauge and matter field configurations which give rise to homogeneous and isotropic observables, such as the energy-momentum tensor. The formalism of symmetric fields is particularly useful in studying homogeneous and isotropic FRW cosmologies in the presence of fields with gauge degrees of freedom[22].

In the next section, we shall review the main ideas contained in refs [18,22] and obtain the Wheeler-DeWitt equation for a radiation dominated Universe. In section 3 we obtain wormhole-type solutions of the euclideanized EYM systems with arbitrary gauge groups SO(N). In section 4, we discuss the (anti-)selfdual solutions of the Yang-Mills equations for an SO(3) gauge group and present our construction for the ground-state wave function fo the Universe. Section 5 contains our conclusions and an outlook on possible implications of our proposal.

2. EYM systems with SU(m) and SO(n) gauge groups ($m \geq 2$, $n \geq 3$).

We are interested in solutions of the EYM equations which are obtained from the action:

$$S = k^2 \int_M d^4x \sqrt{-g}\,(R - 2\Lambda) - 2k^2 \int_{\partial M} d^3x \sqrt{-h}\,K + \frac{1}{8e^2} \int_M d^4x \sqrt{-g}\, \mathrm{Tr}(F_{\mu\nu} F^{\mu\nu}) \qquad (2.1)$$

where $g_{\mu\nu}$ is a Lorentzian metric with signature (-,+,+,+) in the four-dimensional manifold M; R, g and K are the scalar curvature corresponding to $g_{\mu\nu}$, $g = \det [g_{\mu\nu}]$ and the trace $K = K^\mu{}_\mu$ of the extrinsic curvature $K^\mu{}_\nu$ of ∂M respectively. Constant e denotes the gauge coupling constant and $F_{\mu\nu}$ the usual field strength tensor.

The most general form of an SO(4)-invariant metric, ie. a metric which is spatially homogeneous and isotropic in a $M = \mathbb{R} \times S^3$ topology is given by:

$$g = \sigma^2 (- N^2(t)\, dt^2 + a^2(t) \sum_{b=1}^{3} \omega^b \omega^b), \qquad (2.2)$$

where N(t) and a(t) are arbitrary non-vanishing functions of time t; the constant $\sigma^2 = \frac{2G}{3\pi}$ and ω^b are the left-invariant one-forms in $SU(2) \overset{\mathrm{diff}}{\cong} S^3$.

The metric (2.2) gives rise also to an SO(4)-invariant Einstein tensor $G_{\mu\nu}$ which in the gauge N = 1 reads:

$$G = G_{(0)}(t)\, dt^2 + G_{(1)}(t)\, a(t)^2 \sum_{b=1}^{3} \omega^b \omega^b, \qquad (2.3a)$$

where

$$G_{(0)}(t) = 3\frac{\dot{a}^2}{a^2} + 3\frac{1}{a^2} \qquad (2.3b)$$

$$G_{(1)}(t) = -2\frac{\ddot{a}}{a} - \frac{\dot{a}^2}{a^2} - \frac{1}{a^2} \quad , \qquad (2.3c)$$

and the dots denote time derivatives.

Due to the Einstein equations, the energy-momentum tensor (EMT) must share the symmetries of the Einstein tensor (2.3a), that is it must be of the form

$$T = T_{(0)}(t)\, dt^2 + T_{(1)}(t)\, a(t)^2 \sum_{b=1}^{3} \omega^b \omega^b . \qquad (2.4)$$

This corresponds to the EMT of a comoving perfect fluid with energy density $\rho(t) = T_{(o)}(t)$ and pressure $p(t) = T_{(1)}(t)$. There is a large class of fields A_μ and generically of all matter fields Φ which possess the required property of generating SO(4)-invariant EMT's. This class, or in other words this truncation procedure, is also consistent with the corresponding equations of motion. This is the class of the so-called SO(4)-*symmetric* fields A_μ and Φ, meaning that they are SO(4)-*invariant* up to a gauge transformation [18,22].

For the simplest embedding of the isotropy group SO(3) in the gauge group SO(n), the SO(4)-symmetric ansatz for the gauge fields is given by [18,22]:

$$A(t) = \hat{A}(t) + X(t)\,, \tag{2.5a}$$

where

$$\hat{A}(t) = \left(\sum_{k,m=1}^{n-3} \frac{1}{2} \Lambda^{km}(t)\, T_{k+3\ m+3} \right) dt \tag{2.5b}$$

is a gauge field in $\mathbb{R}$ with gauge group SO(n-3),

$$X(t) = \sum_{a=1}^{3} \left(\frac{1}{2}\Big(1 + \sqrt{\frac{2\alpha}{3\pi}}\,\chi_o(t)\Big) \sum_{b,c=1}^{3} T_{bc}\, \varepsilon_{bac} + \sum_{k=1}^{n-3} \sqrt{\frac{2\alpha}{3\pi}}\,\chi_k(t)\, T_{a\ k+3} \right) \omega^a \tag{2.5c}$$

$T_{km} \in \mathfrak{so}(n)$=Lie(SO(n)) are the generators of SO(n), $\chi_o(t)$, $\chi_k(t)$ - $k = 1, \dots, n-3$ - are arbitrary functions and $[\Lambda^{km}(t)]_{k,m=1}^{n-3}$ is for any t an arbitrary antisymmetric matrix. In (2.5c) we have introduced, for convenience, the fine structure constant $\alpha = e^2/4\pi$.

It is shown in refs [18,22] that the energy-momentum tensor generated by the gauge field (2.5) is, as expected, the one of a perfect fluid (2.4) satisfying the equation of state for a radiation fluid:

$$p_{YM}(t) = \frac{1}{3} \rho_{YM}(t) \quad . \tag{2.6}$$

Substituting (2.2) and (2.5) into action (2.1) we obtain a Lagrangian density independent of the space coordinates. The integration over these coordinates, gives just the volume $\mu = 2\pi^2$ of S^3 (for a = 1) as a multiplicative factor. We are then led to an effective action for the SO(4)-symmetric degrees of freedom

$$S_{eff}[a, \chi_o, \vec{\chi}; N, \Lambda^{km}] = \frac{1}{2}\int_{t_1}^{t_2} dt \left(\frac{N}{a}\right)\left[- \left(\frac{a}{N}\frac{da}{dt}\right)^2 + a^2 - H^2 a^4 + \left(\frac{a}{N}\frac{d\chi_o}{dt}\right)^2 + \left(\frac{a}{N}\vec{\mathfrak{D}}_t\chi\right)^2 - 2\, V_{g.f.}(\chi_o, \vec{\chi})\right], \qquad (2.7)$$

where $H^2 = \frac{2G\Lambda}{9\pi}$. In action (2.7) $\vec{\chi} = \{\chi_k\}_{k=1}^{n-3}$, $\hat{\mathfrak{D}}_t$ is the covariant derivative of $\vec{\chi}$ with respect to the gauge connection $\hat{A}$ (t) in $\mathbb{R}$

$$\vec{\mathfrak{D}}_t\chi = \frac{d}{dt}\vec{\chi} + \hat{\Lambda}\vec{\chi} \quad , \qquad (2.8)$$

$\hat{\Lambda} = [\Lambda^{km}(t)]_{k,m=1}^{n-3}$ and

$$V_{g.f.}(\chi_o, \vec{\chi}) = \frac{\alpha}{3\pi}\left[\left(\chi_o^2 + \vec{\chi}^{\,2} - \frac{3\pi}{2\alpha}\right)^2 + 4\chi_o^2\vec{\chi}^{\,2}\right] \qquad (2.9)$$

is the effective potential of self-interaction for χ_o and $\vec{\chi}$.

Notice that, in (2.7), N(t) and $\Lambda^{km}(t)$ are not dynamical variables. These functions play the rôle of Lagrange multipliers. The lapse function N(t) is associated with the invariance of the minisuperspace model under time reparametrizations and $\Lambda^{km}(t)$ with local SO(n-3) symmetry remnant from the original SO(n) gauge symmetry.

It is worth mentioning the striking similarity, due to conformal invariance, between our effective action (2.7) for the EYM system and the one of Hartle and Hawking for the free massless conformally coupled scalar field (eq. (5.5) in ref. [4]). The only differences being that we have more variables (χ_o and $\vec{\chi}$) describing the matter sector and a self-interacting potential which is quartic rather than quadratic.

The absolute minima of the potential (2.9) correspond to pure gauge configurations belonging to three topologically inequivalent vacua [18]. These are the only vacua which survive the condition of SO(4)-symmetry, among the infinitely degenerate vacua of the full Yang-Mills theory.

We stress that the above results for SO(n) gauge groups can be generalized to any arbitrary compact simple Lie groups - SU(n), Sp(n), G_2, F_4, E_6, E_7 or E_8. In particular, the SU(n) case turns out to be quite similar to the one of an SO(n) group (see refs [18] for details).

Let us now turn to the derivation of the Wheeler-DeWitt equation corresponding to the minisuperspace with dynamics described by the effective action (2.7). We start by computing the canonical momenta conjugate to a(t), $\chi_o(t)$ and $\vec{\chi}(t)$:

$$\pi_a = \frac{\partial L_{eff}}{\partial \dot{a}} = -\frac{a}{N}\dot{a} \; ,$$

$$\pi_{\chi_0} = \frac{\partial L_{eff}}{\partial \dot{\chi}_0} = \frac{a}{N}\dot{\chi}_0 \; , \tag{2.10}$$

$$\vec{\pi}_{\vec{\chi}} = \frac{\partial L_{eff}}{\partial \dot{\vec{\chi}}} = \frac{a}{N}\, \vec{\partial}_t \chi \; .$$

The Hamiltonian constraint is obtained by considering the variation of S_{eff} with respect to the lapse function N(t), that is:

$$-\frac{\partial L_{eff}}{\partial N} = 0 \quad \leftrightarrow \quad \mathcal{H} = 0 \; , \tag{2.11}$$

which yields, after using (2.10)

$$\frac{1}{2}\frac{N}{a}\left\{-\; \pi_a^2 - a^2 + H^2 a^4 + \pi^2_{\chi_0} + \vec{\pi}^2_{\vec{\chi}} + 2\, V_{g.f.}(\chi_0, \vec{\chi})\right\} = 0 \, . \tag{2.12}$$

It is relevant to remark that, in obtaining (2.12), we have allowed π_a and a to commute, which, although perfectly legitimate at the classical level, leads to ambiguities at quantum level due to the ordering of operators. Quantization proceeds by promoting the canonical conjugate momenta in (2.12) into operators, in the following way:

$$\pi_a \to -i\frac{\partial}{\partial a} \; , \quad \pi_{\chi_0} \to -i\frac{\partial}{\partial \chi_0} \; , \quad \vec{\pi}_{\vec{\chi}} \to -i\frac{\partial}{\partial \vec{\chi}} \tag{2.13a}$$

Due to the operator ordering ambiguity, it is usual to parametrize π_a^2 as [4]:

$$\pi_a^2 = -\, a^{-p}\frac{\partial}{\partial a}\left(a^p \frac{\partial}{\partial a}\right) \; , \tag{2.13b}$$

where p is an arbitrary real constant. The Wheeler-DeWitt equation is then found to be:

$$\frac{1}{2}\left\{ a^{-p}\frac{\partial}{\partial a}\left(a^p \frac{\partial}{\partial a}\right) - a^2 + H^2 a^4 - \frac{\partial^2}{\partial \chi_0^2} - \frac{\partial^2}{\partial \vec{\chi}^2} + 2\, V_{g.f.}(\chi_0, \vec{\chi})\right\} \psi\,(a, \chi_0, \vec{\chi}) =$$

$$= E\, \psi\,(a, \chi_0, \vec{\chi}), \tag{2.14}$$

where E is an arbitrary constant arising from matter-energy renormalization.

An additional constraint emerging from the effective action (2.9) is the vanishing of the components of the angular momentum corresponding to the motion in the subspace $\mathbb{R}^{n-3}$ of the variables $\vec{\chi}$. This arises from the variation of S_{eff} with respect to $\hat{\Lambda}$:

$$-\frac{\partial L_{eff}}{\partial \Lambda_{km}} = 0 \quad \leftrightarrow \quad \mathcal{M}_{km} = 0\,, \tag{2.15a}$$

which reads

$$\chi_k \pi_{\chi_m} - \chi_m \pi_{\chi_k} = 0\,. \tag{2.15b}$$

Thus, the wave function $\psi\,(a,\chi_0,\vec{\chi})$ has to satisfy besides (2.14), the condition:

$$\left(\chi_k \frac{\partial}{\partial \chi_m} - \chi_m \frac{\partial}{\partial \chi_k}\right)\psi\,(a,\chi_0,\vec{\chi}) = 0\,. \tag{2.16}$$

In (2.12) we can see that the Hamiltonians corresponding to the gravitational and gauge degrees of freedom effectively decouple:

$$\frac{a}{N}\mathcal{H} = \frac{a}{N}\left(\mathcal{H}_{grav} + \mathcal{H}_{gauge}\right). \tag{2.17}$$

Therefore, solutions of the Wheeler-DeWitt equation can be obtained by separating variables in the following way:

$$\psi\,(a,\chi_0,\vec{\chi}) = \sum_n C_n\,(a)\; U_n\,(\chi_0,\vec{\chi})\,, \tag{2.18}$$

where $C_n\,(a)$ and $U_n\,(\chi_0,\vec{\chi})$ are solutions of:

$$\frac{1}{2}\left[-\,a^{-p}\frac{d}{da}\left(a^{p}\frac{d}{da}\right) + a^2 - H^2 a^4\right]C_n(a) = (\lambda_n - E)\,C_n(a) \tag{2.19}$$

and

$$\frac{1}{2}\left[-\frac{\partial^2}{\partial \chi_0^2} - \frac{\partial^2}{\partial \vec{\chi}^2} + 2\,V_{g.f.}(\chi_0,\vec{\chi})\right]U_n(\chi_0,\vec{\chi}) = \lambda_n U_n(\chi_0,\vec{\chi})\,. \tag{2.20}$$

Notice that, eq. (2.19) is the very one obtained by Hartle and Hawking (eq. (5.11)) in ref. [4]. We shall restrict ourselves to the case of a SO(3) gauge group for which the vector $\vec{\chi}$ does not exist and therefore:

$$\psi = \psi\,(a,\chi_0) = \sum_n C_n\,(a)\; Y_n\,(\chi_0) \quad , \tag{2.21}$$

where C_n (a) satisfy eq. (2.19) and $Y_n\,(\chi_0)$ are solutions of:

$$\left[-\frac{1}{2}\frac{d^2}{d\chi_0^2} + V_{g.f.}(\chi_0) \right] Y_n(\chi_0) = \lambda_n Y_n(\chi_0) \ . \tag{2.22}$$

This equation coincides with the Schrödinger equation for an anharmonic oscillator of the double-well type. On general grounds, the quantum anharmonic oscillator has been studied in ref. [23]. More concretely, numerical solutions are known for a long time [24] and perturbation theory at very high order has been analyzed in depth [25]. Moreover, variational methods involving more than one variational parameter prove to be fairly accurate in the description of the energy levels [26]. In the section 4, we shall see that, as the ground-state solution of eq. (2.22) can be obtained with the help of the path integral, in the semiclassical approximation, this solution can be related with (anti-)selfdual solutions of the euclideanized YM equations.

3. Wormhole Solutions

In this section we shall obtain wormhole soutions of the euclideanized EYM equations. The former equations are obtained from the euclideanized version of action (2.1) or as shown in ref. [18] from the euclideanized effective action corresponding to (2.7). The Friedmann equation arises from the variation of effective action with respect to $N(\tau)$ – τ = - it. In the conformal gauge N=a, $\hat{\Lambda} = 0$ this is given by:

$$\frac{1}{2}\frac{da}{d\tau}^2 - \frac{1}{2}a^2 + H^2 a^4 = -\frac{1}{6}\left(\frac{1}{2}\left(\frac{d\chi_o}{d\tau}\right)^2 + \frac{1}{2}\left(\frac{d\vec{\chi}}{d\tau}\right)^2 - V_{g.f}(\chi_o, \vec{\chi})\right) \ . \tag{3.1}$$

The equations for $\chi_o, \vec{\chi}$, i.e. the YM equations are the following:

$$\frac{d^2\chi_o}{d\tau^2} = \frac{\partial V_{g.f.}}{\partial \chi_o} \tag{3.2a}$$

$$\frac{d^2\vec{\chi}}{d\tau^2} = \frac{\partial}{\partial\vec{\chi}} V_{g.f.} \tag{3.2b}$$

$$\chi_k \frac{d\chi_m}{d\tau} - \frac{d\chi_k}{d\tau}\chi_m = 0 \ . \tag{3.2c}$$

Equations (3.2) are the equations of motion for a unit mass particle in the space $\mathbb{R}^{n-2}$ under the action of the inverted potential $-V_{g.f.}$ and with vanishing projection of the angular momentum in the subspace $\mathbb{R}^{n-3}$ of the variables χ_k, $k = 1, \ldots, n-3$. On the other hand since the right hand side of eq. (3.1) is the conserved mechanical energy W_0 of eqs. (3.2) ,

$$W_0 = \frac{1}{2}\left(\frac{d\chi_0}{d\tau}\right)^2 + \frac{1}{2}\left(\frac{d\vec{\chi}}{d\tau}\right)^2 - V_{g.f}(\chi_0, \vec{\chi}) ,$$

the Einstein and Yang-Mills equations decouple and can be solved separately.

We are interested in finite action non-static solutions of equations (3.1) and (3.2), corresponding to the motion between classical turning points where both the scale factor as well as χ_0 and $\vec{\chi} = \chi\,\vec{e}$ - $\vec{e}$ being an arbitrary but fixed unit norm vector - tunnel through the potential barrier. This implies that only bounded motions between the classical turning points of the gauge fields and of the scale factor are relevant. We study only bounded solutions of eq. (3.2) with either $\chi = \tilde{\chi}$ and $\chi_0 = 0$, or $\chi_0 = \tilde{\chi}$ and $\chi = 0$ (Figure I). The solutions $\tilde{\chi}(\eta)$ and $a(\eta)$ are periodic. The period of oscillations of $\tilde{\chi}(\eta)$ between the turning points $\alpha_{1,2} = \pm\sqrt{1 - \sqrt{e^2 W_0/9\pi^2\sigma^2}}$ is given by:

$$T(W_0) = \frac{4}{\sqrt{1+\sqrt{e^2 W_0/9\pi^2\sigma^2}}}\, K\left(\sqrt{\frac{1-\sqrt{e^2 W_0/9\pi^2\sigma^2}}{1+\sqrt{e^2 W_0/9\pi^2\sigma^2}}}\right) \qquad (3.3)$$

where K(x) is a complete elliptic function of the first type [27].

Solutions of eq. (3.1) are bounded between turning points r_{MIN} and r_{MAX} (see Figure II):

$$r^2_{MIN,MAX} = \frac{\sigma^2}{2H^2}\left[1 \pm \sqrt{1 - 64 H^2 W_0/3\sigma^2}\right] . \qquad (3.4)$$

These solutions describe the transition of an S^3 Universe of radius r_{MIN} into another with radius r_{MAX} . The period of oscillation which depends also on the value of the cosmological constant is given by:

$$\tilde{T}(H, W_0) = 2\frac{\sigma}{H}\frac{1}{r_{MAX}}\, K\left(\sqrt{1 - \left(\frac{r_{MIN}}{r_{MAX}}\right)^2}\right) . \qquad (3.5)$$

Wormhole solutions are found if the ratio of periods (3.3) and (3.5) is a rational number

$$\frac{\tilde{T}(\Lambda,W_0)}{T(W_0)}=\frac{m}{n} \qquad \text{m, n integers} \quad . \tag{3.6}$$

Solving (3.6) with respect to W_0 yields that wormhole solutions (with either χ or χ_0 equal to zero) exist only for discrete values of W_0, i.e. for $W_0 = W_0(m,n,\Lambda)$. Thus the wormhole quantization, first encountered in refs [16,17] for the gauge group SU(2), generalizes to arbitrary non-abelian gauge groups.

An interesting feature of our wormhole solutions is that they have a fixed upper bound for their radii ($O(\sqrt{G})$). Thus we have no difficulties such as the large wormhole catastrophe in the $\Lambda \to 0^+$ limit [28]. This can be easily understood by expanding the square root in (3.4) for $\Lambda \to 0^+$. One finds then that r_{MIN} depends essentially on W_0 which is bounded, hence r_{MIN} is also bounded and always of the same order as the Planck length. This is in contrast to the other known wormhole-type solutions which besides being found in rather special theories [29-31], have unbounded wormhole radii.

4. (Anti-)Selfdual solutions and the ground-state wave function of the Universe

In what follows, we shall use the Euclidean path integral (1.3) to obtain the ground-state solution of eq. (2.14) for an SO(3) gauge group.

The evaluation of the ground-state wave function is greatly simplified as the gravitational and gauge functional integrals in (1.3) may be computed separately:

$$\Psi[a,\chi_0] = C(a)U(\chi_0) = \int \mathcal{D} a \; \exp(- S^E_{grav}[a(\tau)]) \int \mathcal{D}\chi_0 \exp(- S^E_{gauge}[\chi_0(\tau)]) \tag{4.1a}$$

where

$$S^E_{grav}[a(\tau)] = \frac{1}{2}\int_{\tau_1}^{\tau_2} d\tau \left(- \left(\frac{da}{d\tau}\right)^2 - a^2 + H^2 a^4 \right), \tag{4.1b}$$

$$S^E_{gauge}[\chi_0(\tau)] = \frac{1}{2}\int_{\tau_1}^{\tau_2} d\tau \left(\left(\frac{d\chi_0}{d\tau}\right)^2 + 2\, V_{g.f.}(\chi_0) \right), \tag{4.1c}$$

τ is the conformal Euclidean time introduced in the previous section and $V_{g.f.}(\chi_0)$ is given by (2.9) with $\vec{\chi} = \vec{0}$.

The functional integrals in the r.h.s. of (4.1a) are evaluated with the following boundary conditions, appropriate for the state of minimal excitation:

$$a(-\infty) = 0, \qquad a(0) = a \tag{4.2a}$$

and

$$\chi_0(-\infty) = \mp\sqrt{\frac{3\pi}{2\alpha}}, \qquad \chi_0(0) = \chi_0 \ . \tag{4.2b}$$

The function C(a) in (4.1a) is given, in the semiclassical approximation, by [4]:

$$C(a) = \exp\left(\frac{1}{2}a^2 - \frac{1}{3H^2}\right) \quad , \text{ for } H\,a << 1 ;$$
$$C(a) = 2\cos\left(\frac{(a^2 H^2 - 1)^{3/2}}{3H^2} - \frac{\pi}{4}\right), \text{ for } H\,a > 1 . \tag{4.3}$$

Let us now compute semiclassically the function $U(\chi_0)$. For that we must find finite action solutions satisfying boundary conditions (4.2b), of the euclideanized YM equations obtained from (4.1c). Action (4.1c) can be seen as the action of an unit mass particle subject to the potential - $V_{g.f.}(\chi_0)$. With the help of this analogy, it is clear that the solutions satisfying (4.2b) are those with vanishing mechanical energy and therefore are given by the solutions of the following first order differential equations:

$$\frac{d\chi_0}{d\tau} = \pm\sqrt{\frac{2\alpha}{3\pi}}\,(\chi_0^2 - \frac{3\pi}{2\alpha}) \ . \tag{4.4}$$

These equations coincide with the equations for (anti-)selfdual YM fields. Indeed, by substituting the ansatz (2.5) into the (anti-)selfduality conditions:

$$F_{\mu\nu} = \pm {}^*F_{\mu\nu} \ , \tag{4.5}$$

where

$${}^*F_{\mu\nu} = \frac{\sqrt{g}}{2}\,\varepsilon_{\mu\nu\rho\lambda}\,F^{\rho\lambda} \quad , \tag{4.6}$$

one obtains precisely eqs. (4.4). Notice that conditions (4.5) on their own ensure that the EMT vanishes [32].

In the present case, the solutions of eqs (4.4), satisfying (4.2b) read:

$$\chi_0(\tau) = \sqrt{\frac{3\pi}{2\alpha}}\ \mathrm{cotanh}(\tau - \tau_0) \ , \qquad \text{for } \chi_0 < -\sqrt{\frac{3\pi}{\alpha}} \tag{4.7a}$$

$$\chi_0(\tau) = \pm\sqrt{\frac{3\pi}{2\alpha}}\ \tanh(\tau - \tau_0) \ , \qquad \text{for } |\chi_0| < \sqrt{\frac{3\pi}{\alpha}} \tag{4.7b}$$

$$\chi_0(\tau) = -\sqrt{\frac{3\pi}{2\alpha}}\ \mathrm{cotanh}(\tau - \tau_0) \ , \qquad \text{for } \chi_0 > \sqrt{\frac{3\pi}{\alpha}} \ . \tag{4.7c}$$

It is important to remark that for $|\chi_0| < \sqrt{\frac{3\pi}{2\alpha}}$ (see (4.7b)), there are two solutions with the same boundary value $\chi_0(0) = \chi_0$, the relevance of each being determined by the relative value of their corresponding actions. In this region, the wave function is approximately given by the even combination of the semiclassical contributions from these two solutions. Therefore, semiclassically, the wave function is the following (see Figure III):

$$U(\chi_0)=A\ \exp\left(-(\chi_0+\sqrt{\frac{3\pi}{2\alpha}})^2+\frac{1}{3}\sqrt{\frac{2\alpha}{3\pi}}(\chi_0+\sqrt{\frac{3\pi}{2\alpha}})^3\right),\ \text{for}\ \chi_0<-\sqrt{\frac{3\pi}{2\alpha}} \quad (4.8a)$$

$$U(\chi_0)=B\ \exp\left(-(\chi_0+\sqrt{\frac{3\pi}{2\alpha}})^2+\frac{1}{3}\sqrt{\frac{2\alpha}{3\pi}}(\chi_0+\sqrt{\frac{3\pi}{2\alpha}})^3\right)+ B\ \exp\left(-(\chi_0-\sqrt{\frac{3\pi}{2\alpha}})^2-\frac{1}{3}\sqrt{\frac{2\alpha}{3\pi}}(\chi_0-\sqrt{\frac{3\pi}{2\alpha}})^3\right),\ \text{for}\ |\chi_0|\leq\sqrt{\frac{3\pi}{2\alpha}} \quad (4.8b)$$

$$U(\chi_0)=A\ \exp\left(-(\chi_0-\sqrt{\frac{3\pi}{2\alpha}})^2-\frac{1}{3}\sqrt{\frac{2\alpha}{3\pi}}(\chi_0-\sqrt{\frac{3\pi}{2\alpha}})^3\right),\ \text{for}\ \chi_0>\sqrt{\frac{3\pi}{2\alpha}}, \quad (4.8c)$$

where $A = \left[1+\exp\left(-\frac{2\pi}{\alpha}\right)\right] B$, and B is a normalization factor. We see that near the two vacua, $\chi_0 = \pm\sqrt{\frac{3\pi}{2\alpha}}$, the wave function has essentially a Gaussian behaviour, as one should expect from the double-well form of potential. This is because (4.8) is an approximate solution of the Schrödinger equation (2.22) with n=0 and around the minima the potential is approximately quadratic. For large χ_0, solutions (4.8) reproduce the expected $\exp\left(-\frac{1}{3}\sqrt{\frac{2\alpha}{3\pi}}|\chi_0|^3\right)$ behaviour of the wave function [24].

5. Conclusions and outlook

We have studied the minisuperspace Wheeler-DeWitt equation for Eintein-Yang-Mills systems. Due to the conformal invariance of the Yang-Mills action we find that, similarly to the case of a free massless conformally invariant scalar field studied in ref. [4], the Wheeler-DeWitt equation can be solved by separating the gravitational and gauge degrees of freedom. We build then the ground-state wave function for the case of an SO(3) gauge group in the semiclassical approximation using (anti-)selfdual solutions of the euclideanized Yang-Mills equations. The wave function constructed in this way accommodates naturally the expectation that the very early Universe - earlier than a conjectured inflationary stage - was dominated by radiation.

We have obtained also wormhole solutions for the EYM systems with SO(n) gauge groups. The wormhole transitions correspond to between a closed Friedmann Universe into a de Sitter Universe. The wormhole radius (r_{MIN}) is always of the order of the Planck lenght. Thus the so-called giant wormhole catastrophe is avoided for EYM theories.

Let us now briefly comment on the implications of our ground-state wave function construction. It is believed that the main effect of the presence of wormholes is the fixing of the values of the constants of Nature [28,33,34]. Moreover, the wormhole influence can be seen as resulting from the interaction of our Universe with a bath of harmonic oscillators in an unknown state. Nevertheless, we have seen that the degenerate structure of the gauge fields vacua leads to a wave function which is picked for the different values of the dynamical variables which correspond to these vacua. Thus, a possible implication of our construction would consist in an indeterminacy in the process of fixing the constants of Nature. This may have consequences on the various phenomenological questions that are related with the breaking of global symmetries as for instance, the violation of the baryon number about which has been already some controversy [35,36]. Of course, any definite statement on this and related questions requires the inclusion of fermions into the EYM system. Modest progress have already been achieved for fermions coupled with an SU(2) gauge group [15,17,37]. The generalization to larger gauge groups will be the scope of future work [38].

References

[1] For a review see for instance S.W. Hawking, in 300 Years of Gravitation, S.W. Hawking and W. Israel, eds. (Cambridge University Press 1989).

[2] B.S. DeWitt, Phys. Rev. 160 (1967) 1113.

[3] J.A. Wheeler, in Battelle Rencontres, C. DeWitt and J.A. Wheeler, eds. (Benjamin, New York 1968).

[4] J.B. Hartle and S.W. Hawking, Phys. Rev. D28 (1983) 2960.

[5] K.V. Kuchar and M.P. Ryan, Jr., Phys. Rev. D40 (1989) 3982.

[6] S.W. Hawking, Nucl. Phys. B244 (1984) 135.

[7] S.W. Hawking and P.C. Luttrell, Nucl. Phys. B247 (1984) 250.

[8] I.G. Moss and W.A. Wright, Phys. Rev. D29 (1984) 1069.

[9] S.W. Hawking and Z.C. Wu, Phys. Lett. 151B (1985) 15.

[10] P.F. Gonzales-Dias, Phys. Lett. 159B (1985) 19.

[11] D.N. Page, in Quantum Concepts in Space And Time, R. Penrose and C.J. Isham eds. (Clarendon Press, Oxford 1986).

[12] G.W. Gibbons and L.P. Grishchuk, Nucl. Phys. B313 (1989) 736.

[13] O. Bertolami, J.M. Mourão, "The Ground-State Wave-Function of a Radiation Dominated Universe" Lisbon Preprint, IFM-19/90.

[14] S.W. Hawking, Mod. Phys. Lett. A5(1990)453.

[15] A. Hosoya and W. Ogura, Phys. Lett. 225B (1989) 117.

[16] Y. Verbin and A. Davidson, Phys. Lett. 229B (1989) 364.

[17] S.J. Rey, Nucl. Phys. B336(1990)146.

[18] O. Bertolami, J.M. Mourão, R.F. Picken and I.P. Volobujev, "Dynamics of Euclideanized Einstein-Yang-Mills Systems with Arbitrary Gauge Groups" Lisbon Preprint, IFM-9/90, to appear in Int. Journ. Mod. Phys. A.

[19] K. Yoshida, S. Hirenzaki and K. Shiraishi, Phys. Rev. D42 (1990) 1973.

[20] N.S. Manton, Nucl. Phys. B158 (1979) 141;
P. Forgacs and N.S. Manton, Comm. Math. Phys. 72 (1980) 15;
F.A. Bais, K.J. Barnes, P. Forgacs and G. Zoupanos, Nucl. Phys. B236(1986)557;
R. Coquereaux and A. Jadcyzk, Riemannian Geometry, Fiber Bundles, Kaluza-Klein Theories and all that ... (World Scientifc Lecture Notes in Physics, vol. 16, Singapore, 1988);
Yu.A. Kubyshin, J.M. Mourão and I.P. Volobujev, Intern. Journ. Mod. Phys. A4 (1989) 151; Theor. Math. Phys. 78 (1989) 41; 78 (1989) 191;
G. Rudolph and I.P. Volobujev, Nucl. Phys. B313 (1989) 95.

[21] Yu.A. Kubyshin, J.M. Mourão, G. Rudolph and I.P. Volobujev, "Dimensional Reduction of Gauge Theories, Spontaneous Compactification and Model Building" (Lecture Notes in Physics vol. 349, Springer-Verlag, 1989).

[22] P.V. Moniz and J.M. Mourão, "Homogeneous and isotropic closed cosmologies with a gauge sector" Lisbon Preprint IFM - 11/90.

[23] M. Reed and B. Simon, in "Methods in Modern Mathematical Physics", Vol. II (Academic Press, London 1975).

[24] R. McWeeny and C.A. Coulson, Proc. Cambridge Phil. Soc. 44 (1948) 413.

[25] C.M. Bender and T.T. Wu, Phys. Rev. Lett. 27 (1971) 461; Phys. Rev. D7 (1973) 1620.

[26] J. Dias de Deus, Phys. Rev. D26 (1982) 2782.

[27] I.S. Gradshteyn and I.M. Ryzhik, "Tables of Integrals, Series and Products" (Academic Press, New York, 1965).

[28] W. Fischler and L. Susskind, Phys. Lett. 217B (1989) 48;
I. Klebanov, L.Susskind and T. Banks, Nucl. Phys. B317 (1989) 665.

[29] S. Giddings and A. Strominger, Nucl. Phys. B306 (1988) 890.

[30] R.C. Myers, Phys. Rev. D38 (1988) 1327.

[31] J.J. Halliwell and R. Laflamme, Class. Quantum Grav. 6 (1989) 1839.

[32] J.M Charap and M.J. Duff, Phys. Lett. 69B (1977) 445.

[33] S. Coleman, Nucl. Phys. B310 (1988) 643.

[34] S.W. Hawking, Phys. Lett. 195B (1987) 337; Phys. Rev. D37 (1988) 907;
G.V. Lavrelashvili, V.A. Rubakov and P.G. Tinyakov, JETP Lett. 46 (1987) 167; Nucl. Phys. B299 (1988) 757;
S. Giddings and A. Strominger, Nucl. Phys. B307 (1988) 854;
S. Coleman, Nucl. Phys. B307 (1988) 867;
S.W. Hawking and R. Laflamme, Phys. Lett. 209B (1989) 39;
I. Klebanov, L. Susskind and T. Banks, Nucl. Phys. B317 (1989) 665.

[35] G. Gilbert, "Wormhole-Induced Proton Decay", Caltech Preprint 68-1524(88).

[36] K. Choi and R. Holman, Phys. Rev. Lett. 62(1989)2575.

[37] O. Bertolami, "Wormhole Solutions of the Einstein-Yang-Mills Systems with the Inclusion of Fermions" Lisbon Preprint IFM - 14/90, to appear in the Proc. of the XVIII International Colloquium on Group Theoretical Methods in Physics, Moscow, June 1990.

[38] O. Bertolami, J.M. Mourão, R.F. Picken and I.P. Volobujev, work in progress.

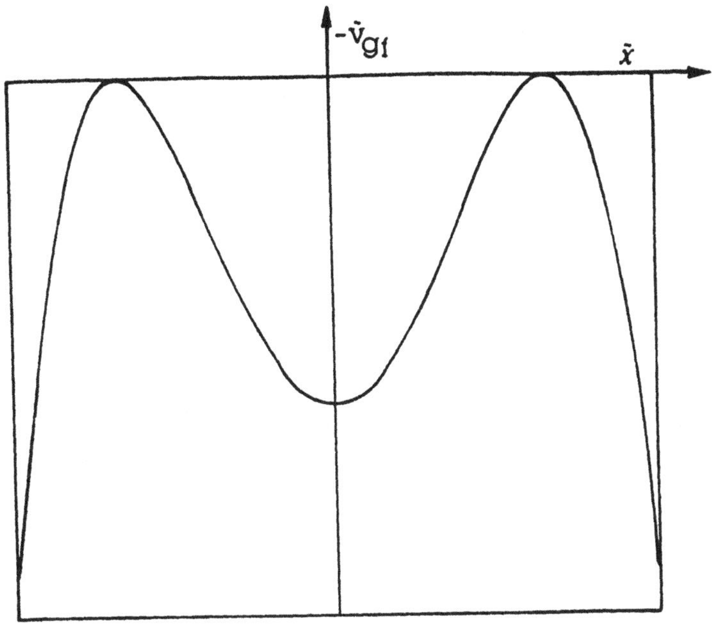

Figure I

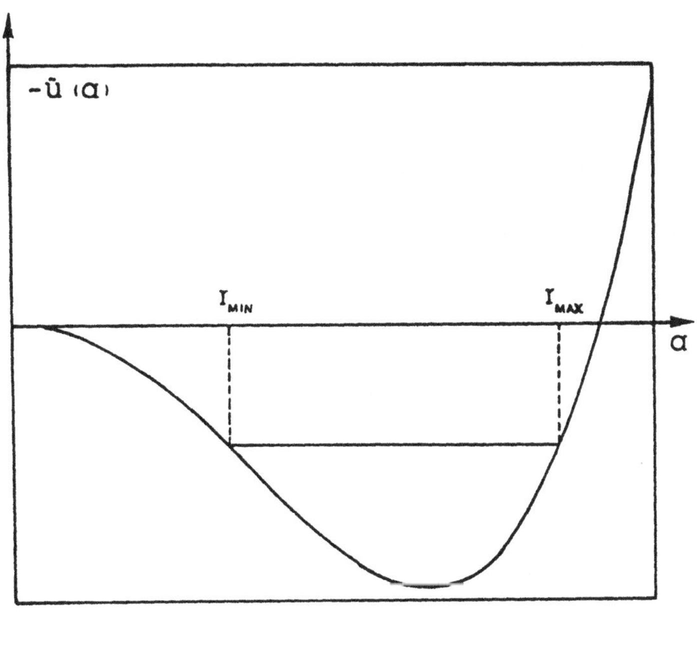

Figure II

Figure III

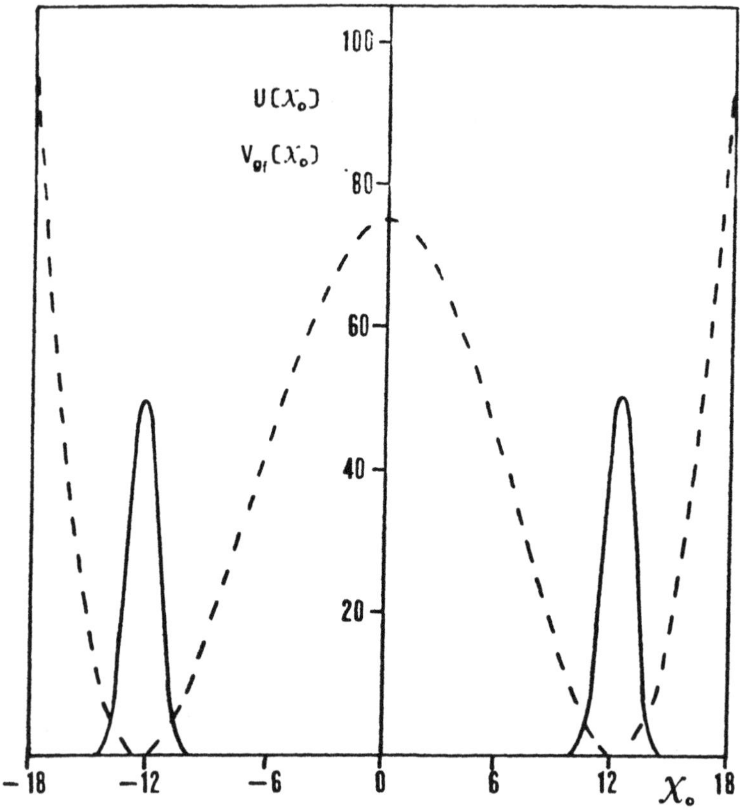

Ground-state wave function for A = 50 and $\frac{\alpha}{\pi} = 0.01$

EXPERIMENTS WITH NEUTRINOS

Paula Bordalo
LIP/IST

Laboratório de Instrumentação e Física Experimental de Partículas
Av. Elias Garcia 14 1^0
1000 Lisboa
PORTUGAL

ABSTRACT

Non-accelerator experiments envolving neutrinos are briefly presented. Accent is put on neutrino mass measurements in β and double β decays, and on the observation of solar neutrinos.

1. Introduction

The questions of neutrino mass and of solar neutrino flux are very important, both in High Energy Physics (HEP) and Cosmology . In fact, dark matter could be explained attributing to the neutrino mass a few eV and in Grand Unified Theories (GUT) the relations between quarks and leptons do impose a small mass for the neutrino. The solar neutrino flux is calculate in the Standard Solar Model (SSM), under the assumption that the energy source of the sun is nuclear fusion; one of the tests of SSM is the measurement of this flux; on the other hand, solar neutrinos properties can be studied in HEP by their interaction with matter.

Neutrino sources can be natural: formed in nuclear reactions inside the sun or due to the explosion of, stars such as supernova SN1987A [1]; or created in accelerators, as refered in these proceedings[2]. Here we will focus on some non-accelerator electron neutrinos experiments.

This contribution does not pretend to review such subjects, but rather use them as a motivation to present a few experiments and describe their detection methods.

In section 2 we refer two types of experiments which purpose is the measurement of electron anti-neutrino mass (m_ν) by β decay and double β decay, and we show the principles of Time Projection Chambers (TPC).

The measurement of solar neutrino flux is discussed in section 3, being ilustrated with neutrino ^{37}Cl capture and neutrino-electron elastic scattering experiments; a brief description on the principles of Cherenkov detectors is presented.

In section 4 we summarise.

2. M_ν measurements

Neutrino interacts with matter via the weak interaction. This means that no direct measurement of the m_ν is possible; up to now all measurements performed are indirect, such as β decay and double β decay.

2.1 β decay

In β decay, the electron spectrum end-point is dependent on the $\bar{\nu}_e$ mass (Fig. 1). All results come from tritium β decay: $^3H \rightarrow ^3 He + e^- + \bar{\nu}_e$, but the major problem of these experiments is the use of tritium in a surrounding environment, such as valine ($C_5H_{11}NO_2$).

An example is the Moscow experiment, which has been running during the last decade, and recently presented the very accurate value (consisten with its previous one) of $m_\nu = 30.3 \pm^2_8 \ eV$ [3].

This value is in contradiction with a very recent result from the Los Alamos experiment which uses molecular tritium ($^3H_2$) and obtained an upper limit of $m_\nu < 9.4 \ eV$ corresponding to 95% confidence level[4].

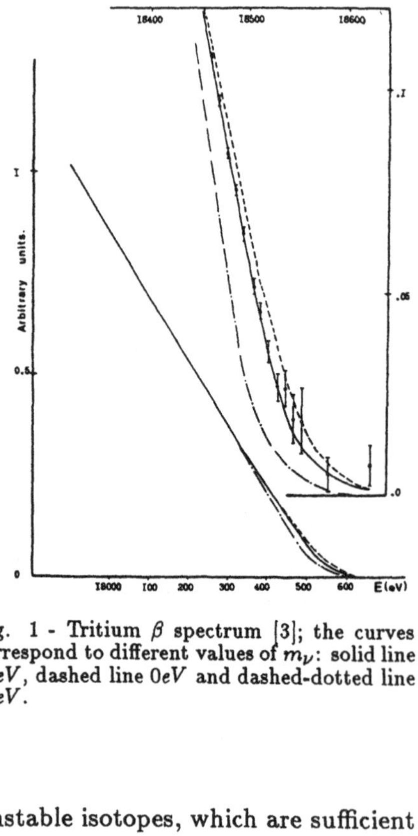

Fig. 1 - Tritium β spectrum [3]; the curves correspond to different values of m_ν: solid line $37eV$, dashed line $0eV$ and dashed-dotted line $80eV$.

2.2 Double β decay

Double β decay is a rare process occuring in unstable isotopes, which are sufficient stable against single β decay; there are several double β decay modes:

a) $2\nu\beta\beta$ – two neutrino emission double β decay mode,

$$(A, Z) \rightarrow (A, Z+2) + 2e^- + 2\bar{\nu} \ ,$$

that occurs via the second order perturbation of the weak interaction within the standard model;

b) $0\nu\beta\beta$ – neutrinoless double β decay mode,

$$(A, Z) \rightarrow (A, Z+2) + 2e^- \ ,$$

which violates lepton number conservation and demands neutrino masses and right handed currents; this is possible if the neutrinos are majorana particles ($\nu_e \equiv \bar{\nu}_e$);

c) $0\nu\chi$ – neutrinoless mode with a majoron emisson,

$$(A, Z) \rightarrow (A, Z+2) + 2e^- + \chi \ ,$$

where the majoron χ is a massless scalar boson arising from a baryon-lepton symmetry breaking.

Fig. 2 shows a two electron energy spectrum concerning each decay mode; a narrow pic due to $0\nu\beta\beta$ should be observed.

The $0\nu\beta\beta$ lifetime is related to the m_ν, to the transition energy between isotopes, and to the matrix element describing the interaction between nucleons in the nucleus. On the other hand, the $2\nu\beta\beta$ lifetime depends on the transition energy and on the matrix element. So, from the lifetime measurement of these two decay, modes one can extract the m_ν value.

Many experiments devoted to these measurements used as unstable isotopes mainly ^{76}Ge, ^{82}Se, ^{100}Mo and ^{136}Xe, but most of them didn't observe these decay modes, altogether leading to lifetime lower limits of $\sim 10^{23}yr$ for $0\nu\beta\beta$ and $\sim 10^{20}yr$ for $2\nu\beta\beta$ decays. These limits can constrain m_ν; in particular, a lifetime $> 8 \cdot 10^{23}yr$ corresponds to $m_\nu < 1eV$.

In Fig. 3 we show a result obtained by the UC Irvine group representing the energy spectrum corresponding to $2\nu\beta\beta$ [5] using ^{82}Se and a time projection chamber (TPC). They measure a lifetime of $(1.1\pm^{0.8}_{0.3}) \cdot 10^{20}yr$ within a 68% confidence level.

Also a $\beta\beta$ decay with majoron emission was claimed with a lifetime $(6\pm1)\cdot10^{20}yr$ in a ^{76}Ge experiment[6], but has not been confirmed by the other germanium experiments.

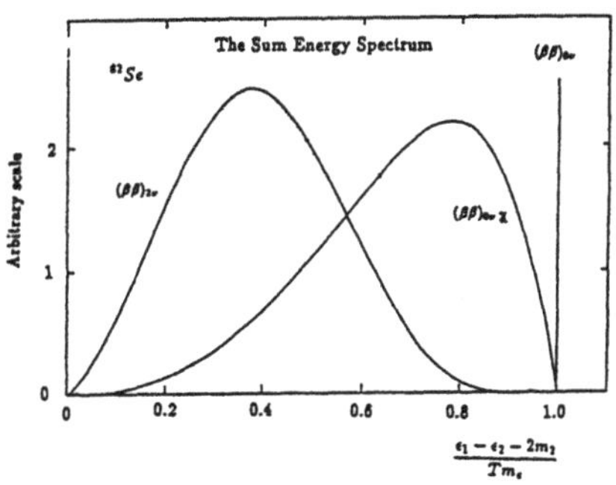

Fig. 2 - $\beta\beta$ energy spectrum.

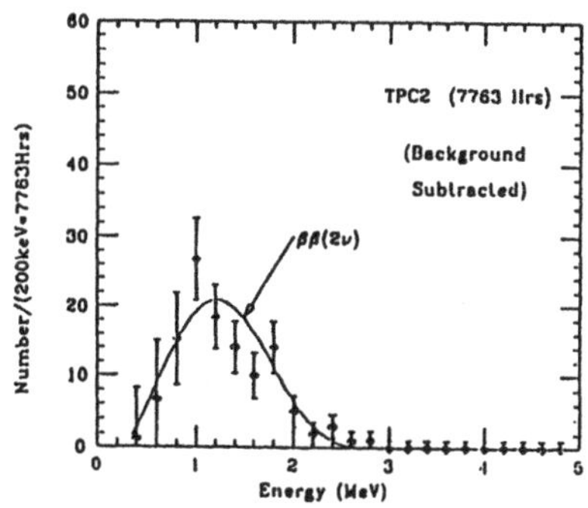

Fig. 3 - UCI sum energy spectrum [5].

2.2.1 Time Projection Chamber

A TPC is a 3-dimensional tracking detector (an electronic-like bubble chamber): it gives the position of many space points of a particle track projected on to the endcap plane.

In Fig. 4a we show a TPC scheme consisting of a tube filled with gas, a high voltage cathode plane placed in the middle, and anode planes formed by multiwire proportional chambers (MWPC) followed by padded cathodes at the endcaps. The resulting electric field must be uniform, so that electrons (created by the gas ionization due to charged particles) drift with a constant velocity.

The 3 coordinates necessary to the definition of a space point, are given by (see Fig. 4b):

x – the firing wire,
y – the signal induced on the raw pads,
z – the drift time of ionization electrons.

The UCI experiment used a 97% enriched ^{82}Se foil as cathode electrode, and measured the energy spectrum of the β rays emitted in the process $^{82}Se \rightarrow\ ^{82}Kr + 2e^- + 2\bar{\nu}_e$ by analysis of the electron trajectory curvature in the magnetic field supplied to the TPC. Its energy resolution is $\Delta E/E\ \sim\ 0.12$.

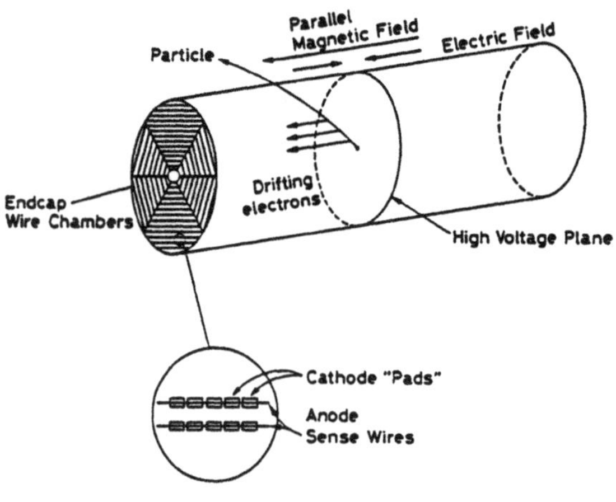

Fig. 4a - TPC scheme.

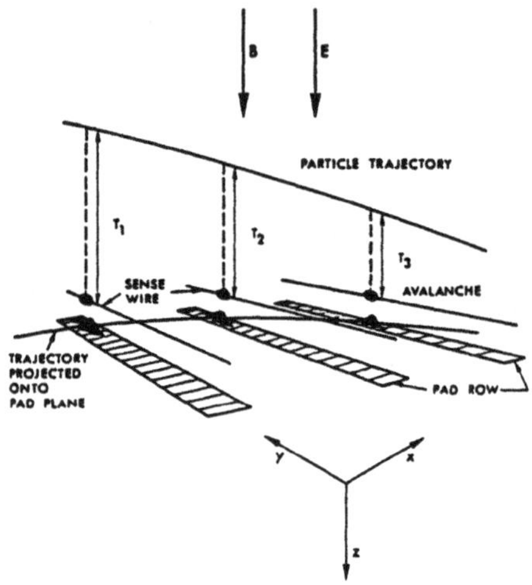

Fig. 4b - The end-cap MWPC scheme.

3. Solar neutrino flux

Solar ν_e flux experimental results are lower by a factor 2-4 than the predicted by the SSM. This ν_e depletion can be explained by one of the three main explanations: the sun temperature is lower than the one used in SSM; neutrinos are massive and ν_e can oscilate to another type (ν_μ *or* ν_τ); or due to the MSW mechanism[7].

Fig. 5 shows the solar neutrino energy spectra from nuclear fusions of various origins, namely the pp chain and the CNO cycle[8].

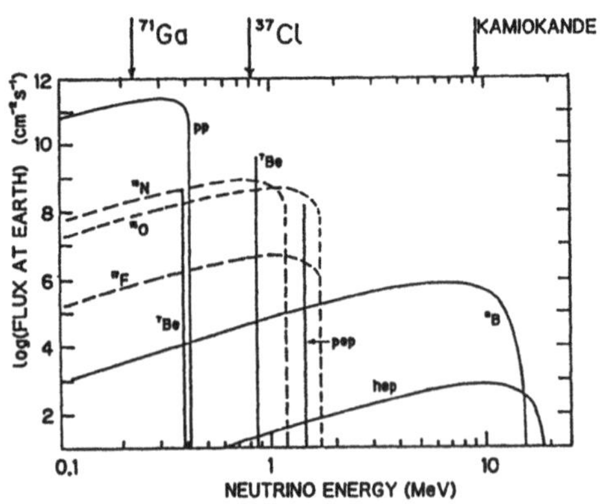

Fig. 5 - Solar neutrinos energy spectra in SSM from various processes; the threshold energies for three experiments are quoted.

The pioneering experiment on solar neutrinos is being performed by R.Davis since 1970 and uses the ν_e induced nuclear reaction of C_2Cl_4: $^{37}Cl + \nu_e \rightarrow^{37} Ar + e^-$ [10].

Its 814 keV threshold energy is specially sensitive to the 860 KeV ν_e coming from $e^- + {}^7Be \rightarrow {}^7Li + \nu_e$ and to the $0 - 14$ MeV ν_e continuous-spectrum issued from two processes: $p + {}^7Be \rightarrow {}^8B + \gamma$ and ${}^8B \rightarrow {}^8Be^* + e^+ + \nu_e$. The results as a function of time are shown in Fig.6. The average event rate of ν_e interactions in this experiment, concerning the total period (from 1970 to 1988), is 2.33 ± 0.25 SNU [11]; the flux derived from SSM is $7.9 \pm 2.6(3\sigma)$ SNU [8] or $5.8 \pm 1.2(1\sigma)$ SNU [9] depending on the model detail calculations.

Recently, data was taken from 1987 to 1990 using the Kamiokanda II Ring Imaging Water Cherenkov detector[12]. This experiment, having a high energetic threshold, is sensitive to the ${}^8B$ solar high energetic ν_e by means of neutrino-electron elastic scattering $\nu_e + e^- \rightarrow \nu_e + e^-$. Background subtracted data are displayed in Fig.7 and compared to the SSM prediction[8]. Their ν_e flux value relative to SSM ones, obtained with two calculations[8,9], is $0.46 \pm 0.05(stat) \pm 0.06(syst)$ and $0.70 \pm 0.08(stat) \pm 0.09(syst)$ respectively.

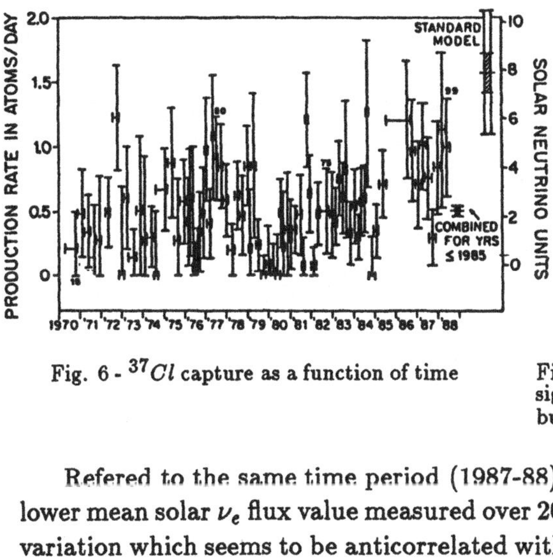

Fig. 6 - ${}^{37}Cl$ capture as a function of time

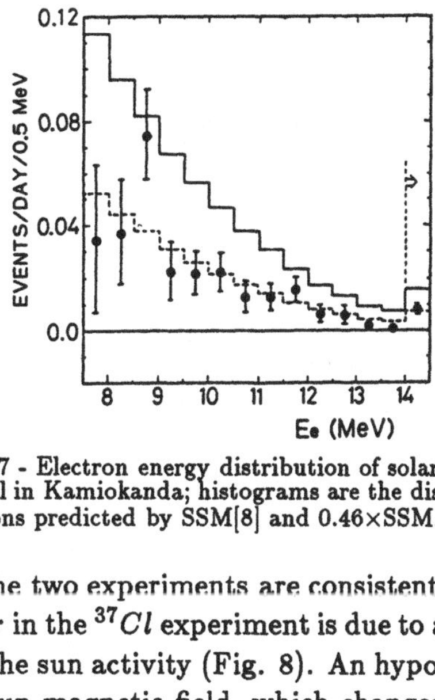

Fig. 7 - Electron energy distribution of solar ν_e signal in Kamiokanda; histograms are the distributions predicted by SSM[8] and 0.46×SSM.

Refered to the same time period (1987-88) the two experiments are consistent. The lower mean solar ν_e flux value measured over $20yr$ in the ${}^{37}Cl$ experiment is due to a time variation which seems to be anticorrelated with the sun activity (Fig. 8). An hypothesis of the ν_e magnetic moment coupling with the sun magnetic field, which changes with the sun activity, was proposed to explain this time evolution[13]; but it fails because the variation affects mainly the low energy neutrinos and day-night effects were not seen. On the other hand, no significant time variation is observed in the Kamiokanda data.

There are other new experiments using namely ${}^{71}Ga$ as neutrino target [14] and having a very low threshold ($E_\nu > 233$ keV), sensitive to ν_e coming from the center of the sun (a part of the pp solar neutrino spectrum: $0 < E_\nu < 420$ keV which production is fixed by the solar luminosity). Preliminary results[15] from SAGE experiment, measuring 90% of ν_e total flux, suggest a lower value than the predicted by SSM[8]: $132\pm^{20}_{17}$ SNU.

The major problem of all the experiments being the radioactive background, one needs a next generation of low background higher sensitive and better energy resolution detectors[16] besides more statistic.

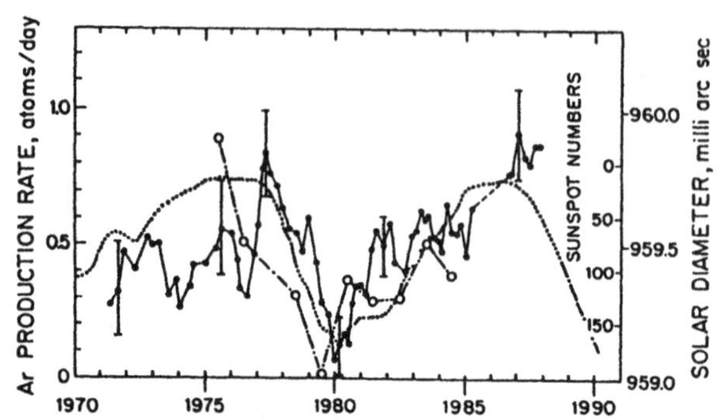

Fig. 8 - Comparasion of the capture rate with sunspot numbers. Solid circle: five-point running average; dotted curve: sunspot numbers; open circles: solar diameter.

3.1 Cherenkov detectors

Cherenkov counters identify charged particles, $e^{\pm}$, $\pi^{\pm}$, $k^{\pm}$, p, $\bar{p}$, ... in a given momentum range. They are based on the Cherenkov effect which is a radiation emission at fixed angle (θ_C) when the velocity of a charged particle (βc) exceedes the velocity of the light (c/n) in the propagation medium[17]. The characteristic angle (θ_C) is derived from the relation:

$$cos\ \theta_C = 1/\beta n \qquad \text{, whith } \beta > 1/n \ .$$

There are many Cherenkov detector types, but in general (see Fig. 9) all of them are made of a tank, filled with gas, with mirrors at the foward edge for the light collection by photomultipliers (PM).

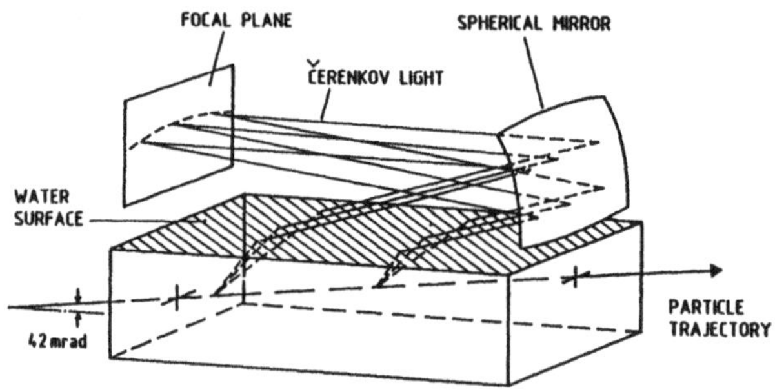

Fig. 9 - Cherenkov detector scheme.

a) Threshold Cherenkov — gives light if $\beta > 1/n$, n being dependent on the nature, pressure and temperature of the radiator; particles can only be identified with a series of them tuned to different thresholds.

b) Differential Cherenkov — measures θ_C by using a diaphragm, and so gives the velocity of the particle in a medium.

c) DISC - Differential Isochronus Self-Collimating: by changing the diaphragm radius and the gas pressure one can perform a scan on the velocity.

d) RICH – Ring Imaging CHerenkov[18] is the Cherenkov used nowadays; it has a large phase-space acceptance and gives high resolution velocity. The ring image formation principle is shown in Fig. 10a. The light emitted, reflected by a mirror, is focalized at the detector (PMs), which forms a ring image (Fig. 10b) of a radius r and is placed at the mirror focal plane:

$$tan\ \theta_D = 2r/R_M\ , \qquad \text{whith}\ \theta_D \simeq \theta_C$$

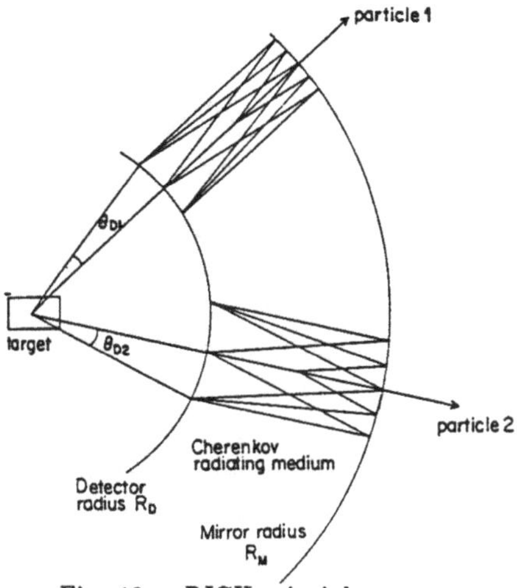

Fig. 10a - RICH principle.

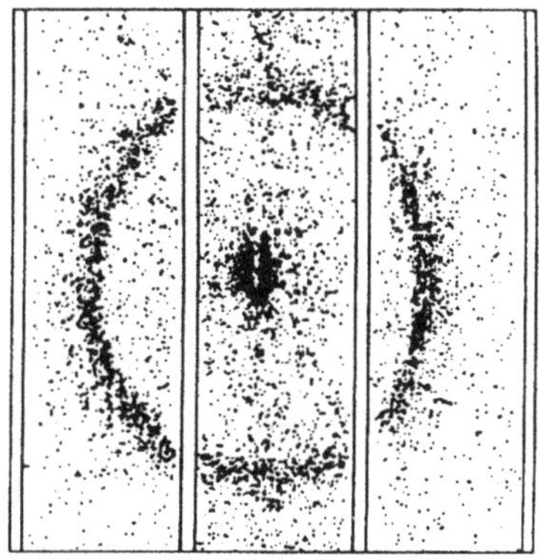
Fig. 10b - Typical ring image.

4. Conclusions

A large number of non-accelerator experiments concerning m_ν and solar ν flux measurements are running in order to give a contribution to HEP and Cosmological problems.

Until now, only one tritium β decay experiment claimed a nonzero mass for the neutrino ($m_\nu = 30.3 \pm^2_8\ eV$), which is in contradiction to the very low upper limit ($m_\nu < 9.4\ eV$) of another experiment; it seems that the discrepancy is due two the difficult treatment of source environment effects.

There is no evidence for $0\nu\beta\beta$ and for $\chi\beta\beta$ decays, and just one experiment, which distinguishes the two e^- by means of a TPC, observed $2\nu\beta\beta$ decay with a lifetime of $(1.1\pm^{0.8}_{0.3}) \cdot 10^{20} yr$.

After 20 years of solar ν flux measurements, statistics are still very poor; nevertheless, three different energy threshold experiments give a ν rate deficit, as compared to SSM, indicating that the problem may be not solved within Cosmology alone.

References

[1] K. Hirata et al., Phys. Rev. Lett. 58 (1987) 1490.
R.M. Bionta et al., Phys. Rev. Lett. 58 (1987) 1494.

[2] See, for example, M. Pimenta "Experiments in HEP ..." in these Proceedings.

[3] V.A. Lubimov et al., Phys. Rev. Lett. 94B (1980) 266.
S. Boris et al., Phys. Rev. Lett. 58 (1987) 2019.

[4] Oral presentation at Int. Conf. on HEP, 1990, Singapour;
J.F. Wilkerson et al., Phys. Rev. Lett. 58 (1987) 2023.
[5] S.R. Elliott et al., Phys. Rev. Lett. 59 (1987) 2020.
[6] F.T. Avignone III et al., presented at Int. Conf. Neutrino 88, 1988, Boston.
[7] S.P. Mikheyev and A.Yu. Smirnov, Nuovo Cim. 9C (1986) 17;
L. Wolfenstein, Phys. Rev. Lett. D17 (1978);
H.A. Bethe, Phys. Rev. Lett. 56 (1986) 1305.
[8] J.N. Bachal and R.K. Ulrich, Rev. Mod. Phys. 60 (1988) 297.
[9] S. Turck-Chièze et al., Astrophys. J. 335 (1988) 415.
[10] R. Davis et al., Phys. Rev. Lett. 20 (1968) 1205.
[11] R. Davis et al., Annu. Rev. Part. Sci., 39 (1989) 467.
[12] K.S. Hirata et al., Phys. Rev. Lett. 63 (1989) 16; 65 (1990) 1297; 65 (1990) 1301.
[13] M.B. Voloshin et al., Sov. Phys. JETP, 64 (1986) 446.
[14] E. Fiorini, Nucl. Phys. B (Proc. suppl.) 16 (1990) 479;
V.N. Gavrin et al., SAGE coll., Nucl. Phys. B (Proc. suppl.) 16 (1990) 483;
T. Kirsten et al., GALLEX coll., in Proc. of 9th Workshop on Grand Unification, 1988, Aix-les-Bains.
[15] R.S. Raghavan, presentation at the 25th Int. Conf. on HEP, 1990, Singapour.
[16] L. Gonzalez-Mestres et al., in Proc. of $24th$. Rencontre de Moriond, March 1989.
[17] see for instance, R.C. Fernow, "Introduction to experimental particle physics", Cambridge Univ. Press, 1986.
[18] see for instance, J. Seguinot, "les compteurs Cherenkov ...", CERN-EP/89-92.

Topological Defects in the Early Universe

Edmund Copeland

Division of Physics and Astronomy
University of Sussex
Brighton BN1 9QH
U.K.

I. The Formation of Defects in the Early Universe

This is an intriguing time to be working in the area of Cosmology. Over the past decade or so we have come to recognise that particle physics could be playing an important role in the evolution of and the formation of structure in our universe (for an excellent up to date review of its impact see the textbook of Kolb and Turner [1]). Two of the most promising particle physics inspired candidates have been the Inflationary Universe Scenario[2](see Barrow's lectures in these proceedings[3]), and the role of Topological defects in the early universe[4, 5, 6]. However the excitement generated by the models that have come out of the particle physics factory has recently been matched by the astronomical community as extremely tight observational constraints are now being placed on many of the theories. They have been systematically probing the very large scale structure of our universe, and have made observations that theorists are now struggling to explain. For example, it is still not clear what causal mechanism could produce the primordial energy density perturbations which are thought of as necessary to seed galaxies and clusters of galaxies. How could the inhomogeneity represented by this large scale structure and the galaxies be reconciled with the observed smoothness of the microwave background radiation?

A recent all-sky redshift survey of galaxies detected by IRAS (the InfraRed Astronomical Satellite) has been used to map out the universe to a distance of $140h^{-1}Mpc$ ($h = H_o/100kms^{-1}Mpc^{-1}$)[7]. Their main conclusions are first that the underlying density field is skewed to high densities even when smoothed on large scales, with the voids larger and less extreme than the superclusters, and second there is more power on large scales than is predicted in the standard Cold Dark Matter theory of galaxy formation, so much so that the authors claim to rule out the standard CDM model to at least the 97% confidence level. This inturn means that the 'standard' scenario of galaxy formation, that of Inflation and CDM is probably ruled out, although there is alot of work going on to keep it alive[8]. In Efstathiou et al.[8] their way to satisfy the new observational bounds and have more structure on large scales is to drop the old caveat of a zero cosmological constant, and instead to have a spatially flat universe, with up to 80% of the critical density provided by a positive cosmological constant. This is not being included as a criticism, but to emphasize that at the moment, there are many puzzling results coming in from the observers which need to be explained.

What is required from theory is some source of the perturbations essential to produce the structures we see today. Amongst the more unusual large scale features are giant 'filaments' (linear overdense regions in the galaxy distribution, about $200h_{50}^{-1}$ Megaparsecs long and $10h_{50}^{-1}$Mpc across[9], large 'voids', empty of bright galaxies, $120h_{50}^{-1}$Mpc in diameter, and galaxies lying on the surface of 'bubbles' 40-$60h_{50}^{-1}$Mpc across[10], (h_{50} is Hubbles constant in units of $50kms^{-1}$). An Abell cluster is defined to be a region smaller than $3h_{50}^{-1}$Mpc in radius containing more than 50 bright galaxies. These clusters appear to be clustered on scales of $100h_{50}^{-1}$Mpc with a mean separation of $110h_{50}^{-1}$Mpc. Just as fascinating are the recent results of Broadhurst et al[11], who with recent pencil beam surveys detect what appears to be periodic structure (peaks in the galaxy distribution) roughly every $130h^{-1}Mpc$, in observations through the north and south poles of our galactic plane. Time and more measurements will determine how characteristic this is of our universe, but it has certainly got the theorists working hard to explain the observations[12].

Whatever it was that produced the primordial density perturbations it had to do so in such a way that at the time that radiation decoupled from matter, there were very small density perturbations in the radiation. The recent results of COBE provide probably the most serious challenge to all the theories of galaxy formation[13]. They show that the microwave background is smooth to at least one part in 10^4 on angular scales of say 10 degrees. Just about every theory of large scale structure formation suggests that anisotropies should show up in the background at or slightly below this level, including the models I will be discussing in these lectures (late time phase transitions which produce light domain walls can avoid the microwave background constraint, as the walls are formed after decoupling [14]). Therefore they could all be in trouble soon, or of course, one of them could turn up trumps!

Now it appears that gravity alone could not have moved galaxies and led to such large scale structure since the big bang. Turok[15] explains this in a succinct argument which I will follow here. Peculiar velocities (velocities relative to the Hubble flow) grow as $t^{\frac{1}{3}}$ in an expanding universe. As we shall see in the third lecture, in the linear regime, $\frac{\delta\rho}{\rho} < 1$, there is a precise relation, $\delta r = H_0^{-1}\delta v$ where δr is the peculiar displacement, δv is the peculiar velocity and the Hubble radius H_0^{-1}, (characterising the expansion rate of the universe)$\equiv 6000h_{50}^{-1}$Mpc. Observational limits placed on the galaxy peculiar velocities are $\delta v_{galaxy} < 600kms^{-1}$, which implies $\delta r \equiv 12h_{50}^{-1}$Mpc, yet structures form on scales some 20 times larger than that. It appears then that by investigating these large scale structures we are looking directly at the primordial density perturbations. The interface of particle physics and cosmology has provided us with one of the most intriguing possible solutions to this problem. The idea that physical processes occurring just 10^{-35} seconds after the initial big bang should directly determine the structures being observed some 15 billion years later is staggering in the extreme.

The hot big bang theory of the early universe successfully predicts the Hubble expansion, the microwave background radiation and the light element abundances. It appears to fit in nicely with ideas of particle physics where, as the energy is increased so is the degree of symmetry used to describe the particle interactions. At high enough energies (i.e the very early universe when the temperature was very hot) we find the universe in the state of maximum symmetry. This is spontaneously broken as the universe expands and cools through some critical temperature. In this lecture we will discuss the ideas behind the formation of topological defects in the early universe. The reader should be aware that there will be quite alot of field theory in this section, which is inevitable as the defects are themselves field theoretic objects. However we will explain at each stage the relevence of each new step. In lecture 2 we will discuss the actual dynamical evolution of the network of cosmic strings and global textures. In lecture 3, we will describe how initial density perturbations grow around strings and global textures emphasizing the Cold Dark Matter (CDM) scenario how it leads to structure formation and importantly to observational tests of the theories. In these lectures we will not be discussing the exciting model due to Witten[16] of superconducting cosmic strings which was proposed as seeds for the large scale structure by Ostriker et al.[18]. For a review of that area see[17]. The possibility of superconducting cosmic strings has generated a great deal of interest in the dynamics of such strings. Some of the nicest recent work is due to Carter[19]. He has used the fact that stable loops of string can be formed due to the balance of the string tension and the current flowing through the string[20, 21]. Infact Carter has speculated that such 'cosmic chumps' could form the dark matter in the universe.

Phase transitions are already known to occur in particle physics. QCD (confinement) occurs at around $1GeV$, the Electroweak phase transition occurs at around $250GeV$. It is generally believed within the particle physics community that the early universe was characterised by a series of such phase transitions. In particular it is thought that Grand Unified Theories (GUT) could have had transitions around $10^{15}GeV$, during which a Higgs field ϕ tended to fall towards the minima of its potential as the universe cooled. As an example, in $\lambda|\phi^4|$ theory, the lagrangian is given by:

$$\mathcal{L} = \frac{1}{2}|\partial_\mu\phi|^2 + \frac{1}{2}m_0^2|\phi|^2 - \frac{\lambda}{4!}|\phi|^4 \qquad , m_0^2 > 0 \tag{1.1}$$

This is the familiar Mexican hat potential for a complex scalar field. At $T > T_c$, the critical temperature, the fields are in the symmetric 'false vacuum' phase with $< |\phi| >= 0$. As the universe expands and cools through T_c, ϕ rolls to the bottom of the potential developing an expectation value $< |\phi| >^2 = \frac{6m_0^2}{\lambda}$, thereby spontaneously breaking the symmetry. In fact it is possible to show[22] that the effective mass of the scalar field vanishes at T_c when the symmetry is once again restored. The partition function for the system in thermal equilibrium at temperature T is

$$Z = e^{-\beta F}$$

$$\propto \int \mathcal{D}\phi\mathcal{D}\phi^\star e^{-I_\beta[|\phi|]}$$

where $\beta = \frac{1}{k_B T}$, $k_B = 1$ is Boltzmanns constant, F is the Helmholtz Free Energy and $I_\beta[|\phi|]$ is the Euclidean form of the action obtained by making the time periodic in imaginary time τ with period β.

$$\begin{aligned} I_\beta[|\phi|] &= \int_0^\beta d\tau \int d^3\underline{x}\left(-\frac{1}{2} \mid \partial_\mu\phi \mid^2 - \frac{1}{2}m_0^2 \mid \phi \mid^2 + \frac{\lambda}{4!} \mid \phi \mid^4\right) \\ \phi(\tau, \underline{x}) &= \sum_n \phi_n(\underline{x}) e^{\frac{2\pi i n \tau}{\beta}} \end{aligned} \tag{1.2}$$

At low temperatures we can expand ϕ in small fluctuations about one of it's $T = 0$ minima, valid because the field is barely excited. The mass of the fluctuations about this $T = 0$ vacuum $\phi_o = \sqrt{\frac{6m_0^2}{\lambda}}$ is just $M^2 \equiv \frac{d^2V}{d\phi^2}|_{\phi_o} = 2m^2$.

In the high temperature regime, we can expand about some constant field solution ϕ, and find when $\frac{m_0}{T} \ll 1$ (see Dolan and Jackiw in[22])

$$\begin{aligned} F(|\phi|, T) &= -\frac{1}{2}m_0^2|\phi|^2 + \frac{\lambda}{4!}|\phi|^4 + \frac{\lambda}{48}T^2|\phi|^2 - \frac{m_0^2}{24}T^2 - \frac{\pi^2}{90}T^4 \\ &\quad + O(\frac{m_0}{T})m_0^2T^2 \end{aligned}$$

The result is a temperature dependent effective potential of the form

$$V_T(|\phi|) = -\frac{1}{2}m^2(T)|\phi|^2 + \frac{\lambda}{4!}|\phi|^4 \tag{1.3}$$

with

$$\begin{aligned} T_c^2 &= 24\frac{m_0^2}{\lambda} \\ m^2(T) &= m_0^2[1 - \frac{T^2}{T_c^2}] \\ < |\phi| >^2 &= 6\frac{m^2(T)}{\lambda} \end{aligned} \tag{1.4}$$

Note the behaviour of the ϕ field as T_c is approached from below, and the symmetry is restored to $< |\phi| >= 0$.

We will now evaluate the spatial correlation of the ϕ field, as it determines the scale of fluctuations in the field, a crucial ingredient in explaining the role of particle physics in the large scale structure of the universe[23] (For a very interesting recent paper which investigates the nature of first and second order phase transitions and the associated correlation regions see Gleiser[24]). Consider first the case of a second order phase transition. The minimum work required to bring the system out of equilibrium for constant pressure and temperature is the difference in the free energy, between the two vacuum states, ΔF, with a corresponding fluctuation probability

$w \propto exp(-\beta\Delta F)$

We concentrate on long wavelength fluctuations in which the ϕ field varies slowly across space. These fluctuations increase anomalously near the transition point. For the case of fluctuations in the symmetric phase, $<\phi>=0, \Delta\phi=\phi$, then to $O(\phi^2)$ we find that the change in the free energy is (from (1.1))

$$\Delta F = \int [\frac{m^2(T)}{2}|\phi|^2 + \frac{1}{2}|\partial_\mu\phi|^2] d^3x \tag{1.5}$$

Now expanding $\phi(r)$ as a Fourier series in a volume V

$\phi = \sum_k \phi_k e^{ik.r} \quad ,\phi_{-k} = \phi_k^*$

we find

$$\Delta F = \frac{V}{2}\sum_k [k^2 + m^2(T)] \mid \phi_k \mid^2 \text{ hence } \qquad <\mid \phi_k \mid^2> \simeq \frac{T}{V(k^2+m^2(T))} \tag{1.6}$$

Note as $T \to T_c$ the long wavelength fluctuations increase. Writing the correlation function as

$$G(r) = <\phi(r_1)\phi(r_2)>, \qquad r = r_1 - r_2 \tag{1.7}$$

we use

$$\begin{aligned} G(r) &= \sum_k <\mid \phi_k \mid^2> e^{ik.r} \\ &= \int <\mid \phi_k \mid^2> e^{ik.r} V \frac{d^3k}{(2\pi)^3} \end{aligned} \tag{1.8}$$

Substituting (1.6) into (1.8) we obtain:

$$\begin{aligned} G(r) &\simeq \frac{T_c}{4\pi r} exp(-\frac{r}{\xi}), \qquad r \gg \xi \\ G(r) &\simeq \frac{T^2}{2\pi^2}, \qquad r \ll \xi \end{aligned} \tag{1.9}$$

where

$$\xi^{-1} = m(T) \simeq \sqrt{\lambda}\,|<\phi>| \tag{1.10}$$

is the correlation length.[1] This implies that domains form of size $\xi \sim m^{-1}$ inside which ϕ is correlated, but outside of which there are no correlations. However as $T \to T_c$, $m \to 0$, $\xi \to \infty$, it appears the domains vanish as all the fields become correlated. There is though an upper bound to the correlation length; from cosmology $\xi < H_0^{-1}(t) \sim t$, the distance over which microscopical forces can establish correlations in one Hubble expansion time. In fact there is a

[1] If gauge fields are present, as in superconductors, there is another relevant length which determines the spatial correlations between the fields. It is the London penetration depth and defines the distance the $\underline{B}$ field penetrates the surface of the superconductor. $\delta^{-1} = e\,|<\phi>|\simeq m_v(T)$. This scale will prove important later on.

tighter constraint which sets the scale over which the domains form. As they form and $< \phi > \sim 0$, there is a possibility that thermal fluctuations in the ϕ field could cause ϕ to return to its false vacuum value, hence wipe out the domains. The free energy associated with such a fluctuation with scale ξ is, using the free energy per unit volume f,

$$\begin{aligned} (2\xi)^3 \Delta f &\simeq m^{-3}(T) \frac{m^4(T)}{\lambda} \\ &\sim \frac{m(T)}{\lambda} \end{aligned} \tag{1.11}$$

The fluctuation has a high probability so long as the free energy required is $\ll$ thermal energy available (T). The two are equal when $\frac{m(T)}{\lambda} \sim T$

$$\frac{1}{T^2} - \frac{1}{T_c^2} \sim \frac{\lambda^2}{m_0^2},$$

(i.e. for small λ, domains set in when $T \sim T_c$). In fact we can then see that at thermal equilibrium: $[\xi \sim [\sqrt{\lambda} < \phi >]^{-1} \sim [\lambda T_c]^{-1}]$ which is the Ginzburg length found in superconductivity.

The second case to consider is that of first order phase transitions such as those proposed originally in the context of Old Inflation[2] and more recently for Extended Inflation[25]. In this case the false and true vacuua are seprated by a potential barrier, causing the release of latent heat during the phase transition from false to true vacuum states. For large barrier heights the tunnelling rate is exponentially suppressed as the field remains stuck in the false vacuum. Eventually the barrier can become flat enough to allow the field to roll down to the true vacuum, and the scenario would proceed as in the previous example. However the phase transition can occur if tunnelling proceeds through the potential barrier, with the field then appearing out the other side of the barrier and rolling down to some point in the true minima. The result is then the formation of bubbles of true vacuum, the value of the field in each bubble being independent of the value of the field in any other bubble. These bubbles then expand at the speed of light, eventually coalescing. Thus again a distribution of domains occurs, the fields being correlated within the domains, but uncorrelated outside of them. This general feature of uncorrelated domains has become known as the Kibble mechanism[4]. It appears generic to most types of phase transition, and has recently been exploited to form topological defects at the end of a period of extended inflation[26].

Hopefully the above arguments are convincing enough to suggest to you that it is worth investigating models of the early universe. One such model was proposed by Kibble[4] in a paper that still remains one of the clearest in the field. At very early times the universe was very hot and the fields describing interactions were in a highly symmetric phase. However as the universe expanded and cooled, symmetry breaking processes would spontaneously occur, occasionally leaving behind remnants of the old symmetric phase, (topological defects), possibly in the form of one dimensional strings or vortex lines[27], two dimensional domain walls or more likely three dimensional monopole configurations[28]. Gibbons article in these proceedings is devoted to some extremely interesting new results concerning magnetic monopoles and global monopoles in gravitational fields [29]. In fact monopoles will always be produced in a GUT symmetry breaking scheme and this is what is commonly referred to as the 'monopole problem'. We don't see any monopoles today, however they would have been produced in a great number density in the early universe. As they are topologically stable, their only means of decay is through annihilation or possibly gravitational radiation[30]. These processes are too slow to rid us of all the monopoles today, so how can we reconcile these apparent discrepancies in the theories? Actually, although I won't go into further details, a possible method of eliminating the monopoles without resorting to inflating them away does exist; inflation would also wipe out any cosmic strings that had been produced prior to the inflationary period, which is bad news for the cosmic strings. It is possible to have a sequence of symmetry breakings which first produce monopoles and then attaches monopole/antimonopole pairs via strings[31]. The monopole flux is confined to exist only on the string, and they rapidly come to annihilate. A novel type of defect which could have interesting cosmological consequences are Global Textures [32, 33]. It is a three dimensional defect formed whenever a global non-abelian symmetry (i.e. SU(2)) is spontaneously and completely broken, preferably at the GUT scale. In theories where the non-abelian global symmetry is only partially broken–these theories produce global monopoles[34, 35] and non-topological textures[33]. Textures differ from conventional defects such as strings and monopoles in three main ways. The Higgs field breaking the symmetry remains in the vacuum everywhere, unlike the string and monopole. Thus the texture is not well localised in the same sense as a string or monopole; in those cases one can point to a locus in spacetime where $|\phi| = 0$ as the location of the defect. Perhaps the most intriguing difference is that a global texture is unstable to collapse in on itself, it unwinds leaving a trivial field configuration.

Returning to the condition for the existence of these topological defects[4], consider a gauge theory with a symmetry group G, this is the group whose elements leave the full potential V, invariant when acting on it. In the phase transition this group is broken to a sub group H as the fields acquire expectation values. H contains the elements of the original group G which when acting on the fields ϕ, leave them with their expectation value, $H \equiv \{g : g\phi = \phi\}$. In fact the manifold of degenerate vacuum states (the manifold corresponding to the state of least energy in the theory), is identified with the coset space: $M = \frac{G}{H}$ (i.e. $g' = gh$, $h \epsilon H$. Identify elements related by right multiplication by H).[2] What then does the topology of these various coset spaces look like? The defects arise by mapping M_0 into space in a non-trivial way, this is homotopy theory. The condition for the existence of strings is that the first homotopy group $\Pi_1(\frac{G}{H}) \equiv \Pi_1(M)$ be non trivial. The vacuum manifold M must contain non contractable loops.

[2]For example, if
(1) $G \equiv Z_2$, (i.e $V(\phi)$ invariant under $\phi \longrightarrow -\phi$), $H = 1, \Longrightarrow V_{min} = \frac{G}{H} = Z_2$
(2) $G = U(1), (i.e\phi \longrightarrow e^{i\theta}\phi), H = 1, \Longrightarrow V_{min} = \frac{G}{H} = U(1)$
(3) $G = SO(3), (i.e\phi \longrightarrow O_{ab}\phi_b), H = SO(2), \Longrightarrow V_{min} = \frac{G}{H} = S_2$

We will look at a specific example later, a more general class of models which contain non contractable loops has been established including some based on superstring theories[36]. Strings are not the only defects that form of course. In fact if $\frac{G}{H}$ is disconnected, then $\Pi_0(M)$ is non trivial and wall like defects form where $V \neq V_{min}$ inside the wall. If $\frac{G}{H}$ contains non contractable 2- spheres then $\Pi_2(M) \neq 1$ and the resulting defect is a monopole. If $\frac{G}{H}$ contain non-contractable three-spheres, then $\Pi_3(\frac{G}{H}) \neq 1$ and Texture defects can result[4, 32].

Apart from the possibility of forming domain walls during late time phase transitions, along with monopoles, domain walls are generally thought to be disastrous for cosmology. Considering first the domain wall solution. A real scalar field theory which permits wall like solutions is of the form

$$\mathcal{L} = \frac{1}{2}(\partial_\mu \phi)^2 + \frac{1}{2}m_0^2\phi^2 - \frac{\lambda}{4!}\phi^4 \tag{1.12}$$

This possesses the symmetry $(\phi \leftrightarrow -\phi)$. The Euler-Lagrange equation is

$$(\Box - m_0^2)\phi + \frac{\lambda}{6}\phi^3 = 0 \tag{1.13}$$

A static wall solution in the z-y plane is of the form

$$\phi_W(x, y, z) = \eta \tanh(\frac{m_0 x}{\sqrt{2}}), \qquad \eta^2 = \frac{6m_0^2}{\lambda} \tag{1.14}$$

The wall has thickness $\Delta_x \sim \frac{1}{m_0}$, hence a surface tension, $\sigma \sim \frac{m_0^3}{\lambda}$. The problem with GUT era domain walls arises from their contribution to the microwave background. In lecture 3, we will see that

$$\frac{\delta T}{T} \sim \frac{\delta M}{M_{HOR}} \tag{1.15}$$

where δT is the fluctuation in the background temperature of the universe, caused by a mass perturbation δM, and M_{HOR} is the mass in our horizon volume. For the domain wall if there is of order one wall per horizon, and the horizon is today of size t, then we obtain that

$$\frac{\delta T}{T} \sim \frac{\sigma t}{m_{pl}^2}$$

where m_{pl} is the Planck mass. The tight observational constraint that $\frac{\delta T}{T} < 10^{-4}$ then implies that symmetry breakings which make walls above an energy scale of order 10MeV are ruled out. GUT monopoles can also cause major problems for cosmology[30]. The number density of monopoles divided by the entropy density at the GUT era is

$$\frac{n_m}{s} \sim \frac{1}{\xi^3 T^3} |_{T_{GUT}} \tag{1.16}$$

$T_{GUT} \sim 10^{15} GeV$ corresponds to roughly one monopole per horizon volume. There is however a bound on the correlation size from causality, $\xi < t_{GUT} \sim \frac{m_{pl}}{T^2}$. This inturn then constrains $\frac{n_m}{s} \geq (\frac{T_{GUT}}{m_{pl}})^3 \sim 10^{-13}$. Rewriting this in terms of energy densities we obtain $\frac{\rho_{mon}}{\rho_{TOT}} \sim \frac{n_m \times m_{mon}}{T^4} \sim 10^{-13}(\frac{a}{a_{GUT}})$. The monopoles will rapidly come to dominate the energy density of the universe. This is why a mechanism for getting rid of them is required, or possibly of avoiding forming them in the first place.

For the rest of this and the other lectures we concentrate on the cosmologically more interesting cases of cosmic strings and global textures.

a) Cosmic Strings.

The most familiar strings, i.e flux tubes in superconductors correspond to the complete breaking of an Abelian group G=U(1)[27]. The lagrangian for the theory is:

$$\mathcal{L} = -\frac{1}{4}F_{\mu\nu}F^{\mu\nu} + \frac{1}{2} \mid D_\mu\phi \mid^2 - \frac{\lambda}{4!}(\mid \phi \mid^2 - \eta^2)^2 \tag{1.17}$$

where $F_{\mu\nu} = d_{[\mu}A_{\nu]}$, $D_\mu = \partial_\mu + ieA_\mu$, A_μ is the gauge field, e the gauge coupling constant, λ is the self coupling of the Higgs field and η is the value of the symmetry breaking Higgs field. The manifold M of ground states is a circle

$M = [\phi \mid \phi = \eta e^{i\theta}, 0 \leq \theta \leq 2\pi]$

i.e V is minimised by $\phi = \eta e^{i\theta}$ with θ arbitrary, corresponding to the winding number which is an integer. We have already seen how domains could be formed during a phase transition. Now as the system cools below the critical temperature, ($T_c \sim \eta$), the ϕ field begins to fall to the minima of it's potential. Domains form of size $\xi \sim \eta^{-1}$ due to the thermal fluctuations of the ϕ field. In these domains ϕ points in arbitrary directions in M, but match smoothly at the boundary, with θ varying so as to cause defects to form on the edges common to certain domains. This is easily seen. Consider one such edge where θ varies by 2π in encircling the edge, i.e all around the edge we continuously encircle V_{min}. This implies that ϕ must vanish on that edge for it corresponds to a region where it is not in V_{min}. Such regions line up to minimise the spatial gradient energy, forming a defect line or cosmic string. It corresponds to a thin tube of vacuum energy, V(0), being stored in there. These lines where $\phi = 0$ are either in the form of closed loops or infinitely long, for if they had ends, then it would be possible to move the circle (corresponding to V_{min}) beyond the end of the string and then shrink it continuously to a point without having to encounter the $\phi = 0$ region. Hence $\Pi_1(M) = 1$ in that region, there would then be contractable loops and the strings wouldn't exist [3].

Returning to the string solutions of (1.17), we look for z- independent static solutions to the field equations[27]. Figure (1) shows the resulting solutions for $\mid \phi \mid$ and $\underline{B} = \underline{\nabla} \wedge \underline{A}$ as a function of the radial distance from the string. The width of the string is roughly $m_\phi^{-1} \sim (\sqrt{\lambda}\eta)^{-1}$ where m_ϕ is the Higgs mass. The string tension, or mass per unit length:

[3] Strings could be finite in length, by connecting them to monopoles or domain walls[31]

$$\mu = \int d^2r[\frac{1}{2} \mid (\underline{\nabla} + ie\underline{A})\phi \mid^2 + \frac{\lambda}{4!}(\mid \phi \mid^2 - \eta^2)^2 + \frac{1}{2}\underline{B}^2] \tag{1.18}$$

hence from $V(\phi)$:

$$\begin{aligned} \mu &\sim (\sqrt{\lambda}\eta)^{-2}\lambda\eta^4 \\ &\sim \eta^2 \end{aligned} \tag{1.19}$$

For example if the symmetry breaking scale is during the GUT era: $[\eta \sim 10^{15}$ or $10^{16}Gev]$ then the dimensionless parameter $G\mu$, (G is Newtons constant), lies between 10^{-7} and 10^{-5}. As we shall see observational constraints place a tight upper bound, (not lower) on $G\mu$ of: $[G\mu \leq 10^{-5}]$. There is a big difference between global and local strings. The latter, as the name implies possess a local gauge field whose presence results in no long range interactions between the Higgs fields: $[\lim_{r\to\infty}(\nabla + ie\underline{A})\phi = 0]$. Also the magnetic flux in such strings is quantised in units of $\frac{2\pi}{e}$: $[\int \underline{B}.d\underline{S} = \oint \underline{A}.d\underline{l} = \frac{2\pi}{e}]$ where $d\underline{S}$ and $d\underline{l}$ are the area and line elements surrounding a portion of string. If a global symmetry is broken, there are no local gauge fields present, resulting in Goldstone bosons, long range forces and an infinite string mass per unit length.

Strings are formed via the Kibble mechanism[4]. A lattice of domains is constructed typically of size $\sim \xi$, (the length scale above which the orientation of the Higgs fields are uncorrelated). In each domain a value of $\phi \in V_{min}$ is randomly chosen, as this reflects ϕ choosing a minimum energy configuration as the phase transition is passed through. After each domain has a value in it, look at each link on the lattice and using a prescription to smoothly vary the phases from one domain to the next, decide if there is a net winding number, anti winding number or no net winding number, hence is there a string, anti string or no string passing along that link. Numerical tests of this mechanism[37] indicate that after the phase transition, about 80% of the string is in long 'infinite' string as long as the box in which the simulation is performed. The rest is in a scale invariant distribution of closed loops where the number of loops between radius r and r + dr,

$$n(r) \propto \frac{dr}{r^4}$$

(i.e independent of ξ). At high densities both the infinite and closed loops of string are in the form of Brownian walks of length $L \sim \frac{r^2}{\xi}$. Analytical approaches have placed these results on firmer ground. Mitchell and Turok[38], by counting the density of states in the quantised closed bosonic string, demonstrated all the above results. An alternative approach uses finite temperature field theory to investigate the distribution of the Higgs field around the phase transition, also deriving the same behaviour[39]. We are unable to derive a precise number for the amount of infinite string length (i.e 80%), because that is a phenomenon out of equilibrium where the canonical ensemble breaks down.

The simplest theory to possess vortex solutions is scalar QED

$$\mathcal{L}[\phi, A] = -\frac{1}{4}F_{\mu\nu}F^{\mu\nu} + \frac{1}{2} \mid (\partial_\mu + ieA_\mu)\phi \mid^2 + \frac{1}{2}m_0^2 \mid \phi \mid^2 - \frac{\lambda}{4!} \mid \phi \mid^4 \tag{1.20}$$
$$m_0^2 > 0$$

The partition function is:
$Z \propto \int D\phi D\phi^* DA(detM)exp(-I_\beta[\phi, A])$
where det M describes the gauge fixing, and
$I_\beta[\phi, A] = -\int_0^\beta d\tau \int d^3x \mathcal{L}_E[\phi, A]$
with $\mathcal{L}_E$ the Euclidean form of the lagrangian, ($\phi[x,\tau] = \sum_n \phi_n(x)exp(\frac{2\pi in\tau}{\beta})$) [4]. Now to $O(\lambda, e^2)$, for $\frac{m}{T} \ll 1$ we find as previously discussed the one loop correction to the free energy, having integrated out the massive modes:
$V(\phi) = -\frac{1}{2}m_0^2 \mid \phi_0 \mid^2 [1 - \frac{T^2}{T_c^2}] + \frac{\lambda}{4!} \mid \phi_0 \mid^4$
where $T_c^2 = \frac{m_0^2}{\frac{\lambda}{18} + \frac{e^2}{4}}$. This is then substituted back into the partition function leaving only the functional integration over the zero modes remaining.
$Z = \int D\phi_0 DA_{0\mu} exp(-\beta I[\phi_0, A_{0\mu}])$
The statistical properties of strings around the phase transition are found by evaluating Z with string like configurations. The dominant contributions to the integral come from field configurations satisfying the stationary equations $\frac{\partial I}{\partial \phi} \mid_{\phi=\phi_{saddle}, A^\mu=A^\mu_{saddle}} = 0$, $\frac{\partial I}{\partial A_\mu} \mid_{\phi=\phi_{saddle}, A^\mu=A^\mu_{saddle}} = 0$.

$$\partial^i F_{ji} = \frac{1}{2}ie(\phi^*\partial_j\phi - \phi\partial_j\phi^*) - e^2 A_j \mid \phi \mid^2$$
$$(\partial_i + ieA_i)^2\phi = -m_0^2(1 - \frac{T^2}{T_c^2})\phi + \frac{\lambda}{3!} \mid \phi \mid^2 \phi$$

The contribution of any solution to Z is found by substitution. The solution $\phi = const, A = 0$ is a solution of minimum energy, so gives the maximum contribution. Well away from T_c, Z is well represented by this term. However as $T \to T_c$ we must sum all maxima to the functional because the total contribution due to the large number of non constant field configurations becomes larger. This is easy to see, as $T \to T_c$, the mass, $m(T) \to 0$, so it becomes possible to form strings at no energetic cost, and there is a second order phase transition. There are many string configurations and we must sum over them all,[5] here though the solutions vary with temperature.[6] In particular
$\lim_{r\to\infty} \mid \phi(r) \mid \longrightarrow \eta, \quad \eta^2 = \frac{m_s^2}{\lambda}, \quad m_s^2 = m_0^2(1 - \frac{T^2}{T_c^2})$
At the core of the string $\mid \phi \mid$ vanishes, it's thickness $\sim m_s^{-1}$. The magnetic field is determined by m_v^{-1},

$$m_v = e\eta = \frac{e}{\sqrt{\lambda}} m_s \tag{1.21}$$

[4]It is important to note that we are not using the mean field approach here, in that case we would expand about constant solutions for ϕ and A

[5]In principle we should sum over all the maxima of the functional, not just the string contributions.

[6]These solutions are strictly speaking infinite string solutions. In dealing with loops of string, if we are to use these solutions, then we must work in the regime where the radius of curvature is large compared to the width of the string, so the string is approximately straight in any given region

and the string has an energy per unit length $\sigma = \sigma_s + \sigma_v$

$$\begin{aligned}
\sigma_s &= O(\eta^2(T)) \text{ scalarfield} \\
\sigma_v &= O(\frac{e^2\eta^4}{m_v^4}) \\
&= O(\eta^2(T)) \text{ vectorfield}
\end{aligned}$$

We are in a position to discuss the thermodynamics of a gas of non-interacting strings. For a recent review with detailed references see[40]. Historically the problem was first discussed in[41] and [42], whose interest was stimulated by a desire to understand hadronic matter at high density. We no longer believe that the string model is a good description of hadronic matter at high density, but their work is still of interest because they reached some important conclusions about the statistical properties of free strings. As they pointed out, the statistical properties of free strings are somewhat different to those of 'normal' systems and care is needed to correctly derive their thermodynamic properties. We shall start our discussion, therefore, by briefly reviewing some basics of statistical mechanics.

The principle idea behind statistical mechanics is that an isolated macroscopic system specified by its energy E and volume V has an enormous number $\Omega(E, V)$ of different microstates, i.e. different ways of distributing the energy of the system between the 'particles' of the system. The fundamental postulate of equilibrium statistical mechanics is that the system is equally likely to be found in any of these microstates. Thus the most probable state of the system will be that which maximizes the density of states. This is the the microcanonical approach. Rather than work directly using this microcanonical approach it is often easier, for large systems, to use the less fundamental canonical ensemble approach.

In the canonical ensemble the isolated system is assumed to include a large 'heat bath', in thermal contact with a smaller part considered to be the system under investigation. Assuming homogeneity, the properties derived for the small system will be the same as those of the system as a whole. A second assumption made is that $T_{bath}C_v^{bath}V_{bath} >> \epsilon_i$ for all states i of the system of interest. T is the temperature of the heat bath, V its volume and C_v its heat capacity. The fundamental quantity that needs to be evaluated in the canonical approach is the partition function Z, defined by

$$Z[V] = \int dE\,\Omega[E, V]e^{-\beta E} \tag{1.22}$$

$$\beta = \frac{1}{T} \tag{1.23}$$

$\Omega[E, V]$ the density of states must therefore be determined. This entails evaluating the number of configurations of a string of length L. Classically an infinitely thin string (of any length) has infinitely many different configurations because it is able to 'wiggle' on any scale. To be able to count the number of configurations it is necessary to impose a lower cutoff to the scale at which strings can bend. This can be done classically 'à la Boltzmann' by introducing an arbitrary cutoff scale which then has to be determined by experiment. Alternatively, the string can be quantized, the lower cutoff then being related to the string tension. For 'fat' strings, quantization is unnecessary, as the width of the string provides the lower cutoff. Our approach to counting the number of string states will be to treat all the different types of strings as classical strings with a cutoff 'a' on the scale at which strings are allowed to bend.

As a tactical device to make the string configurations denumerable, we require the strings to lie on the links of a lattice. We consider the strings to be *orientable*, to form *closed* paths within the lattice and to have a constant intrinsic energy/length, independent of the orientation of the string (it is possible to also account for rigidity in strings[43]). For simplicity we consider the underlying lattice to be a simple cubic lattice of spacing a. All the strings that we shall consider have a real (or effective) non-zero thickness, and we identify this core diameter with a.

On a lattice a string consists of a sequence of hinged straight segments, each joining two lattice sites. (For the cubic lattice above each segment has minimum length a). The configuration of a string of length $L > a$ can be specified by enumerating the lattice sites $\mathbf{r}^{(i)}$ $(i = 1, 2, \ldots, L/a)$ of the segment ends, with respect to a fixed origin. At thermal equilibrium (temperature $T = 1/k_B\beta$), the partition function for a system containing N strings of lengths $L_1, L_2, \ldots, L_N$ is

$$Z_N(L_1, \ldots, L_N) = \sum_{\{\mathbf{r}_p\}} e^{-\beta E[\mathbf{r}_1, \mathbf{r}_2, \ldots, \mathbf{r}_N]} \tag{1.24}$$

where $p = 1, 2, \ldots, N$. Here $\mathbf{r}_p = (\mathbf{r}_p^{(1)}, \mathbf{r}_p^{(2)}, \ldots, \mathbf{r}_p^{(n)}; n = L_p/a)$ denotes the configuration of the pth string, length L_p. The sum is taken over all N-string configurations, the energy for the configuration $\{\mathbf{r}_1, \mathbf{r}_2, \ldots, \mathbf{r}_N\}$ being $E[\mathbf{r}_1, \mathbf{r}_2, \ldots, \mathbf{r}_N]$. The general N-string partition function is obtained by summing over all string lengths,

$$Z_N = \sum_{\{L_p\}} Z_N(L_1, L_2, \ldots, L_N). \tag{1.25}$$

When the string $\mathbf{r} = (\mathbf{r}^{(1)}, \mathbf{r}^{(2)}, \ldots, \mathbf{r}^{(n)}; n = L/a)$ has $n >> 1$ steps it can be approximated by a continuous curve, parameterised by $\mathbf{r}(s)$, $0 \le s \le L = na$. In this approximation the sum (1.24) is re-expressible as a functional integral over paths $\mathbf{r}_p(s_p)$ $(0 \le s_p \le L_p)$ as

$$Z_N(L_1, \ldots, L_N) = \oint \prod_1^N d\mu_{L_p}[\mathbf{r}_p]\, e^{-\beta E[\mathbf{r}_1, \ldots, \mathbf{r}_N]} \tag{1.26}$$

where $d\mu_L[\mathbf{r}]$ is a measure for the number density of configurations in which the lattice string is close to the continuum string $\mathbf{r}(s)$. By definition

$$\oint d\mu_L[\mathbf{r}] \equiv n(L) \tag{1.27}$$

where $n(L)$ is the number of closed loops of length L on the lattice.

There are two separate contributions to $E[\mathbf{r}_1, \mathbf{r}_2, \ldots, \mathbf{r}_N]$. The first is the intrinsic core energy of the strings

$$E_o[\mathbf{r}_1,\ldots,\mathbf{r}_N] = \sigma \sum_{p=1}^{N} L_p \tag{1.28}$$

where σ is the core energy-density.

The second, and much more complicated term, describes string interactions. We will give examples shortly of such terms relevant for cosmic strings. Concentrating initially on the case of free strings, we restrict the position of the string segments to lie on a cubic lattice of coordination number z.

Since we are assuming that the only role of string interactions is to enable the phase space to be explored, it is immediately apparent that the string trajectories will be Brownian. Thus the number of configurations of a closed string of length L is (assuming $\frac{L}{a} >> 1$):

$$\rho(l) = \left(\frac{A}{\{2\}}\right)\left(\frac{V}{a^3}\right)\left(\frac{L}{a}\right)^{-3/2}\left(\frac{L}{a}\right)^{-1} e^{\frac{L}{a}\ln z} \tag{1.29}$$

The factor $\left(\frac{V}{a^3}\right)$ accounts for the number of possible starting points for the string on the lattice; the factor $e^{\frac{L}{a}\ln z}$ is the number of possible configurations of a string of length L having no restriction placed on whether or not it is closed; the factor $A\left(\frac{L}{a}\right)^{-3/2}$ is the probability of a Brownian walk returning to the origin after $\frac{L}{a}$ steps; the factor of $\frac{1}{\{2\}}$ is included if the loop is non-orientable and the final factor accounts for the $\frac{L}{a}$ possible starting points for a given string. For open strings the factor $A\left(\frac{L}{a}\right)^{-3/2}\left(\frac{L}{a}\right)^{-1}$ is excluded.

Whether the strings are open or closed depends on the string type. For example, QCD flux tubes are typically open whilst cosmic strings are closed or infinite in length. The difference is not, however, of major interest, as only the fine details of the thermodynamics of free strings depend on whether or not the strings are closed. Therefore, for the rest of this review we shall assume that the strings are closed.

The energy of a single non interacting string of length L is $E(L) = \sigma L$. Thus the partition function describing a system containing a single closed string of arbitrary length is

$$Z_{1,closed} = \frac{AV}{2a^3}\sum_{\frac{L}{a}=1}^{\infty}\left(\frac{L}{a}\right)^{-5/2} e^{-\beta L\sigma_{eff}} \tag{1.30}$$

where

$$\sigma_{eff} = \sigma - \frac{\ln z}{\beta a} \tag{1.31}$$

σ_{eff} plays the role of an effective string tension. It is the vanishing of σ_{eff} that causes the system to undergo a phase transition. This occurs at a temperature

$$T_{st} = \frac{\sigma a}{\ln z} \tag{1.32}$$

We shall return shortly to discuss further the significance of this temperature. First let us consider a system containing a 'gas' of string. For many types of string, interactions enable a string to split up and rejoin so that the number and length of each string is not conserved. Therefore, the appropriate partition function is that for a system containing an arbitrary number of strings, i.e.

$$\begin{aligned} Z &= 1 + Z_1 + \frac{1}{2!}Z_1^2 + \frac{1}{3!}Z_1^3 + \ldots. \\ &= exp(Z_1) \end{aligned} \tag{1.33}$$

where the factor of $1/n!$ accounts for the indistinguishability of loops. We have restricted ourselves to the 'classical regime' in which the probability for two strings to be of the same length is very small.

From this partition function all the interesting thermodynamic quantities can be derived. For example the mean energy density is

$$\frac{\overline{E}}{V} = -\frac{1}{V}\frac{\partial}{\partial\beta}\ln Z = \frac{A\sigma}{\{2\}a^2}\sum_{\frac{L}{a}=1}^{\infty}\left(\frac{L}{a}\right)^{-\frac{3}{2}} exp(-\beta L\sigma_{eff}) \tag{1.34}$$

and the pressure is

$$P = T\left(\frac{\partial(\ln Z)}{\partial V}\right)_T = \frac{A}{\{2\}a^3}\sum_{\frac{L}{a}=1}^{\infty}\left(\frac{L}{a}\right)^{-5/2} e^{-\beta L\sigma_{eff}} \tag{1.35}$$

Any other thermodynamic quantity may be determined by an appropriate application of Maxwell's relations.

Other potentially interesting quantities may also be evaluated. For example the mean number density of loops can be evaluated and is found to equal P. The mean number of loops $R(L)$ of size L can also be determined:

$$R(L) = \frac{AV}{\{2\}a^3}\left(\frac{L}{a}\right)^{-\frac{5}{2}} exp(-\beta L\sigma_{eff}) \tag{1.36}$$

It can be seen immediately that for $T << T_{st}$ long strings are exponentially suppressed while for $T \approx T_{st}$ the loops have a scale invariant distribution. These results have been confirmed for non-interacting Nambu strings[44, 45].

The unusual property of a system of free strings that is not shared by more 'normal' systems is that the partition function diverges for all temperatures $T > T_{st}$. Thus canonical thermodynamics does not make sense at temperatures above T_{st}. T_{st} defines the maximum temperature of the system; this temperature is often referred to as the Hagedorn Temperature[46]. Now T_{st} is reached while the energy density E_{st} is still finite. Evidently, systems with mean energy density greater than this are thermodynamically unstable. With the present approach the nature of the instability is not clear. We can guess, however, that this instability must be due to the system being characterized by some large internal inhomogeneity or by a state of energy that is not small in comparison to $T_{bath}C_vV_{bath}$, or both, as these are the only obvious assumptions that we have made. This is indeed the case, the system becomes unstable to the formation of an infinite string.

To see this we need to return to the more fundamental microcanonical ensemble. Following Turok[47] we examine the divergence of the mean square energy[47]. The mean square energy diverges as $T \rightarrow T_{st}$ like:

$$\overline{E^2} = \overline{E}^2 + \frac{\sigma^2 AV}{\{2\}a} \sum_{n=1}^{\infty} n^{-1/2} e^{-n(\beta\sigma a - \ln z)} \tag{1.37}$$

$$\approx \frac{C\Gamma(\frac{1}{2})}{\sqrt{\beta\sigma a - \ln z}} \tag{1.38}$$

From

$$\overline{E^2} = \int dE E^2 P(E) \tag{1.39}$$

where $P(E)$ is the probability of finding the subsystem with energy E, it follows that at $T = T_{st}$ and for large E

$$P(E) \approx \frac{C\Gamma(\frac{1}{2})}{E^{\frac{5}{2}}} \tag{1.40}$$

But

$$P(E, T_{st}) = \frac{\Omega(E) e^{-\beta_{st} E}}{Z(T_{st})} \tag{1.41}$$

and so

$$\Omega(E) \approx Z(T_{st}) C \frac{e^{\frac{E}{\sigma a} \ln z}}{E^{\frac{5}{2}}} \tag{1.42}$$

The number of states available to a single string of energy E is (from (1.29)):

$$\Omega(E) = \frac{AV}{2a^3} \left(\frac{E}{a\sigma}\right)^{-\frac{5}{2}} exp\left(\frac{E}{\sigma a} \ln z\right) \tag{1.43}$$

Thus we see that the last term in (1.42) is simply the number of states of a single string of energy E. As we increase the energy density of our isolated system the excess energy above E_{st} goes into a long string. We can also see that increasing the energy density further does not cause an increase in temperature. Using equation (1.29) and (1.43) we find

$$\frac{1}{T} = \frac{-5}{2E} + \frac{1}{T_{st}} \tag{1.44}$$

But our analysis is only valid for $E >> 1$ and therefore to a good approximation $T = T_{st}$.

Let us now consider the implications of this result. Consider the effect of compressing a box of fixed energy E. At first, as the energy density increases the energy remains in loops and the temperature of the system increases. After a period of time, the temperature of the box becomes T_{st}. At this temperature, the loops have a scale invariant distribution and the system is unstable to loop energy densities higher than those present. Increasing the energy density further will no longer cause the energy density in loops to increase. Instead the extra energy density floods into an infinite string[7] and in this way the energy density in the box can become arbitrarily large. If string interactions for fundamental strings can be ignored at high densities the temperature T_{st} is the maximum temperature of a system in thermal equilibrium with a string network. T_{st} can be considered to be a transition temperature of the system as it marks the point at which the energy density of the system changes from being dominated by loops to being dominated by infinite strings.

Recall that in terms of the fields T_c is the mean field critical temperature. How is it related to T_{st}? As $T \to T_c$, $\sigma \to 0$, $m_s^{-1} \to \infty$. As we have just said, think of T_{st} as the temperature at which our strings are formed,

$T_{st} \simeq \sigma(T)l(T) \simeq \gamma\eta^2 m_v^{-1} \quad \gamma \sim O(1) \qquad \lambda \gg e^2$

Then it follows $T_{st} < T_c$ with

$$\begin{aligned} 1 - \frac{T_{st}^2}{T_c^2} &= O(\lambda), \qquad m = m_s \\ &= O(e^2) \qquad , m = m_v \end{aligned}$$

The width of the strings at formation can now be calculated ($\sim$ mean separation)

$$\begin{aligned} m_s(T_{st}) &= O(em_s(T=0)) \qquad , m = m_v \\ &= O(\sqrt{\lambda} m_s(T=0)) \qquad , m = m_s \end{aligned}$$

That is the network of strings at the phase transition has the separation of the centers of the flux tubes scaled up by a factor $O(\frac{1}{e^2})$, $(e^2 < \lambda)$, a result obtained by Kibble[4], and discussed earlier in this lecture. This is a useful confirmation of the validity of the Kibble mechanism. What has become clear from the analysis is that as we approach T_{st}, the mean field approach does not give the correct value for the critical temperature at which the strings form. We predict a second order phase transition at $T_{st} < T_c \mid_{meanfield}$. As the symmetry is restored, overlapping strings fill the whole of space. Above $T_{st} \sim O(T_{Ginzburg})$, our model is unable to recognise string like configurations. Below T_{st} the system evolves to a state with an exponentially suppressed distribution of large loops.

For composite strings, string interactions can definitely *not* be ignored. As the system approaches T_{st} the strings become fat, interactions between strings become important and the free string picture breaks down. Both of the above effects reduce the number of available states making the number of states available less than the number available to free strings. The entropy can no longer balance the energy and as a consequence the system does not possess a maximum temperature. The system can exist in a new phase that is no longer describable in terms of 'free' strings at temperatures above T_{st}.

Assuming string interactions are pairwise it can be written as

[7] We are considering the thermodynamic limit as the volume of the system tends to infinity.

$$E_{int}[\mathbf{r}_1,\ldots,\mathbf{r}_N] = \frac{1}{2}\sum_{\alpha,\beta} E^{(2)}_{int}[\mathbf{r}_\alpha,\mathbf{r}_\beta] \tag{1.45}$$

where the general form for $E^{(2)}_{int}[\mathbf{r}_1,\mathbf{r}_2]$ depends on the system under investigation. For example, when considering dilute *global cosmic strings*, the relevant form for $E^{(2)}_{int}[\mathbf{r}_1,\mathbf{r}_2]$ is

$$E^{(2)}_{int}[\mathbf{r}_1,\mathbf{r}_2] = \frac{\gamma}{4\pi}\oint \frac{d\mathbf{l}_1.d\mathbf{l}_2}{r_{12}} \tag{1.46}$$

where r_{12} is the separation between the two incremental string sections $d\mathbf{l}_1$ and $d\mathbf{l}_2$ and γ measures the strength of the interaction. The underlying physics fixes γ (e.g. for local cosmic strings $\gamma = 8\pi^2\sigma$).

For crystal dislocations a very similar form

$$E_{int}[\mathbf{r}_1,\mathbf{r}_2] = \frac{\delta\mu}{4\pi}\oint \frac{(\mathbf{b}_1.d\mathbf{l}_1)\,(\mathbf{b}_2.d\mathbf{l}_2)}{r_{12}} \tag{1.47}$$

can be used to describe the interaction energy between two dislocations in a crystal, where the vectors $\mathbf{b}_1$ and $\mathbf{b}_2$ are the Burgers vectors associated with the two dislocation sections $d\mathbf{l}_1$ and $d\mathbf{l}_2$. The expression (1.47) is a simplified version of the general form for $E_{int}[\mathbf{r}_1,\mathbf{r}_2]$ given by Blin [48]. The factor δ in (1.47) was introduced in [49] in an attempt to take into account the other terms in Blin's description.

Further, it is possible to describe short-range forces of the type experienced by *local* cosmic strings and superconducting flux tubes. This is much more difficult to handle properly because there are *two* length scales in these cases. A simple modification of (1.46) which allows for this is

$$E^{(2)}_{int}[\mathbf{r}_1,\mathbf{r}_2] = \frac{\gamma}{4\pi}\oint \frac{d\mathbf{l}_1.d\mathbf{l}_2}{r_{12}} e^{-mr_{12}} \qquad , m > 0, \tag{1.48}$$

which is enough to display some of the general features of short range forces. The expressions (1.46,1.47,1.48) are only valid in the dilute string approximation, and this should be kept in mind throughout.

In [50, 51] we were able to incorporate the type of interaction (1.48) and determine how in the dilute string approximation it affected the thermodynamic properties and distribution of the string network. For example depending on the strength of the short range interaction (the value of m) previously second order phase transitions can become first order. Complete details can be found in [51]. A substantial amount of literature concerning interactions in condensed matter syetems, helium vortices and their relations to field theories already exists. Perhaps the most comprehensive survey of the field is to be found in the monographs by Kleinert[52].

b) Global Textures.

Just as broken abelian [U(1)] global symmetries lead to global strings, it is possible to construct Grand unified models which permit the formation of global textures[32, 33]. The simplest example involves the breaking of a global SU(2) symmetry (although broken non abelian global symmetries always lead to the formation of texture). The SU(2) global symmetry can be broken by a complex doublet Φ (which we shall write as a four component real field);

$$\Phi = \frac{1}{\sqrt{2}}(\phi_1 + i\phi_2, \phi_3 + i\phi_4), \qquad \vec{\phi} \equiv (\phi_1 ... \phi_4) \tag{1.49}$$

There has ben much work into non-abelian gauge fields (for a thorough review see [53]). Let G be the gauge group of order r, not necessarily semi-simple, and let the gauge field be coupled to a multiplet of scalar fields Φ, transforming according to some irreducible n-dimensional representation. The matter Lagrangian is

$$\mathcal{L}_m = -\frac{1}{4}F_{\mu\nu a}F_a^{\mu\nu} + [D_\mu\Phi]^\dagger[D^\mu\Phi] - V(\Phi) \tag{1.50}$$

where

$$D_\mu = \partial_\mu - \sum_{a=1}^{r} g_a A_{\mu a} T^a, \tag{1.51}$$

the anti-hermitian T matrices are the infinitesimal generators of the representation, and the coupling constants may depend upon the simple component of the group. Consider the case where $V(\Phi) = \lambda(\Phi^2 - \eta^2)^2$. Dropping the gauge field makes this a global non abelian field theory and we are left with n-component scalars. The case when a global SU(2) is broken by a complex doublet with the potential $V(\Phi)$ is equivalent in the pure scalar sector to the theory where SO(4) is broken by a four vector to SO(3) with Φ^a acquiring a vacuum expectation value. Then since $\Pi_3(SO(4)/SO(3)) = S_3$ we will have non-trivial solutions of the field Φ^a - global textures with $\Phi^2 = \eta^2$. The action in this pure scalar sector giving rise to textures is

$$S = \int d^4x\sqrt{-g}[\partial_\mu\Phi^a.\partial^\mu\Phi_a - \lambda((\Phi^a\Phi_a - \eta^2)^2 - \frac{1}{16\pi G}R] \tag{1.52}$$

where R is the scalar curvature, (-g) is the metric determinant and Φ^a is a four component real field[54].

Texture 'knots' are regions where the Higgs field winds around a three-sphere in a non-trivial way, generally the case in regions larger than the horizon. As mentioned earlier the global textures are unstable to collapse, as may be understood from Derricks theorem[55]. We can write the energy of a static configuration schematically as

$$E = \int d^3\underline{x}[(\nabla\Phi(\underline{x}))^2 + V(\Phi(\underline{x}))] \tag{1.53}$$

Replacing $\Phi(\underline{x})$ by $\Phi(\lambda\underline{x})$ with $\lambda > 1$, thereby shrinking it, the gradient term scales as λ^{2-D}, the potential term as λ^{-D} in D spatial dimensions. For $D > 1$ any localised configuration of scalar fields is unstable to shrinking. As Turok points out, because global strings or monopoles are not localised (their energy diverges at infinity) this argument fails in those cases[33].

The formation picture developed by Turok and his collaborators is as follows. The universe cools through a temperature of order the GUT scale, and the initial conditions for Φ are specified by its being in thermal equilibrium. The resulting random domain picture is one where causality demands that the symmetry breaking gives rise to vacuum configurations which are uncorrelated on scales above the horizon. In the next lecture we will discuss the actual texture solutions and their evolution.

Finally in this lecture we address the issue of what does a texture look like, how would you know if one was beside you? Three dimensional textures are hard to visualise, they live on a three sphere vacuum manifold. The simplest textures exist in one spatial dimension, and although they are not unstable, they are useful to at least gain a feel for how the textures can be defined. Recently we have investigated the initial conditions for the formation of textures[56]. In one space dimension textures arise if the vacuum manifold has the topology S_1. In general, a texture can form in d spatial dimensions if the d-th homotopy group of the vacuum manifold is non-vanishing.

The simplest texture we can have is where, as we move along the real line from $-\infty$ to $+\infty$, the direction in which the symmetry breaks starts at some angle (which we can choose to be $-\pi$), twists round the potential and then tends to $+\pi$ at $+\infty$ (we identify $+\pi$ and $-\pi$). There is no change in potential energy as we have remained in vacuum throughout, but there is some energy associated with the gradient energy of the field as we wind around, and this energy cannot be removed while we remain in vacuum. If the number of dimensions is greater than one, however, Derrick's theorem dictates that the texture will collapse, and be able to unwind when the texture size becomes comparable to the symmetry breaking length scale. Such an event is schematically illustrated in Figure 2 for the case of the spherically symmetric texture described in lecture 2. Note how the gradient energy increases (the slope of the curve increases) with time. Eventually this energy is so large, it can lift the field off the vacuum manifold and unwind the texture.

In order to set up initial conditions for our one dimensional texture, we discretise both space and the vacuum manifold. Following Vachaspati and Vilenkin [37], the circular manifold can be replaced by a triangle with vertices labelled **0**, **1** and **2** in a clockwise direction. Then the sequence **0120** corresponds to winding once around the vacuum manifold (we assume the shortest path between two vertices is always taken). Space is divided up into evenly spaced discrete points, where the spacing is taken as the correlation length of the field at symmetry breaking. A configuration representing the conditions after symmetry breaking is simply a random sequence of numbers such as **...02102202100120....** We shall use this specific sequence to illustrate some points.

A random sequence of numbers will include windings around the vacuum manifold. For instance, between the first two occurrences of **0** in the above sequence we wind around once (backwards, hence an antitexture). However, that antitexture is not localised, because we cannot say that it is definitely between two adjacent numbers. Indeed, although we have identified it as being between the two **0**s, we could equally well say that the winding is between the first two instances of the number **2**. Clearly the notion of where the texture is is a bit fuzzy. Textures can also exhibit varying degrees of diffuseness; the texture knot **0120** is the minimum possible size, but one can have a longer texture where partial windings and unwindings occur. An example is the texture between the first two occurences of **1** in the sequence above, where there is a single winding in a sequence of seven numbers. Further, there are pathological sequences, of which **...1201021201021...** is an example, where if we take **0** as our reference point we see no textures at all, whereas if we take **1** as our reference point we see a sequence of textures and antitextures. What is well defined is the overall texture charge q between any two selected points; this charge gives the difference in texture and antitexture numbers[33]. Between randomly chosen points, it need not be an integer, but may be in units of 1/3. If the two endpoints are chosen to be the same number then the charge will be an integer. Note though that during dynamical evolution texture unwindings violate conservation of the texture charge, with a net change of ± 1. A useful concept is that of partial winding. Associate a value with the link between the points of the spatial lattice as follows; if the numbers at each end are the same leave the link blank (signalled by $\bullet$), whereas if the numbers are different, then label the link by the third number which is not present at the ends. Further, give this third number a sign + or − according to whether the rotation between the endpoints is clockwise or anticlockwise. Thus the sequence **...0210220...** gives rise to the link labelling ...1^- 0^- 2^- 1^- $\bullet$ 1^+.... These link values are the basic building blocks of texture; they are the partial windings. A full texture corresponds to a collection of the three numbers all with a + sign. Thus in the above sequence the 1^- 1^+ cancel out, leaving exactly one antitexture. The $\bullet$s play a passive role, making the textures more diffuse, and the order of the building blocks is irrelevant.

This picture also works in two spatial dimensions and appears to encompass all the main features of the full three dimensional case while remaining relatively easy to illustrate and visualise. In that case the vacuum manifold is a two sphere. The archetypal texture corresponding to that of Figure 2 is the spherically symmetric texture as illustrated in Figure 3. On the 2-sphere, label the north pole as $\theta = 0$ and the south pole as $\theta = \pi$; $\phi \in [0, 2\pi]$ is the angle around the globe. To form the spherically symmetric texture, map the 2-sphere into space by opening out the north pole and taking it to infinity (exactly as in the mapping of the Riemann sphere onto the plane). At constant radius ϕ just winds round from 0 to 2π, while θ varies from 0 at infinity up to π at the origin and then decreases again as we move away from the origin. As before, this winding around the 2-sphere cannot be continuously removed while remaining on the vacuum manifold.

S_2 can be discretised as a tetrahedron, with the vertices numbered **0** to **3**. It is most convenient to find a tesselation of the plane which respects as far as possible the symmetries of this discretisation, so we choose to tesselate the plane with equilateral triangles, on the vertices of which the phases shall be randomly thrown. Figure 4 represents the discretised version of the spherically symmetric texture. Having thrown the phases at random on the vertices of the triangle,we use the partial building blocks to locate the textures. To see this choose a triangle from the tesselation. If two or three of the phases are the same, the triangle corresponds just to an edge or vertex of the tetrahedron, in either case there is no net winding. If the three numbers are different, associate with the triangle the fourth, missing, number. To define whether it is a partial winding or antiwinding- write the sequence $n_1n_2n_3n_4$ where n_4 is the missing number and the others are the vertices in clockwise order. Then the block is assigned $+$ if $n_1n_2n_3n_4$ is a positive permutation of **0123** and $-$ otherwise. Again, the partial windings are the basic building blocks of texture, and by gathering a complete set of $+$ or $-$ triangles we obtain a unit texture charge- a knot. The presence of blank triangles of no net winding can make a texture more diffuse. In this case one can calculate that the probability of any given triangle being blank is $5/8$, compared with only $1/3$ in the one dimensional case, thus favouring more diffuse textures. In three dimensions the probability of a blank goes up to about $4/5$.

Figure 5 illustrates a two dimensional initial configuration, alternately labelled via the random vertices and then by the partial winding technique. Note the clustering of the partial windings, and also the ubiquity of partial winding-antiwinding pairs.

The machinery for the full three dimensional version is exactly that just illustrated. The vacuum manifold now has the topology of a 3-sphere, and is discretised by a 3-tetrahedron, made up from 5 tetrahedra with vertices which are assigned a phase from **0** to **4**. The most convenient tesselation of space-time is via tetrahedra. The phases are randomly thrown on the vertices of the tesselating tetrahedra, and partial windings associated in a similar manner to above (of course we now have $n_1n_2n_3n_4n_5$).

Blank tetrahedra are now by far the likeliest result, with a probability of $101/125 \simeq 4/5$. Hence one expects the typical texture to be more and more diffuse as the number of dimensions increases. However, as is also true in the two dimensional case, there are correlations between neighbouring tetrahedra; given a tetrahedron containing a partial winding each neighbour has a probability of $2/5$ of also containing one as it is known to have three common vertices which must be different. $1/5$ of that probability is that it will be the antiwinding of the first. Similarly, there is an enhanced probability of a blank adjacent to a blank. Hence this distribution of partial windings is more clustered than if they were Poisson distributed.

Our method differs from those suggested by Srivastava [57] but we reach similar conclusions. If he demands that the texture should be completely surrounded by vertices where the Higgs fields are aligned (thus ensuring integer windings), he rightly concludes that such a configuration is extremely unlikely, and translates this conclusion into a heavy suppression on the number of textures. However, we would conclude that the presence of aligned boundary conditions is not necessary for a region to possess a winding which will be developed by the dynamics into a texture; all we need is a close collection of an appropriate set of partial windings with no particular boundary conditions. Hence we believe that the suppression of textures is not as great as suggested by that requirement. A rather different method for examining formation of a new kind of texture defect, associated with the second homotopy group when in three spatial dimensions, has been demonstrated by Perivolaropoulos [58].

II. The Evolution of Defects in the Early Universe

In this lecture we will be concentrating on the evolution of both cosmic strings and global textures in the early universe.

a). Cosmic Strings.

For a string of width W and radius of curvature R, if $R \gg W$ then the action for a string is approximately a locally boosted version of the straight static solution. This is the Nambu action[59]:

$$S = -\mu \int d\sigma d\tau [-det g^{(2)}_{ab}]^{\frac{1}{2}} \tag{2.1}$$

where $g^{(2)}_{ab}$ is the world sheet metric,(a,b = 0,1). In terms of the string coordinates $X^{\mu}(\sigma, \tau)$ and the background spacetime metric we have

$$g^{(2)}_{ab} = X^{\mu}_{,a} X^{\nu}_{,b} g^{(4)}_{\mu\nu} \tag{2.2}$$

The Nambu action (2.1) also has the interpretation of being the area of the two dimensional worldsheet. Corrections to the action and hence to the solutions to the equations of motion, are of $O(\frac{W}{R})$. Since for typical cosmic strings, $W \sim 10^{-30}m, \quad R \sim kpc$ the corrections are small everywhere except near a cusp where $R \sim W$[60]. Recently Haws et al. have demonstrated that the $O(\frac{W}{R})$ correction to the Nambu action which arises from expanding the full field theory action for the string (1.17) vanishes exactly[61]. Thus to first order it is impossible to say whether a local cosmic string is rigid or anti-rigid.

A vital issue for string theory concerns what happens when two strings collide? The Nambu action breaks down here, so the full non linear field equations must be solved (1.17). This was first studied numerically for the case of global strings by Shellard[62], who demonstrated that strings nearly always intercommute or exchange partners, for relative velocities below ·9. This is a very important result, as it is the only process by which loops can form in sufficient amounts from an initial configuration which contains nearly 80% of its length in infinite string. Analytical approaches to this problem, including introducing gauge fields have been tried, but only for special cases[63], although the full analytical problem remains unsolved. Matzner has studied the case of local strings numerically and concluded they are even more likely to intercommute when they cross than the global case[64].

From (2.1), which is a purely geometrical object, we can write down the string equations of motion in say a flat Friedmann Robertson Walker universe (FRW). One important point to note is that the equations are independent of μ, the scale of the symmetry breaking. Reparameterisation invariance of S under $\sigma \rightarrow \overline{\sigma}(\sigma,\tau)$, $\tau \rightarrow \overline{\tau}(\sigma,\tau)$ enables a suitable gauge to be chosen:

$$x^0 = \tau \qquad\qquad \dot{\underline{x}}.\acute{\underline{x}} = 0 \tag{2.3}$$

where $\dot{\underline{x}} \equiv \partial_\tau \underline{x}$, $\acute{\underline{x}} \equiv \partial_\sigma \underline{x}$. Then the equations of motion become, in terms of the scale factor $a(\eta)$ (where $dt^2 = a^2(\eta)d\eta^2$),

$$\ddot{\underline{x}} + 2\frac{\dot{a}}{a}(1-\dot{\underline{x}}^2)\dot{\underline{x}} = \frac{1}{\varepsilon}\frac{\partial}{\partial\sigma}(\frac{\acute{\underline{x}}}{\varepsilon}) \tag{2.4}$$

with

$$\varepsilon = [\frac{\acute{\underline{x}}^2}{1-\dot{\underline{x}}^2}]^{(\frac{1}{2})} \tag{2.5}$$

the string energy per unit sigma. This implies

$$\dot{\varepsilon} = -2\frac{\dot{a}}{a}\dot{\underline{x}}^2\varepsilon \tag{2.6}$$

The energy in a string is

$$E = \mu a \int d\sigma\varepsilon \tag{2.7}$$

In general it isn't possible to analytically solve these equations, but in certain regimes it can be done.

1]. If $R \ll$ damping term, then $\dot{\underline{x}} \sim 0$, and the string is simply conformally stretched with a radius of curvature $\gg H_0^{-1}$.

2].Vachaspati and Garfinkle have recently obtained the exact solution for a travelling wave moving along a straight string[65].

3]. If $R \ll \frac{a}{\dot{a}} = H_0^{-1}$, then the string doesn't notice the curvature of spacetime and acts as if it was in flat space. We can set $\varepsilon = 1$, and the equations become

$$\begin{aligned} \frac{\partial^2 \underline{x}}{\partial\tau^2} &= \frac{\partial^2 \underline{x}}{\partial\sigma^2} \\ \dot{\underline{x}}.\acute{\underline{x}} &= 0 \qquad (2.8) \\ \dot{\underline{x}}^2 + \acute{\underline{x}}^2 &= 1 \end{aligned}$$

(2.9)

The general solution is made up of right and left moving modes:

$$\underline{x}(\tau,\sigma) = \frac{1}{2}[\underline{a}(\sigma-\tau) + \underline{b}(\sigma+\tau)] \tag{2.10}$$

with the gauge condition implying:

$$\underline{\acute{a}}^2 = \underline{\acute{b}}^2 = 1$$

in the centre of mass frame of the string. Thus $\underline{\acute{a}}$, $\underline{\acute{b}}$ are closed curves on the unit sphere, generally intersecting if they are continuous[66]. In fact the picture for $R \ll t$, is that loops of string break off the network of long string and oscillate as if in flat space. At an intersection $\underline{\acute{x}} = 0$, $\underline{\dot{x}}^2 = 1$ and such a point is a cusp (defined by the vanishing of $\det g^2_{ab}$). However as $\underline{\acute{x}}$ need not be continuous, such points where it isn't are called kinks. Four kinks are always produced when a string intersects. This will be an important feature in determining the fate of the string network. The kinks travel along the string at the speed of light. In flat space, no gravitational radiation is emitted, the kinks remain infinitely sharp with constant amplitude. However in curved space the kinks are damped due to both stretching caused by the universe expanding, and by gravitational radiation.

Before discussing the evolution of the network we address the energy loss mechanism from cosmic strings. As the loops oscillate, due to their tension they decay primarily by emitting gravitational radiation[67]. For a loop of size R it will oscillate with a frequency $w \sim R^{-1}$. The quadrupole formula provides an order of magnitude estimate for the lifetime of the loops. The rate of loss of energy E is $\frac{dE}{dt} \sim -GE^2R^4w^6 \sim -G\mu^2$ where the energy of the loop is $E = \mu R$. This gives a lifetime of order $\tau \sim \frac{E}{|\dot{E}|} \sim \frac{R}{G\mu}$. Thus a loop undergoes about 10^6 oscillations before it decays. Of course we expect gravitational backreaction to play a role towards the end of the loop lifetime. Recently this issue has been addressed by a number of authors investigating the small scale structure on the long strings. Quashnock and Spergel[68] have argued from numerical simulations that a kink lifetime is of order $\tau_k \sim 50(G\mu)^{-1}\bar{d}$, where $\bar{d}$ is the average separation of like moving kinks on the string. Hindmarsh in an elegant letter reaches a similar conclusion, analytically solving for the distribution of kinks on an infinite kinky string[69]. In particular he obtains $\tau_k \sim (G\mu)^{-1}\bar{d}$. These results tend to validate the cruder but more straightforward quadrupole approximation, but as we shall see the kinks have serious implications for the fate of the string network in their own right.

In the evolution of the string network loop production plays an essential role. If the rate of loop formation from infinite strings is too small, then the strings quickly dominate the total energy of the universe. As we have seen, energy in strings longer than H_0^{-1}, (several per horizon volume), scales as a(t), whereas energy in radiation scales as $a^{-1}(t)$. Thus $\frac{\rho_\infty}{\rho_{rad}} \propto a^2(t) = t$. These strings must chop off a constant fraction of their length into loops each expansion time, to avoid this problem. The way this works is that the correlation length, $\xi(t)$ of an infinite string at time t, is $\sim$ t. This is the condition for the scaling solution which is vital if strings are not to dominate the universe. To see this we know at formation

$$\rho_{string} = \frac{\mu\xi}{\xi^3} = \frac{\mu}{\xi^2}.$$

Thus

$$\frac{\rho_{string}}{\rho_{rad}} = \text{const if } \xi \propto \text{t}.$$

In a scaling solution the only length scale is H_0^{-1}. If events occurring 10^{-35} seconds after the big bang are to explain events some 10 billion years later, we need some sort of scaling if we are to be able to predict anything. Fortunately as we shall now see the numerical results do show long string scaling[70, 71, 72], and analytically we can explain why small loops would form rather than infinite string. It is due to the increased amount of phase space available in flat space for loops[38, 39].

Recall that at formation of order 80% of all the string is infinite in length. As the network evolves under the effects of the string tension and the expansion of the universe, the long strings which are moving with velocities a fraction of the speed of light begin intercommuting. This leads to: the formation of loops as a long string intersects itself; long string intercommutations which doesn't effect the length of long string in the network but which does increase the number of kinks present; reconnection of loops back onto the long string network.

The cosmic string picture of galaxy formation has undergone something of a revolution over the past few years. Early analytic[73] and numerical studies[74] suggested that the loops self-intersected until a class of non-self intersecting loop solutions were reached. These loops were to be identified with the galaxies and clusters of galaxies as matter accreted around them, the amount of matter in proportion to the size of the loop. The studies also suggested that the configuration of strings approached a scaling regime in which the total energy density in loops was much larger than that of long strings

$$\rho_{loops} \sim \sqrt{G\mu}, \qquad\qquad \rho_{ls} \sim G\mu \tag{2.11}$$

Therefore the loops played the dominant role in the early models. However these conclusions have been called into question by recent work. All recent simulations agree that the long strings reach a scaling solution where the persistence length of the long string network is substantially smaller than the horizon size[70, 71, 72]. Having said that there still appears to be significant differences between the various authors. Albrecht and Turok[70], [AT], find a much larger density of long string than that found by Bennett and Bouchet[71], [BB], or Allen and Shellard[72], [AS]. They all agree with the one-scale model $\rho_{ls} = A\mu t^{-2}$, but they differ in their values of A. In particular in the radiation era we find

$$\begin{aligned} A &= 50 \pm 25, && [AT] \\ &= 13 \pm 2.5, && [BB] \\ &= 16 \pm 4.0, && [AS]. \end{aligned}$$

More significantly the small scale structure on the network appears to be playing a vital role, leading to differences in the solutions. In BB and AS, the long strings appear very kinky on scales much less than the overall persistence length. The kinks build up on the long strings, controlling the size distribution of the stable loops produced by the long string network. Loop production is then dominated by very small scales and loops are produced with an essentially constant physical size, close to their numerical cut-off. Thus the small scale structure prevents stable non-self-intersecting loops being produced except at very small sizes. In contrast to this AT find that their distribution of loops does reach a scaling solution, with the typical size of a loop formed being a fixed fraction of the horizon distance. Such disagreements need to be resolved, first because of the implications for structure formation, and secondly for the observational tests on the scenario, especially the limit on gravitational radiation derived from bounds on the millisecond pulsar. The dearth of large loops lowers the bound on $G\mu$, and the loops are no longer the dominant feature for string seeded structure formation.

The simulations of BB and AS are not necessarily inconsistent with the idea that the loop distribution would eventually reach scaling. It could be that the typical scale of loops would ultimately be a fixed fraction of the horizon size, but a much smaller fraction than the persistence length of long strings. Indeed this has recently been proposed by a number of authors. Allen and Caldwell[75], [AC], examined the evolution of the linear density of kinks on a cosmic string network, under the assumptions:

1. A spatially flat FRW cosmological model with power law scale factor.
2. Scaling of long string energy density.
3. Uniform kink distribution on the long strings.
4. Kink lifetime proportional to the average distance between kinks.
5. New loops are formed with an average length proportional to the mean distance between kinks.

Under these assumptions, they claim that gravitational radiation from and the resultant back-reaction on the kinks causes them to decay faster than they would due just to the stretching of the kinks in an expanding universe and consequently determines the minimum scale of the structure. The result is that the kink density on long strings grows rapidly at early times, but the mean distance between the kinks becomes proportional to the horizon length at late times, with of order 10^5 kinks present on a horizon sized segment. The explaination for the fact that the simulations of AS and BB do not show scaling then follows; a typical radiation-dominated era simulation runs for $t_f < t < 25t_f$ where t_f is the time the network is formed. Scaling of the kink density would not occur until $10^4 t_f$, so they expect to see a build up in kink density in the simulations. As mentioned earlier Hindmarsh calculated the mean separation of kinks on a long string as $G\mu t$[69].

In a recent paper[76] we addressed the possibility of there being two scales to the cosmic string network. Explicitly introducing persistence lengths $(\xi,\bar{\xi})$ for the long string and small scale structure on the network respectively, we derived equations governing the evolution of the small scale structure on the strings. Intriguingly we found that the question of whether or not both the loop and the long string distribution approached scaling depended critically on the value of a parameter q describing the relative kinkiness of a loop compared to a section of long string of the same length.

As in BB we take the spacetime metric to be Robertson-Walker with conformal time τ:

$$ds^2 = R^2(\tau)(d\tau^2 - d\underline{x}^2) \tag{2.12}$$

Choosing string world sheet coordinates u,v to satisfy

$$x_u^2 = x_v^2 = 0 \tag{2.13}$$

where $x_u \equiv \partial x/\partial u$ etc..., variation of the Nambu-Goto Action leads to equations of motion:

$$\begin{aligned} \underline{p}_v &= -h_0 x_v^0(\underline{q} - \underline{p}\,\underline{p}.\underline{q}) \\ \underline{q}_u &= -h_0 x_u^0(\underline{p} - \underline{q}\,\underline{p}.\underline{q}) \end{aligned} \tag{2.14}$$

in terms of the unit vectors $\underline{p} = x_u/x_u^0$, $\underline{q} = x_v/x_v^0$ and $h_\mu = \partial_\mu R/R$. The importance of (2.14) lies in the coupling of the effective left $(\underline{p})$ and right $(\underline{q})$ movers on the string in curved space. Returning to the orthogonal gauge, with coordinates (τ, σ) defined by

$$x^0 = \tau, \quad \dot{x}.x' = 0, \quad \dot{x} = \partial x/\partial\tau, \quad x' = \partial x/\partial\sigma \tag{2.15}$$

it can be shown that

$$\dot{\underline{x}}^2 = \frac{1}{2}(1 + \underline{p}.\underline{q}) \tag{2.16}$$

Defining a parameter α equal to the average value of $-\underline{p}.\underline{q}$

$$\alpha = \frac{-\int du x_u^0 \underline{p}.\underline{q}}{\int du x_u^0} \tag{2.17}$$

it is possible to show that the increase in energy in the network comes
from the work done in stretching $E \propto R^\alpha$, whereas the momentum of the strings redshifts in the usual way $\underline{P}R$ =constant.

The evolution equations can be written in discrete form, which proves useful as it allows directly for the division of a loop into left and right moving loops characterised by x_u and x_v

with the expansion coupling the two. Dividing the loops into N segments, each of length l, beginning at $u = ml$, or $v = nl$, $(m, n = 0, 1, ..., N-1)$ we define:

$$\begin{aligned} a_m(v) &= \int_{ml}^{(m+1)l} du\, x_u(u,v) = x((m+1)l, v) - x(ml, v) \\ b_n(u) &= \int_{nl}^{(n+1)l} dv\, x_v(u,v) = x(u, (n+1)l) - x(u, nl) \end{aligned} \tag{2.18}$$

The unit vectors $(\underline{p}, \underline{q})$ in discretised form become

$$\underline{p}_m = \frac{\underline{a}_m}{a_m^0}, \qquad \underline{q}_n = \frac{\underline{b}_n}{b_n^0} \tag{2.19}$$

and in general $\underline{p}_m^2 < 1$, $\underline{q}_n^2 < 1$, tending to unity as the universe expands. We assume (cf. (2.17)) that

$$< \underline{p}_m \cdot \underline{q}_n >= -\alpha \tag{2.20}$$

which in (2.16) corresponds to

$$< \dot{\underline{x}}^2 >= \frac{1}{2}(1-\alpha). \tag{2.21}$$

Later we will see how α is a function of the overall type of configuration, it depends on the amount of kinkiness on small scales. It is possible to show how the angle between successive $\underline{p}$ (and $\underline{q}$) segments decreases as the universe expands, leading to the decay of the kinks[71]. Infact

$$< 1 - \underline{p}_m \cdot \underline{p}_{m+1} > \propto R^{-2\alpha} \tag{2.22}$$

If θ is the angle between neighbouring segments we obtain $sin(\frac{\theta}{2}) \propto R^{-\alpha}$ so for small angle kinks $\theta \propto R^{-\alpha}$. In principle α is derivable from the equations governing the string dynamics. It is related to the mean square velocity of the string segments by

$$\alpha = 1 - 2 < \dot{\underline{x}}^2 > \tag{2.23}$$

From the numerical simulations of BB and AT we obtain in the radiation and matter eras respectively

$$\begin{aligned} \alpha_r &= 0.12 \pm .04 \qquad \alpha_m = 0.20 \pm .04 \quad [AT] \\ \alpha_r &= 0.14 \pm .04 \qquad \alpha_m = 0.26 \pm .04 \quad [BB]. \end{aligned} \tag{2.24}$$

Assuming that at any given time the correlation between any two segments $\underline{q}_n$ and $\underline{q}_{n-r}$ depends only on the distance between them measured along the string, then using the comoving distance

$$s_{n,n-r} = \sum_{k=1}^{r} b_{n-k}^0 \tag{2.25}$$

we specifically assume

$$< \underline{q}_n \cdot \underline{q}_{n-r} >= f(s_{n,n-r}). \tag{2.26}$$

The function f could in principle vary slowly with time. The instant $\underline{p}_m$ and $\underline{q}_n$ meet can be analysed. Given that the correlation between $\underline{p}$ and $\underline{q}$ is weak, in the continuum limit we obtain[76]

$$\alpha = - < \underline{p}_m . \underline{q}_n >= \frac{2}{3} h_0 \int ds f(s) \tag{2.27}$$

In (2.27) the integral runs from zero to an upper cutoff at a point where the correlation may be assumed to vanish. It represents a persistence length along the left or right moving string measured in comoving coordinates. We define the true persistence length of the left or right moving string as

$$\bar{\xi} = R \int dv x_v^0 < \underline{q}(0).\underline{q}(s) >= R \int ds f(s) \tag{2.28}$$

If as the simulations suggest, the long strings approach a scaling solution, then $\bar{\xi}$ should be a fixed fraction of the horizon distance

$$\bar{\xi} = \frac{d_{hor}}{\bar{\gamma}} = \frac{R\tau}{\bar{\gamma}} \tag{2.29}$$

From (2.27,2.28) we then have a prediction for α

$$\alpha = \frac{2m}{3\bar{\gamma}} \tag{2.30}$$

where m is defined by $R \propto \tau^m$ i.e. m=1 in the radiation dominated era and m=2 in the matter era. From (2.24) we obtain the result that if scaling is achieved for $\bar{\xi}$ then when this occurs

$$\begin{aligned} \bar{\gamma}_r &= 5.6 \pm 1.8 \qquad \bar{\gamma}_m = 6.7 \pm 1.3 \quad [AT] \\ \bar{\gamma}_r &= 4.8 \pm 1.4 \qquad \bar{\gamma}_m = 5.1 \pm 0.8 \quad [BB]. \end{aligned} \tag{2.31}$$

It is encouraging that within the uncertainties the <u>same</u> value of $\bar{\gamma}$ gives good agreement in both eras.

As indicated earlier, the main aim is to study whether the system of strings will approach scaling in a model with two distinct length scales. The scale ξ characterises the long string density, ρ_{ls}, and is defined by

$$\rho_{ls} = \frac{\mu}{\xi^2} \tag{2.32}$$

where μ is the string tension (assumed constant). From (2.28) we see that $\bar{\xi}$ is the persistence length describing the small scale structure on the strings. Relating both scales to the horizon distance

$$\xi = \frac{R\tau}{\gamma}, \qquad \bar{\xi} = \frac{R\tau}{\bar{\gamma}} \tag{2.33}$$

the scaling solution corresponds to γ and $\bar{\gamma}$ approaching constant values. It is this condition we will look for.

The equation for the long-string density is

$$\frac{d\gamma}{d\tau} = \frac{\kappa}{\tau}\gamma - \frac{c}{2\tau}\gamma\bar{\gamma} \tag{2.34}$$

where

$$\kappa = 1 - \frac{(1-\alpha)}{2}m, \qquad \alpha = \frac{2m}{3\bar{\gamma}} \tag{2.35}$$

In (2.34) the first term on the right hand side represents the stretching of the long string network, and the final term represents the loss of string due to loop production. The scale relevant to the production of small loops is the persistence length along the string $\bar{\xi}$. The parameter c measures the efficiency of loop production. We expect there to be at least two contributions to c, from the small and large loops produced[76]. Therefore we write

$$c = c_{sh} + c_{lo} \tag{2.36}$$

c_{sh} is difficult to calculate, but we expect it to be approximately constant. This is not the case for the large loops, we can estimate the efficiency of producing these loops off the network[76] using the fact that large loops follow Brownian paths. The result is that in (2.36) we obtain

$$\begin{aligned} c &= c_{sh} + c_m \frac{\bar{\gamma}^{\frac{1}{2}}}{\gamma} \\ c_m &= \frac{3}{8\pi}\left(\frac{3(1+m)}{2\chi}\right)^{\frac{1}{2}} \end{aligned} \tag{2.37}$$

Turning to the question of the rate equation for $\bar{\xi}$, the small scale structure, we obtain

$$\frac{d\bar{\gamma}}{d\tau} = \frac{\bar{\kappa}}{\tau}\bar{\gamma} - (q-1)\frac{c}{\tau}\bar{\gamma}^2 + \frac{\chi}{\tau}\gamma^2 \tag{2.38}$$

where

$$\bar{\kappa} = 1 + (1-3\alpha)m \tag{2.39}$$

The first term on the right hand side of (2.38) is due to the string stretching caused by the expansion of the universe. The third term is the effect of long string intercommutation which creates four new kinks on the network (two left moving and two right moving), $\chi \sim .24$ and is constant. The second term arises out of loop production. As before c is the efficiency of chopping loops off the network, but we have introduced the quantity q to try and account for the fact that when a loop is excised off a long string, it is more likely to have more than the average number of kinks on that loop as it carries away all the kinks on that section of string. Loops are not a random sample of the string within the volume V. It is reasonable to expect that a section of string containing more than the usual number of kinks is more likely to be incorporated into the loop. In principle q should be a function of ξ, $\bar{\xi}$ but it is difficult to calculate. For the time being we have kept it constant, but expect it to be greater than one.

Rewriting (2.34) and (2.38) using (2.35), (2.37) and (2.39) we obtain the rate equations in their full form[77]

$$\frac{d\gamma}{d\tau} = \frac{1}{\tau}\left[\left(1-\frac{m}{2}\right)\gamma + \frac{m^2}{3}\frac{\gamma}{\bar{\gamma}} - \frac{c_{sh}}{2}\gamma\bar{\gamma} - \frac{c_m}{2}\bar{\gamma}^{\frac{3}{2}}\right] \tag{2.40}$$

$$\frac{d\bar{\gamma}}{d\tau} = \frac{1}{\tau}\left[(1+m)\bar{\gamma} - 2m^2 - (q-1)c_{sh}\bar{\gamma}^2 - (q-1)c_m\frac{\bar{\gamma}^{\frac{5}{2}}}{\gamma} + \chi\gamma^2\right] \tag{2.41}$$

We need to look for scaling solutions of the coupled rate equations (2.40) and (2.41). Setting them both equal to zero we obtain the critical values of γ and $\bar{\gamma}$. From (2.40)

$$\gamma_c = \frac{3c_m\bar{\gamma}_c^{\frac{5}{2}}}{3(2-m)\bar{\gamma}_c + 2m^2 - 2c_{sh}\bar{\gamma}_c^2} \tag{2.42}$$

The resulting equation for $\bar{\gamma}$ is messy and not very revealing. From the rate equations we can solve for c and q if we know $\gamma, \bar{\gamma}$, or for $\gamma, \bar{\gamma}$ if we know c and q. This is currently under investigation. The sort of questions we hope to address are given the numerical values estimated for c and q from the simulations, when (if at all) would we expect the simulations of BB and AS to reach scaling of γ and $\bar{\gamma}$? Currently we require $q > 1.5$, $c \sim 0.1$ in order to have stable scaling solutions. The results are however particularly sensitive to c, q and will reported in detail elsewhere. We note though that it is possible to reach scaling solutions of the long and short string network without having to introduce the effect of gravitational backreaction. Whether backreaction is the dominant effect remains to be seen but we would be surprised if the small scale structure did not reach scaling before the backreaction scale was reached. This analytical approach is certainly enabling these important issues to be addressed, and hopefully soon the source of the differences will be pinned down.

b). Global Textures.

Recall the action in the pure scalar sector that gives rise to global textures is (1.52) [54]

$$S = \int d^4x\sqrt{-g}[(\partial_\mu\Phi^a)(\partial^\mu\Phi_a) - \lambda(\Phi^a\Phi_a - \eta^2)^2 - \frac{1}{16\pi G}R]$$

Writing the solution for Φ in terms of real scalar field

$$\Phi = \frac{1}{\sqrt{2}}(\phi_1 + \imath\phi_2, \phi_3 + \imath\phi_4), \qquad \vec{\phi} = (\phi_1 \ldots \phi_4) \tag{2.43}$$

the vacuum manifold V_o corresponds to a three sphere with $\vec{\Phi}^2 = \eta^2$. Variation of the action gives the wave equation for ϕ in terms of the conformal time $d\tau \equiv \frac{dt}{a(t)}$ and the scale factor $a(\tau)$

$$\frac{\partial^2\Phi^b}{\partial\tau^2} + \frac{2}{a}\left(\frac{\partial a}{\partial\tau}\right)\left(\frac{\partial\Phi^b}{\partial\tau}\right) - \nabla^2\Phi^b = -a^2\frac{\partial V}{\partial\Phi^b} \tag{2.44}$$

Here ∇ is with respect to comoving coordinates. This wave equation has a Hubble damping term $(\frac{\partial\Phi}{\partial\tau})$ and a forcing term $(\frac{\partial V}{\partial\Phi})$ which drives Φ into the minimum of its potential. The formation of textures was discussed in lecture 1. At temperatures above the critical temperature $(T_c \sim \eta)$

the symmetry is restored and Φ is localised about $\Phi = 0$. For $T < T_c$ the global symmetry is broken, a phase transition occurs and the Higgs field falls to V_0 in correlation regions $< t$. In this broken symmetry phase the Higgs field is made up of three massless Goldstone bosons which move tangential to S_3 and one massive radial Higgs ($w_o^{-1} \equiv m_\phi \sim \sqrt{\lambda}\eta$). The massive mode is negligibly excited as long as the length scales we are investigating are $\gg w_o$. Then there are two equivalent descriptions we can use to follow the collapse of the textures. The first is to use the full four component field in $V(\Phi)$ and solve the wave equation exactly. This can be done numerically and is the approach adopted in the numerical simulations[78]. The second approach which is useful in that it allows analytical solutions to be obtained, is to impose the condition $\Phi^2 = \eta^2$ as a constraint on the free-massless field action. This can be done using λ in the potential as a Lagrange multiplier. The effect is to have the three massless modes interacting in a non-linear sigma model with the equation of motion

$$\begin{aligned} \nabla^\mu \nabla_\mu \Phi^a &= -\frac{\nabla^\mu \Phi^b \nabla_\mu \Phi_b}{\eta^2} \Phi^a \\ \Phi^2 &= \eta^2 \end{aligned} \tag{2.45}$$

where ∇^μ is the covariant derivative operator and Φ^a, $(a = 1 \ldots 4)$. This approach breaks down when $|\nabla| \sim m_\phi$ (i.e. at w_o), because then it is possible for the gradient energy in Φ to cause the field to leave the vacuum manifold and unwind. We must therefore ensure that these solutions are used only in the regime $\gg w_o$, otherwise we must use the full field theory.

The initial conditions for Φ in the numerical simulations are to take $\Phi = 0$ everywhere, but have $\frac{d\Phi}{d\tau}$ in a random direction with fixed magnitude at each grid point. This imposes an initial uniform distribution of Φ simulating the phase transition from $\Phi = 0$ to $|\Phi| = \eta$. The unwinding of Φ around the vacuum manifold S_3 can be characterised by a conserved topological current $j^\mu = \frac{1}{12\pi\eta^4}\epsilon^{\mu\nu\alpha\beta}\epsilon^{abcd}\Phi^a\partial_\nu\Phi^b\partial_\alpha\Phi^c\partial_\beta\Phi^d$ if Φ remains on the 3-sphere. There is an associated conserved charge $Q = \int d^3x j^0$. This is an integer if Φ is everywhere parallel on the boundary. Thus j^0 is a winding density. A knot is a region where $Q \simeq 1$, and an anti-knot has $Q \simeq -1$. Of course there are partial windings with $Q < 1$, if $Q > .5$ the partial winding texture will also collapse, this is what we introduced in lecture 1.

We now turn to the solutions of the equation of motion. Turok and Spergel[79], and Notzold[80] have obtained exact flat space solutions. With the spherically symmetric ansatz

$$\Phi^a = \eta(cos\chi,\ sin\chi sin\theta cos\phi,\ sin\chi sin\theta sin\phi,\ sin\chi cos\theta) \tag{2.46}$$

where θ, ϕ are the usual polar angles, the flat spacetime equation of motion becomes

$$\ddot{\chi} - \frac{2}{r}\chi' - \chi'' = -\frac{sin(2\chi)}{r^2} \tag{2.47}$$

where $\dot{\chi} = \frac{d\chi}{dt}$, $\quad \chi' = \frac{d\chi}{dr}$. The exact self-similar solution is

$$\chi(t,r) = 2\arctan(-\frac{r}{t}), \qquad t < 0$$

$$= \quad 2\arctan(\frac{r}{t}) + \pi, \qquad t > r > 0$$
$$= \quad 2\arctan(\frac{t}{r}) + \pi, \qquad r > t > 0 \tag{2.48}$$

the π was included by Turok and Spergel to simulate the texture when it collapsed to a very small size and unwound. In the next lecture this solution will be used to calculate T^{μ}_{ν} in the weak field approximation, hence as a source for metric perturbations. There has been a great deal of interest in obtaining texture solutions in curved space. This is important in its own right, but in our context it is useful as it provides a direct test on the accuracy of using the weak field approximation. Barriola and Vachaspati found the metric of a self-similar global texture upto a function that could be given in closed form in terms of the texture field configurations[81]. The metric reduces to the weak field result of Turok and Spergel and Notzold in the relvent limit, and is canonical even in the strong field case. In the next lecture we will derive the weak field result. Barriola et al[54] and Durrer et al[82] have independently found approximate analytical solutions for the line element and the scalar field of a self similar global texture by considering Einstein's equations. Durrer et al. numerically integrate their solutions to make them exact[82].

The evolution picture developed by Turok and his collaborators is as follows[33]. Assuming that the initial conditions for Φ are specified by its being in thermal equilibrium and that after symmetry breaking, vacuum configurations of the Φ field exist which are uncorrelated on scales above the horizon, then as the universe expands the scale of spatial variation of Φ grows at the speed of light. A texture knot (region where the Higgs field winds around a three sphere in a non-trivial way - usually in regions larger than the horizon) enters inside the horizon, and collapses at the speed of light down to the inverse GUT size ($w_o \equiv m_{\phi}^{-1} \sim \lambda^{-0.5}\eta^{-1}$), the width of the texture whereupon they unwind themselves. Aslong as its size is larger than w_o the field Φ should remain close to the vacuum manifold M throughout space, but when the knot shrinks to $O(m_{\phi}^{-1})$, the field gradients are strong enough to pull Φ over the potential barrier and unwind the knot, leaving behind a lump of Goldstone bosons. The shrinking of knots correlates the Φ field on larger and larger scales, so the correlation length should grow to order the horizon scale and keep growing with it. As still larger scales come across the horizon, knots constantly form anew as Φ points in different directions on M in different horizon volumes. The result should be a number of knots per unit comoving volume n collapsing per unit conformal time τ of the form $\frac{dn}{d\tau} \sim \frac{c}{\tau^4}$ where c is a constant- the texture scaling solution.

It is useful to estimate the density perturbations induced by a collapsing knot[33]. Because they are constantly falling inside our horizon, they are always making perturbations on horizon scales. The density in a knot as it forms is $\sim (\nabla\phi)^2 \sim \frac{\eta^2}{t^2}$. Thus the density perturbation induced is given in terms of the background density ρ_b is $\frac{\delta\rho}{\rho} \sim \frac{\eta^2}{\rho_b t^2} \sim 20G\eta^2$ independent of the time in the matter or radiation eras. In other words the fluctuation is given by $\frac{\delta\rho}{\rho} \sim (\frac{m_{GUT}}{m_{pl}})^2 \sim 10^{-6}$, well within observational bounds. To a first approximation we have a scale-invariant Harrison-

Zeldovich spectrum. Numerically the scaling solution gives $c \sim .04 \sim \frac{1}{25}$. If a scaling solution occurs early as it appears to do, then it should be possible to relate c to the initial distribution of textures. We have been able to do this from our simulation of the phase transition, and are able to calculate how many knots fall within the horizon per unit Hubble time. This agrees well with observed value of c from the full numerical simulations[56].

In the next lecture we will use the known string and texture solutions to investigate how the network of strings and textures provide a source of density perturbations, hence act as seeds for the formation of large scale structure.

III. Defects and the Formation of Large Scale Structure

The hot big bang is incomplete without a source of perturbations, necessary to seed structures on large scales, especially as the universe was essentially very isotropic and homogeneous early on. There are recent models which suggest the possibility of late time phase transitions. These allow for the formation of light domain walls, and can then act as the seeds for large scale structure[14]. Initially we will investigate the density perturbations due to an infinite straight string[83]. Recall the action (2.1)

$$S = -\mu \int d\tau d\sigma [-det g_{ab}^2]^{\frac{1}{2}} \tag{3.1}$$

we can write the stress-tensor:

$$\begin{aligned} T^{\mu\nu}(x) &= -\frac{2}{\sqrt{-g^4(x)}} \frac{\delta S}{\delta g_{\mu\nu}(x)} \mid_{g=\eta} \\ &= \mu \int d\sigma \begin{pmatrix} 1 & \dot{x}^i \\ \dot{x}^i & \dot{x}^i \dot{x}^j - \acute{x}^i \acute{x}^j \end{pmatrix} \delta^3(x - x(\sigma, \tau)) \end{aligned} \tag{3.2}$$

For a straight static string along the z-axis, $\underline{x}(\sigma, \tau) = (0, 0, \sigma)$, so:

$$T^{\mu\nu}(x) = \mu \delta^2(x) diag(1, 0, 0, -1) \tag{3.3}$$

The negative pressure term gives the string its' tension. What is the effect of this string on the background metric? Writing,

$$g_{\mu\nu}(x) = \eta_{\mu\nu} + h_{\mu\nu}(x),$$

which is valid provided $\mid h \mid \ll 1$, then in the Harmonic gauge, $(g^{\alpha\beta}\Gamma^{\mu}_{\alpha\beta} = 0)$, Einsteins' equations become

$$\partial^2 h_{\mu\nu} = -16\pi G (T_{\mu\nu} - \frac{1}{2}\eta_{\mu\nu} T^{\lambda}_{\lambda}). \tag{3.4}$$

The time independent solutions have no Newtonian potential for h_{00}, h_{33}, (i.e $\nabla^2 h_{00} = \nabla^2 h_{33} = 0$). However we do obtain

$$h_{11} = h_{22} = 8G\mu In(\frac{r}{r_0}) \tag{3.5}$$

where r_0 is a suitable lower cut off introduced into the integral (i.e. the width of the string). Thus we find the line element becomes:

$$ds^2 = dt^2 - dz^2 - (1 - 8G\mu \ln(\frac{r}{r_0}))(dr^2 + r^2 d\phi^2) \tag{3.6}$$

Redefining
$d\acute{r} = (1 - 4G\mu \ln(\frac{r}{r_0}))dr$
$d\acute{\phi} = (1 - 4G\mu)d\phi$
we obtain

$$ds^2 = dt^2 - dz^2 - (d\acute{r}^2 + \acute{r}^2 d\acute{\phi}^2). \tag{3.7}$$

The effect of the backreaction of the string on the spacetime is to leave the metric flat everywhere except at the vertex, with an accompanying missing angle, $0 < \acute{\phi} < 2\pi - 8\pi G\mu$[83]. The spacetime is conical shaped with the string passing through the vertex of the cone. However the gravitational attraction due to the string on a static test particle is zero, as the particle effectively experiences a flat spacetime. If the string is moving relative to the test particles, they will feel the conical spacetime and be drawn in behind the string over an angle $2\Delta\phi = 8\pi G\mu$, forming small wakes where $\rho_{wake} = 2\rho_{background}$ as matter falls in on either side of the wake. I will mention observational consequences of this missing angle at the end of the lecture.
Turning our attention to the density perturbations generated by global textures, we follow the work of Turok and Spergel[79]. The stress-energy tensor is calculated from

$$T_{\mu\nu} = \nabla_\mu \Phi^a \nabla_\nu \Phi_a - \frac{1}{2} g_{\mu\nu} \nabla_\beta \Phi^a \nabla^\beta \Phi_a \tag{3.8}$$

Using the flat spacetime solution (2.46,2.48) we obtain

$$\begin{aligned} T_{00} &= \frac{2(r^2+3t^2)}{(r^2+t^2)^2}\eta^2 \\ T_{0i} &= -x^i \frac{4t}{(r^2+t^2)^2}\eta^2 \\ T_{ij} &= 2\delta_{ij}\frac{(r^2-t^2)}{(r^2+t^2)^2}\eta^2 \end{aligned} \tag{3.9}$$

Note the solution has a mass which diverges at large r: $M_{<r} \equiv 4\pi \int_0^r r^2 T^{00} dr \simeq 8\pi\eta^2 r$. This should not be a problem as there is a natural cutoff which we can introduce of size $O(t)$, the size of a knot when it collapses. Now $G\eta^2 \ll 1$, so to calculate the metric perturbations produced by $T_{\mu\nu}$, we use the weak field approximation–a limit which has recently been vindicated by analysis of the full general relativistic collapse of textures[54, 82]. Working in the gravitational Coulomb gauge specified by its conditions $h_{0i,i} = \frac{1}{2}h_{ii,0}$ and $h_{ik,k} = \frac{1}{2}h_{kk,i}$ and using spherical symmetry, $h_{\mu\nu}$ may be evaluated completely from the remaining radial integrals which can be performed analytically. Parameterising $h_{\mu\nu}$ by four functions of r and t, $h_{00} = H$, $h_{0i} = Ix^i$, $h_{ij} = Jx^i x^j + K\delta^{ij}$ one

solves for H, I and h_{kk} from Einsteins equations, and determines J, K from h_{kk} using the second gauge condition above.

$$
\begin{aligned}
H_C &= -2\epsilon[\ln(1+\frac{r^2}{t^2}) + 3(\frac{t}{r}\arctan(\frac{r}{t}) - 1)] \\
I_C &= -2\epsilon t[-\frac{1}{r^2} + \frac{t}{r^3}\arctan(\frac{r}{t}) + \frac{1}{3t^2}] \\
J_C &= -\epsilon[\frac{1}{3r^2} + \frac{t^3}{r^5}\arctan(\frac{r}{t}) - \frac{t^2}{r^4}] \\
K_C &= \epsilon[-\frac{2}{3}\ln(1+\frac{r^2}{t^2}) + \frac{1}{9} - \frac{t^2}{3r^2} + \frac{t^3}{3r^3}\arctan(\frac{r}{t})] \qquad (3.10)
\end{aligned}
$$

where $\epsilon \equiv 16\pi G\eta^2$ and the subscript denotes the Coulomb gauge. For finite radius and large positive or negative time the metric is the Minkowski metric, so particle and photon trajectories are then easy to interpret. The singular behaviour of H_C, I_C and K_C at $t = 0$ is a coordinate singularity, easily seen by a suitable change of coordinates. Recalling that the sigma model approximation upon which this analysis is based breaks down for $r = r_c \sim w_o$, of order the inverse GUT scale, we use r_c in the new coordinates $x'^\mu = x^\mu + \xi^\mu$, with $\xi_0 = 2\epsilon(t\ln|\frac{t}{r_c}| - t)$, $\xi_i = \frac{2}{3}\epsilon x_i \ln|\frac{t}{r_c}|$. The new metric is given by $h'_{\mu\nu} = h_{\mu\nu} - \xi_{\mu,\nu} - \xi_{\nu,\mu}$. This metric diverges logarithmically at large r and large t. Fortunately though, $h'_{\mu\nu}$ is small in the region of interest (inside the horizon). The large r divergence leads to a nice geometrical interpretation similar to the case of cosmic strings. The spatial hypersurfaces have a metric

$$
(1 - \frac{4}{3}\epsilon\ln(\frac{r}{r_c}))(dr^2 + r^2 d\Omega^2) \qquad (3.11)
$$

Under a change of variables $dr' = (1 - \frac{2}{3}\epsilon\ln(\frac{r}{r_c}))dr$, i.e. $r' = r(1 - \frac{2}{3}\epsilon\ln(\frac{r}{r_c}) + \frac{2}{3}\epsilon)$, this yields (to order ϵ),

$$
dr'^2 + r'^2(1 - \frac{4}{3}\epsilon)d\Omega^2 \qquad (3.12)
$$

which is just flat space minus a solid angle $\frac{8\pi\epsilon}{3} = \frac{128\pi^2 G\eta^2}{3}$. Notzold[80] obtains a deficit angle of $2\pi\epsilon$, a result confirmed by Barriola and Vachaspati[81], although as they point out the discrepancy could just be due to a different time slicing. They also demonstrate the interesting result that the metric is conical even in the strong field case. The geometry is exact at $t = 0$, reminiscent of the result for a static global monopole (the actual result obtained by Turok and Spergel[79]; is $\frac{4}{3}$ that of Barriola and Vilenkin[34]). We will later return to this missing angle.

Now though we turn our attention to just how matter can accrete around objects. We will concentrate on the case of cosmic strings and quote the results of global textures. The latter is still in its early period of research so the results we will be quoting are to some degree preliminary should be noted as such.

We saw in the previous lecture how a good fraction of string ends up in loops. It is worth while looking at the fall of matter onto loops as it provides a very good testing area for the types of calculation that are done when investigating structure formation. By the time matter starts accreting around these loops they should be well inside the horizon, $r \ll H_0^{-1}$, so gravity should then be weak inducing small peculiar velocities, $\delta v \ll 1$. In this regime we can use linear perturbation theory[84]. Also on scales large compared to their size, it is a good approximation to treat the loops as Newtonian point particles[85]. Thus in the Newtonian approximation to Einstein's equations we have for a particle at $\underline{r}(t)$,

$$\ddot{\underline{r}} = -\underline{\nabla}_r \Phi \tag{3.13}$$

where the Newtonian potential obeys

$$\underline{\nabla}^2 \Phi = -4\pi G \rho \tag{3.14}$$

For cold dark matter, (CDM), mass is conserved, so in a comoving volume

$$\int \rho dV_{comoving} = \text{const} \tag{3.15}$$

The solution in a homogeneous background, $\rho_B(t)$ only is

$$\begin{aligned} \Phi &= \frac{4\pi G \rho_B \underline{r}^2}{6} \\ \ddot{\underline{r}} &= -\frac{4G\rho_B \underline{r}}{3} \\ \rho_B r^3 &= \text{const} \end{aligned} \tag{3.16}$$

In an $\Omega = 1$ universe, we find on integrating (3.16),

$$\begin{aligned} r &\propto t^{\frac{2}{3}} \\ \rho_B &= \frac{1}{6\pi G t^2} \end{aligned} \tag{3.17}$$

which is the standard FRW matter dominated universe. Now we want to do linear perturbation theory about this solution. Define $a(t) = (\frac{t}{t_i})^{\frac{2}{3}}$ where t_i is the initial time. Then

$$r(t) = a(t) r_i \tag{3.18}$$

Thus $\underline{x} = \underline{r}_i$ is the initial comoving coordinate of the particle. Giving each particle a small comoving displacement $\underline{\psi}(x,t)$ we write

$$\underline{r}(t) = a(t)[\underline{x} + \underline{\psi}(\underline{x},t)] \tag{3.19}$$

Mass conservation implies,

$$\rho_i d^3 x_i = \rho(r) d^3 r \tag{3.20}$$

so using $\rho_i = \rho_B a^3$, this implies

$$\rho(\underline{r}) = \frac{a^3 \rho_B}{\mid \frac{d\underline{r}}{dx} \mid}$$

From (3.19) we have:

$$\rho(\underline{r}) = \rho_B(1 - \underline{\nabla}_x.\psi) + O(\psi^2) \tag{3.21}$$

Defining

$$\frac{\delta\rho}{\rho} \equiv \frac{(\rho(r) - \rho_B)}{\rho_B}$$

we can write

$$= -\underline{\nabla}_x.\underline{\psi} \tag{3.22}$$

for the fractional density perturbation.

To solve for r(t), hence for $\psi(\underline{x}, t)$, substitute (3.21) into (3.13,3.14)

$$\begin{aligned} \nabla_r^2\Phi &= \frac{1}{r^2}\frac{d}{dr}(r^2\frac{d\Phi}{dr}) \\ &= -4\pi G\rho(r) \end{aligned}$$

with

$$(\underline{\nabla}_r = \frac{\underline{\nabla}_x}{a})$$

and obtain

$$r^2\frac{d\Phi}{dr} = -4\pi G\int r^2 dr\rho_B(1 - a\underline{\nabla}_r.\underline{\psi}) \tag{3.23}$$

Hence we find

$$\underline{\nabla}_r\Phi = 4\pi G\rho_B(\frac{\underline{r}}{3} - a\underline{\psi})$$

which when substituted in (3.13) gives

$$\ddot{\underline{\psi}} + \frac{2\dot{a}}{a}\dot{\underline{\psi}} = 4\pi G\rho_B\underline{\psi} \tag{3.24}$$

Taking the homogeneous solution we find $\dot{\underline{\psi}} \propto a^{-2}$. Now the peculiar velocity of a test particle (the physical velocity minus the Hubble flow) is,

$$\begin{aligned} v_p &= \dot{\underline{r}} - \frac{\dot{a}}{a}\underline{r} \\ &= \dot{\underline{r}} - \dot{a}(\underline{x} + \underline{\psi}) \\ &= a\dot{\underline{\psi}} \end{aligned}$$

or

$$v_p \propto \frac{1}{a},$$

so a test particle slows down in comoving coordinates. We interpret the right hand side of (3.24) as the gravitational instability, the feedback of the perturbation on itself. For the case of the matter dominated FRW universe, i.e $\rho_B = \frac{1}{6\pi G t^2}$, the solution to (3.24) is

$$\underline{\psi}(x,t) = \underline{A}(x)t^{\frac{2}{3}} + \underline{B}(x)t^{-1} \tag{3.25}$$

with $\underline{A}$ *and* $\underline{B}$ arbitrary functions. Recalling (3.22) we see that $\frac{\delta\rho}{\rho}$ obeys the same equation as ψ, it has the same time behaviour. Thus

$$\frac{\delta\rho}{\rho} \propto a(t)$$

in the growing mode.

We have seen

$$v_p = \dot{\underline{r}} - (\frac{\dot{a}}{a})r = a\dot{\underline{\psi}}.$$

For the solution (3.25), in the growing mode, $v_p \propto t^{\frac{1}{3}}$. Now (3.25) also tells us that in the growing mode, $\psi \propto a$, whereas from (3.19) we know

$$\underline{\psi} = \frac{(\underline{r} - \underline{x})}{a}.$$

Hence

$$\underline{v}_p = (\frac{\dot{a}}{a})[\underline{r} - \underline{x}] = Hd\underline{r} \tag{3.26}$$

the equation used in lecture 1. The usefulness of (3.26) is that in particular geometries v_p is related to the magnitude of $\frac{\delta\rho}{\rho}$. Defining the Hubble velocity, $v_H = Hr$, write

$$\frac{\underline{v}_p}{v_H} = \frac{d\underline{r}}{r} \tag{3.27}$$

In CDM scenarios we know mass is conserved in a comoving volume (3.15). Equation (3.22) tells us that

$$\frac{\delta\rho}{\rho} = \frac{\int \underline{\psi}.d\underline{S}}{V_i}.$$

For planar collapse over a length scale L, $V_H = HL$, so

$$\frac{\delta\rho}{\rho} = \frac{\psi}{L} = \frac{v_p}{V_H}.$$

For cylindrical collapse it is easy to see, (because $V \propto r^2$)

$$\frac{v_p}{v_H} = \frac{\delta\rho}{2\rho},$$

and for spherical collapse, ($V \propto r^3$),

$$\frac{v_p}{v_H} = \frac{\delta\rho}{3\rho}.$$

The recent observations of peculiar velocities give a unique window on $\frac{\delta\rho}{\rho}$[86]. In particular for the 'great attractor', $v_p \sim 500kms^{-1}$, *at* $L \sim 100h_{50}^{-1}Mpc$ gives $\frac{\delta\rho}{\rho} \sim \cdot 3$.

So far we have thought only about linear collapse in a particular background. What happens when we include sources, (e.g loops of cosmic string)[15]? On scales large compared to the size of the loop, we can treat the loops as Newtonian point particles. Equations (3.14,3.24) become

$$\nabla^2\Phi = -4\pi G(\rho_B + \rho_{source}) \tag{3.28}$$

$$\ddot{\underline{\psi}} + \frac{2\dot{a}}{a}\dot{\underline{\psi}} - 4\pi G\rho_B\underline{\psi} = -\frac{1}{a}\underline{\nabla}\Phi_{source} \tag{3.29}$$

A nice trick used to solve (3.29) is found by recalling that we really require $\frac{\delta\rho}{\rho}$. Using Gauss's theorem to rewrite (3.22)

$$\frac{\delta\rho}{\rho} = \frac{\int \underline{\nabla}.\underline{\psi}dV}{V_i} \tag{3.30}$$

we define $\delta \equiv \underline{\nabla}.\underline{\psi}$, and take the divergence of (3.29)

$$\ddot{\delta} + \frac{4}{3t^2}\dot{\delta} - \frac{2}{3t^2}\delta = 4\pi G\delta\rho_{source}(\underline{x}, t) \tag{3.31}$$

This has a Green's function solution,

$$\delta(\underline{x}, t) = \int_{t_i}^{t} dt' G(t, t') 4\pi\delta\rho_{source}(\underline{x}, t)$$

where

$$G(t, t') = \frac{3}{5}(t^{\frac{2}{3}}(t')^{\frac{1}{3}} - t^{-1}(t')^2).$$

Now for a loop with comoving trajectory $Z(t, \sigma)$,

$$\delta\rho_{source}(t) = \mu \int d\sigma \frac{\delta^3(\underline{x} - \underline{z}(t, \sigma))}{a^3(t)}$$

which leads to the trumpet shape trajectories as the matter falls onto a loop and the loop shrinks in comoving coordinates[87].

We will not describe the models of non - linear collapse and Hot Dark Matter. A thorough treatment can be found in [15], or in Brandenberger's excellent review article[88].

For the case of global textures we note that the possibility of a texture seeded universe has only recently been addressed in any detail. Park et al.[89] have studied the formation of large scale structure by global textures in a flat universe dominated by cold dark matter. Combining a code which evolves the texture field with an N-body code for evolving the dark matter, they claim to have promising results. With only one-free parameter (the symmetry breaking scale) they reproduce the observed galaxy-galaxy correlation function, find significant galaxy clustering on scales of $20 - 50h^{-1}Mpc$ and coherent structures of over $50h^{-1}Mpc$ in the galaxy distribution. Large scale streaming velocities compare favourably with observation, but potential problem areas are that the clusters exhibit too high velocity dispersions, and voids are not as empty as the observed voids. Gooding et al[90] have recently analysed the growth of density fluctuations in CDM induced by the unwinding of a texture knot, and claim that the non-gaussian perturbations produced by global textures lead to the early formation of stars, quasars and galaxies, consistent with some recent observations.

These results are intriguing, they provide good tests for the model. However the cleanest signatures for theories of the origin of perturbations are the temperature fluctuations induced in the microwave background. The form of the fluctuations and their magnitude depend on the form of the metric perturbations causing them, hence on the candidate seed, in our case cosmic strings and global textures. The fluctuation is obtained from the Sachs-Wolfe effect, the energy shift experienced by photons travelling by a source of metric perturbations[91]. In conformal coordinates

$$g_{\mu\nu} = a^2(\tau)(\eta_{\mu\nu} + h_{\mu\nu}) \tag{3.32}$$

The energy shift experienced by a photon traversing a metric $g_{\mu\nu}$ is determined by the geodesic equations

$$\ddot{X}^\mu + \frac{1}{2}\eta^{\mu\nu}(h_{\nu\beta,\alpha} + h_{\alpha\beta,\nu} - h_{\nu\alpha,\beta})\dot{X}^\nu \dot{X}^\alpha = 0 \tag{3.33}$$

The unperturbed photon trajectory is taken to be $x_o^\mu = n^\mu \lambda$, $p_o^\mu = E n^\mu$ with E the energy, $n^0 = 1$, $\vec{n}^2 = 1$ and λ the affine parameter. Under a perturbation $p^\mu = p_o^\mu + \delta p^\mu$, we obtain $\delta p^\mu = E_o \frac{d\delta x^\mu}{d\lambda}$, hence the change in the geodesic equation is

$$\frac{d}{d\lambda}\left(\frac{\delta p^0}{E_o}\right) + \frac{1}{2}(-2h_{\nu 0,\alpha} + h_{\nu\alpha,0})n^\nu n^\alpha = 0 \tag{3.34}$$

or

$$\left.\frac{\delta p^0}{p_o^0}\right|_i^f = -\frac{1}{2}\int_{\lambda_i}^{\lambda_f} h_{\mu\nu,0} n^\mu n^\nu d\lambda + \left[h_{0\mu} n^\mu\right]_i^f \tag{3.35}$$

Imagine that the photon is emitted/received by a massive particle nearly at rest with four momentum $k^\mu(i) = k_o^\mu + \delta k^\mu(i)$, $/k^\mu(f) = k_o^\mu + \delta k^\mu(f)$, where $k_o^\mu = (m, \underline{0})$ and m is its rest mass. The emitted photon energy is $E = -\frac{k.p(i)}{m}$, and the final photon energy measured by the absorbing particle is $E(f) \equiv E + \delta E = -\frac{k.p(f)}{m}$ which implies

$$\left.\frac{\delta E}{E}\right|_i^f = -\frac{1}{2}\int_{\lambda_i}^{\lambda_f} h_{\mu\nu,0} n^\mu n^\nu d\lambda + \frac{1}{2}\left[h_{00}\right]_i^f - \left[\frac{\delta\vec{k}.\vec{n}}{m}\right]_i^f \tag{3.36}$$

The three terms in the Sachs-wolfe effect are a path-dependent redshift, a part from the local gravitational potential and a Doppler shift. This is the full expression and all that remains is to insert the particular source of $h_{\mu\nu}$.

For cosmic strings, the anisotropy was first calculated by Kaiser and Stebbins[92]. They realised that strings produced a background temperature with steplike discontinuities on curves in the sky. As with the lensing, this relies on the canonical structure of spacetime near a string. Light rays travelling on either side of the string produce a Doppler shift, $\Delta\nu = 8\pi G\mu v_\perp$ in their frequency, $v_\perp$ is the transverse velocity of the string. Comoving frames on either side of the string move towards each other with a velocity, $v_\perp \sim \cdot 4$. Thus we find

$$\begin{aligned} \frac{\delta T}{T} &= 8\pi G\mu v_\perp \\ &< 2.10^{-5} \end{aligned}$$

which is almost at the level of accuracy we can now observe. Recently temperature maps of a 'stringy' universe have been produced to show typical angular distributions of the discontinuities, hence what the temperature distribution may typically look like[93].

Turok and Spergel have used the expression derived for $h_{\mu\nu}$ in (3.10) in (3.36) to obtain the Sachs - Wolfe effect for global textures[79]. Ignoring the third term as the Doppler effect is small in this case, they obtain an intriguing result. Photons crossing $z = 0$ before the texture collapses are redshifted, those crossing after collapse are blue shifted. The former photons fall into a cloud of collapsing texture, the latter ones climb out and the net effect is that texture knots produce blue or red discs on the microwave background, with $\frac{\delta T}{T} < 8\pi^2 G\eta^2$. Such a distribution is quite distinctive and quite different from that produced in gaussian scenarios. The distribution of these red and blue discs on the microwave background can be obtained from the scaling solution. With $\Omega = 1$, there appear to be of order 10 spots covering a scale of 10 degrees across the microwave background sky[79]. This angular distribution is almost certainly within the current sensitivity of COBE. Such a distinctive signal, if it does fall within the sensitivity of the COBE experiment, and some simulations suggest that it should now be there, should soon rule out or confirm the texture scenario.

Finally in this lecture we will discuss some of the other observational tests of cosmic string theory, tests on which the model will also either succeed or fail. They all require a determination of $G\mu$. The first test I have already mentioned is the gravitational lensing by strings[83], where an observer could see a double image of a quasar or galaxy behind a cosmic string. The images would be separated by an angle of $4\pi G\mu \sim 5''$, (the missing angle), for $G\mu \sim 10^{-6}$. There are five or so reported cases of lensing between $2.5'' - 7''$. However a line of double images would be strong evidence for the existence of a string. Recently a potential string candidate gas been observed[94, 95]. It consists of a group of four 'twin' galaxies with characteristic pair separations of order 2". They all appear to be genuine binaries with three of the redshifts near $z \sim 0.4$ while the fourth is at $z \sim 0.2$. Further observations of the fine details of the binaries will help confirm or more likely rule out this particular candidate.

One of the tightest constraints on cosmic strings comes from the gravitational radiation from loops of string[96, 70, 71]. This is one reason why it is necessary to reconcile the differences between the various simulations of the string network (see lecture 2). Typically the energy density of radiation from a loop, in terms of the logarithmic spectrum,

$$\Omega_g(w) = w\rho(w)\rho_0^{-1},$$

($w = R^{-1}$ is the main frequency of radiation, ρ_0 is the background energy density), is

$$\Omega_g(w) = 2.10^{-7} h_{50}^{-2}.$$

Now variations in the observed frequency of the millisecond pulsar, places the severest restriction on $G\mu$,

$$G\mu \leq 10^{-5}(\frac{\alpha}{T_{yr}})^8$$

where $\alpha \sim 2\pi$, T_{yr} is the observation period in years. In principle we will soon be in a position to confirm or rule out cosmic strings. However in practice we are reaching the limits of reliability with the pulsar timing, in our knowledge of the dynamics and radiation contribution of the contents of the solar system aswell as with the the accuracy of the atomic clocks we use to standardise things with. The latter problem could be overcome by using a few pulsars and comparing their relative times, as they are the most accurate clocks we know of. It is therefore extremely difficult to pick out any one contribution as the main source of the radiation, hence we still are not in a position to rule out or confirm the cosmic string scenario. This point was recently emphasised by Taylor[97] who is actually performing the measurements.

As we discussed in lecture 1 currently a great deal of observation time is going into investigating the large scale structure of the universe, including things like 3pt correlation functions between clusters, the distribution of bubbles, voids, filaments. The next few years will be an exciting time in comparing the numerical simulations with observation. At the moment the simulations (including those of comic strings) appear to be falling short. Work is currently going on to investigate the growth of matter around a network of string, at least in the linear regime. The full N-body problem is also being investigated. The current picture for structure formation around cosmic strings is as we have seen earlier radically different to that proposed a few years ago. The conical nature of spacetime around the long strings causes matter to accrete in wakes behind them. The role of loops for the accretion of matter has basically disappeared. It is still an open question as to whether the wake picture of structure formation can generate the observed large scale structure, in particular the huge voids. There is also currently both concern and excitement over microwave background limit arising from global textures. They should soon be ruled out or confirmed.

What then can we make of the role of particle physics in cosmology? Particle physics has been extraordinarily successful up to energies of $250GeV$ the Electroweak era, where the ideas of spontaneous symmetry breaking (SSB) have been confirmed (the reader should be aware that there are also models of dynamical symmetry breaking where an explicit Higgs is not required, and which can give the same features as spontaneous symmetry breaking- note also that the Higgs has yet to be discovered). Actually textures have recently been invoked as a possible mechanism whereby baryogenesis could occur at the electroweak rather than the GUT scale, these gauged textures are the ones that arise naturally in the Weinberg-Salam model[99].

The success of the electroweak theory has prompted a surge of interest into GUT models, the inclusion of the strong force and the effects such models produce. Such effects would be found around 10^{-35}seconds into the universe's lifetime. If the early universe was characterised by a series of SSB, then one effect from the GUT models would have been the production of massive topological defects. Irrespective of their usefulness for cosmology, we would need to know their role in the universe (i.e. the monopole problem arises out of one such process). It is remarkable that the objects created at such early times could still be playing an important role today in the evolution of the universe, and that they can produce realistic observable consequences which are currently being tested for. It isn't really the success or failure of any one model which is vital, it is the recognition that particle physics plays an important role in cosmology and moreover cosmology could provide the first direct evidence for GUT models. It is this thought which provides some of the motivation to study this fascinating subject.

All through these lectures we have discussed topological defects in the early universe. There is no reason why they should not be found at much lower energy scales and indeed they are. For an extremely nice review of defects in condensed matter systems Mermim is to be recommended[98]. Recently Turok and his collaborators have been investigating the formation of defects in liquid crystals and have come up with some extremely nice results, including confirmation of the one-scale model discussed in lecture 2[100]. Infact they claim to be able to see strings, monopoles and textures in their experiments - the world really could be encapsulated under the microscope!

Acknowledgements

I would like to thank the organisers of the Lisbon meeting especially A. Henriques for all of his help and support. The meeting was one of the nicest I have had the pleasure of attending and it was in no small measure due to the people of Lisbon–it also helped that one of the worlds largest chocolate manufacturers was hosting an international meeting at the same time as ours, and breaks were spent down at the chocolate bars. I would also like to thank Jose Mimoso for his kindness throughout the stay. I am extremely grateful to the friends and colleagues with whom I have had the pleasure of collaborating. They include J. Barrow, J. Borrill, D. Garfinkle, D. Haws, S. Holbraad, T.W.B. Kibble, E.W. Kolb, A. Liddle, R. Rivers, and N. Turok. I was supported by an SERC Advanced Fellowship.

References

[1] E.W. Kolb and M.S. Turner, *The Early Universe*, Addison-Wesley (1990)

[2] A. Guth, *Phys. Rev. D* **23** (1981) 347.

[3] J.D. Barrow *in this volume*

[4] T.W.B. Kibble, *J. Phys. A* **9** (1976) 1387

[5] Ya. B. Zel'dovich, *Mon. Not. R. Astron. Soc* **192** (1980) 663

[6] A. Vilenkin, *Phys. Rev. Lett.* **46** (1981) 1169; *Phys. Rep.* **121** (1985), 263.

[7] W. Saunders, C. Frenk, M. Rowan-Robinson, G. Efstathiou, A. Lawrence, N. Kaiser, R. Ellis, J. Crawford, X.Y. Xia and I. Parry, *Nature* **349** (1991) 32.

[8] G. Efstathiou, W.J. Sutherland and S.J. Maddox, *Nature* **348** (1990) 705;
(see Barrow in this volume);
P.J. Steinhardt, *Nature* **345** (1990) 47.

[9] R. Giovanelli and M.P. Haynes, *Astron. J* **87** (1982)

[10] V. de Lapperent, M. Geller and J. Huchra, *Ap. J. Lett.* **302** (1986) 11

[11] T.J. Broadhurst, R.S. Ellis, D.C. Coo and A.S. Szalay, *Nature* **343** (1990) 726

[12] P. Coles, *Nature* **346** (1990) 47
J. Barrow and P. Coles, *Mon. Not. R. Astron. Soc* **224** (1990) 188

[13] J. Barrow and P. Coles, *Mon. Not. R. Astron. Soc* **248** (1991) 52

[14] C. Hill, D.N. Schramn and J. Fry, *Comments on Nucl. Part. Phys.* **19** (1989) 25

[15] N. Turok, *Lectures presented at Cern/Eso Winter School on Cosmology and Particle Physics, Erice* (1987)

[16] E. Witten, *Nucl. Phys. B* **249** (1985) 557.

[17] E. Copeland, *Cosmic Strings and Superconducting Cosmic Strings.* Lectures presented at Second Erice Summer School on Dark Matter, 1988, Fermilab-Pub-88-108A

[18] J. Ostriker, C. Thompson and E. Witten, *Phys. Lett. B* **180** (1986) 231.

[19] B. Carter, *Phys. Lett. B* **238** (1990) 166;
Cosmic Rings as Dark Matter Candidates C.N.R.S.-Observatoire de Paris, Meudon preprint (1990)

[20] E. Copeland, M. Hindmarsh and N. Turok, *Phys. Rev. Lett.* 58 (1987) 1910

[21] R.L. Davis and E.P.S. Shellard, *Phys. Lett. B* **209** (1988) 485; R.L Davis, *Phys. Rev. D* **38** (1988) 3730

[22] D. Kirzhnits, *JETP Lett.* **15** (1972) 529; D. Kirzhnits and A. Linde, *Phys. Lett. B* **42** (1972) 471; L. Dolan and R. Jackiw, *Phys. Rev. D* **9** (1974) 3320; S. Weinberg, *Phys. Rev. D* **9** (1974) 3357

[23] E. M. Lifshitz and L. P. P. Pitaevskii, *'Statistical Physics' 3rd Edition Part 1* [Pergamon Press] Chapter 146.

[24] M. Gleiser, *Phys. Rev. D* **42** (1990) 3350

[25] D. La and P.J. Steinhardt, *Phys. Lett. B* **220** (1989) 375

[26] E. Copeland, E. Kolb and A Liddle, *Phys. Rev. D* **42** (1990) 2911

[27] H. Nielsen and P. Olesen, *Nucl. Phys. B* **61** (1973) 45

[28] G. 't Hooft, *Nucl. Phys. B* **79** (1974) 276, A. Polyakov, *JETP. Lett.* **20** (1974) 194

[29] G. Gibbons, *in this volume*

[30] J. Preskill, *Phys. Rev. Lett.* **43** (1979) 1365

[31] A.Everett, T. Vachaspati and A. Vilenkin, *Phys. Rev. D* **31** (1985) 1925; E. Copeland, D. Haws, T.W.B. Kibble, D. Mitchell and N. Turok, *Nucl. Phys. B* **298** (1988) 445

[32] R. Davis, *Phys. Rev. D* **35** (1987) 3705;**36** (1987) 997.

[33] N. Turok, *Phys. Rev. Lett.* **63** (1989) 2625.

[34] M. Barriola and A. Vilenkin, *Phys. Rev. Lett.* **63** (1989) 341.

[35] D. P. Bennett and S. H. Rhie, *Phys. Rev. Lett* **65** (1990) 1709.

[36] T.W.B. Kibble, G. Lazarides and Q. Shafi, *Phys. Lett. B* **113** (1982) 237; D. Olive and N. Turok, *Phys. Lett. B* **117** (1982) 193.

[37] T. Vachaspati and A. Vilenkin, *Phys. Rev. D* **30** (1984) 2036.

[38] D. Mitchell and N. Turok, *Phys. Rev. Lett.* **58** (1987) 1577; *Nucl. Phys. B* **294** (1987) 1138.

[39] E. Copeland, D. Haws and R. Rivers, *Nucl. Phys. B* **319** (1989) 687.

[40] E. Copeland, D. Haws, S. Holbraad and R. Rivers, *The Statistical Properties of Strings. I Free Strings* in 'The Formation and Evolution of Cosmic Strings' eds. G. Gibbons, S. Hawking and T. Vachaspati (C.U.P. 1990).

[41] S. Frautchi, *Phys. Rev. D* **3** (1971) 2821.

[42] R. Carlitz, *Phys. Rev. D* 5 (1972) 3231.

[43] E. Copeland, D. Haws, S. Holbraad and R. Rivers, *Phys. Lett. B* **236** (1990) 49.

[44] G. Smith and A. Vilenkin, *Phys. Rev. D* **36** (1987) 987.

[45] A. Vilenkin and M. Sakellariadou, *Phys. Rev. D* **37** (1988) 885.

[46] R. Hagedorn, *Nuov. Cim. Supp.* **3** (1965) 147.

[47] N. Turok, *Phase Transitions as the Origin of Large Scale Structure in the Universe* in 'Particles, Strings and Supernovae (TASI88)' eds. A. Jevicki and C.I. Tan (World Scientific, Singapore 1989).

[48] J. Blin, Acta Metall. 3,199(1955).

[49] S. F. Edwards and M. Warner, Phil. Mag. A40,257(1979).

[50] E. Copeland, D. Haws, S. Holbraad and R. Rivers, *The Statistical Properties of Strings. II Interacting Strings* in 'The Formation and Evolution of Cosmic Strings' eds. G. Gibbons, S. Hawking and T. Vachaspati (C.U.P. 1990).

[51] E. Copeland, D. Haws, S. Holbraad and R. Rivers, *The Statistical Mechanics of Interacting Strings* Imperial preprint TP/89-90/05

[52] H. Kleinert, *Gauge Fields in Condensed Matter Physics*, Vols. I and II, World Scientific Press, Singapore (1990)

[53] C. Itzykson and J.B. Zuber, *Quantum Field Theory* McGraw-Hill

[54] M. Barriola, T. Vachaspati and A. Vilenkin, *Approximate analytic solutions to self-similar global texture metric* Tufts preprint (1990)

[55] G.H. Derrick, *J. Math. Phys.* **5** (1964) 1252

[56] J. Borrill, E. Copeland and A. Liddle, *Initial Conditions for Global Textures* SUSSEX-AST 91/1-1, *Phys. Lett. B* in press

[57] A. M. Srivastava, "Importance of Boundary Conditions for Topological Production of Textures and Skyrmions", University of Minnesota preprint TPI-MINN-90/42-T, (1990)

[58] L. Perivolaropoulos, "Primordial Perturbations by π_2 Textures", Brown University preprint BROWN-HET-775, (1990)

[59] Y. Nambu *Lectures for the Copenhagen Summer Symposium* (1970) unpublished

[60] K. Maeda and N. Turok, *Phys. Lett. B* **202** (1988) 376; R. Gregory, *Phys. Lett. B* **206** (1988) 199

[61] R. Gregory, D. Haws and D. Garfinkle, *Phys. Rev. D* **42** (1990) 343

[62] E.P.S. Shellard, *Nucl. Phys. B* **283** (1987) 624

[63] P.J. Ruback, *Nucl. Phys. B* **296** (1988) 669; E.P.S. Shellard and P.J. Ruback, *Phys. Lett. B* **209** (1988) 262

[64] R. Matzner, *Computers in Physics* **1** (1988) 51

[65] T. Vachaspati, *Nucl. Phys. B* **277** (1986) 593
D. Garfinkle, *Santa Barbara preprint* (1990)

[66] T.W.B. Kibble and N. Turok, *Phys. Lett. B* **116** (1982) 141

[67] T. Vachaspati and A. Vilenkin, *Phys. Rev. D* **31** (1985) 3052; N. Turok, *Nucl. Phys. B* 242 (1984) 520; C. Burden, *Phys. Lett. B* **164** (1985) 277

[68] J. Quashnock and D. Spergel, *Phys. Rev. D* **42** (1990) 2505

[69] M. Hindmarsh, *Gravitational Radiation from Kinky Infinite Strings* Newcastle preprint NCL-90 TP11 *Phys. Lett. B* in press

[70] A. Albrecht and N. Turok, *Phys. Rev. D* **40** 973 (1989)

[71] D.P. Bennett and F.R. Bouchet, *Phys. Rev. Lett.* **60** 257 (1988);**63**, 2776 (1989)

[72] B. Allen and E.P.S. Shellard, *Phys. Rev. Lett.* **64** 119 (1990); E.P.S. Shellard and B. Allen, MIT preprint 1989

[73] T.W.B. Kibble, *Nucl. Phys. B* **252** (1985) 227
D.B. Bennett, *Phys. Rev. D* **33** (1986) 872; **34** (1986) 3592

[74] A. Albrecht and N. Turok, *Phys. Rev. Lett* **54** (1985) 1868

[75] B. Allen and R. Caldwell, *Phys. Rev. Lett.* **65** (1990) 1705; *Kinky structure on strings* Milwalkee preprint (1990)
J. Quashnock and T. Piran, Fermilab-Pub 90/179-A Sept (1990)

[76] T.W.B. Kibble and E. Copeland, *Evolution of small-scale structure on strings* (to appear in the 'Proceedings of the Nobel Symposium 79 on The Birth and Early Evolution of our Universe' eds B. Scagerstam)

[77] E. Copeland and T.W.B. Kibble, *Scaling solutions in cosmic string networks* Imperial preprint, February (1991)

[78] D. N. Spergel, N. Turok, W. H. Press and B. S. Ryden, *Global Texture as the Origin of Large Scale Structure: Numerical Simulations of Evolution* Princeton PUPT-90-1182,(1990)

[79] N. Turok and D. Spergel, *Phys. Rev. Lett* **64** (1990) 2736

[80] D. Notzold, Fermilab-Pub-90-64/A (1990)

[81] M. Barriola and T. Vachaspati, *Strong Gravity of a Self Similar Global Texture* Tufts preprint *Phys. Rev. D* in press

[82] R. Durrer, M. Heusler, P. Jetzer, N. Straumann, *General Relativistic Collapse of Texture* Princeton PUPT-1222 (1990)

[83] A. Vilenkin, *Phys. Rev. D* **23** (1981) 852; J.R. Gott, *Ap. J.* **288** (1985) 422

[84] P.J.E. Peebles, *'The Large Scale Structure of the Universe' Princeton University Press* (1980)

[85] N. Turok, *Phys. Lett. B* **126** (1983) 437; *Phys. Rev. Lett.* **55** (1985) 1801; N. Turok and R. Brandenberger, *Phys. Rev. D* **33** (1986) 2175; H. Sato, *Prog. Theor. Phys* **75** (1986) 1342; A. Stebbins, *Ap. J. Lett.* **303** (1986) L21

[86] D. Lynden-Bell, S. M. Faber, D. Burstein, R. Davies, A. Dressler, R. J. Turlevich and G. Wegner, *Ap. J.* **236** (1988) 19

[87] N. Turok, *in 'Nearly Normal Galaxies: from the Planck Time to the present'* proceedings of the Santa Cruz Summer Workshop, Eds. S. Faber and J. Primack, Springer-Verlag. (1987)

[88] R. Brandenberger, *Inflationary Universe Models and Cosmic Strings* in 'Physics of the Early Universe' (Proceedings of the 36th SUSS in Physics 1989) eds. J.A. Peacock, A.F. Heavens and A.T. Davies.

[89] C. Park, D. Spergel and N. Turok, *Large Scale Structure in a Texture Seeded CDM Cosmogony* Princeton preprint (1990).

[90] A. Gooding, D. Spergel and N. Turok, *The Formation of Galaxies and Quasars in a Texture Seeded CDM Cosmogony* Princeton preprint (1990).

[91] R. Sachs and A. Wolfe, *Ap. J.* **147** (1967) 73.

[92] C. Hogan and M. Rees, *Nature* **311** (1984) 109; N. Kaiser and A. Stebbins, *Nature* **310** (1984) 391.

[93] A. Stebbins, *Ap.J.* **327** (1988) 584; F.R. Bouchet, D.P. Bennett and A. Stebbins, *Nature* **335** (1988) 410

[94] L.L. Cowie and E.M. Hu, *Ap. J.* **318** (1987) L33

[95] E.M. Hu, *Ap. J.* **360** (1990) L7

[96] A. Vilenkin, *Phys. Lett. B* **107** (1981) 47; F. Accetta and L. Krauss, *Nucl. Phys. B* **319** (1989)

[97] J. Taylor, Talk presented at Texas Symposium, Brighton, December 1990

[98] N.D. Mermin, *Rev. of Mod. Phys.* **51** (1979) 591

[99] N. Turok and J. Zadrozny, *Phys Rev Lett* **65** (1990) 2331
L. McLerran, M. Shaposhnikov, N. Turok and M. Voloshin, "Why the Baryon Asymmetry of the Universe is $\sim 10^{-10}$", Princeton University preprint PUPT-90-1216, 1990

[100] I. Chuang, R. Durrer, N. Turok and B. Yurke, "Cosmology in the Laboratory: Defect Dynamics in Liquid Crystals", Princeton University preprint PUP-TH-1208, 1990

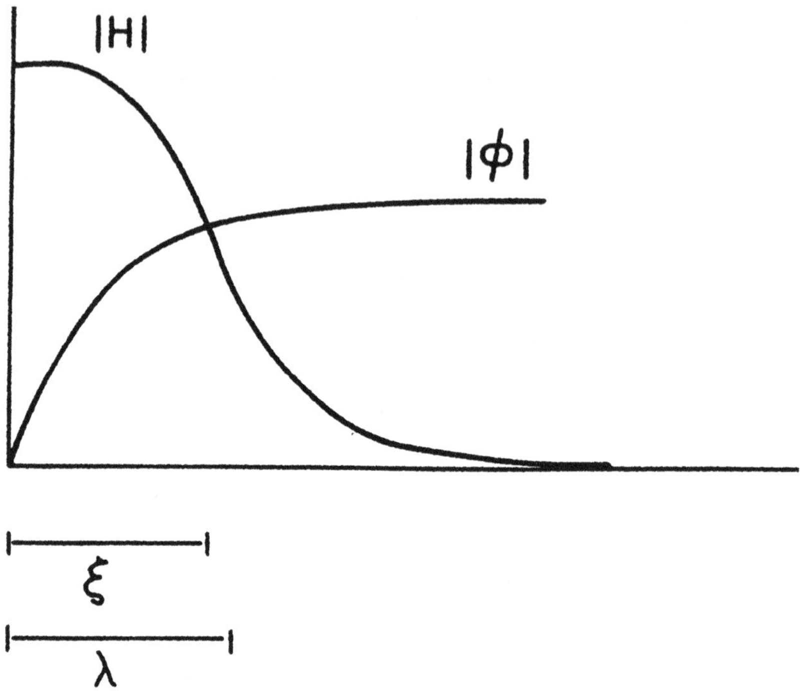

Figure 1

Field configuration for a Nielsen-Olesen vortex solution.

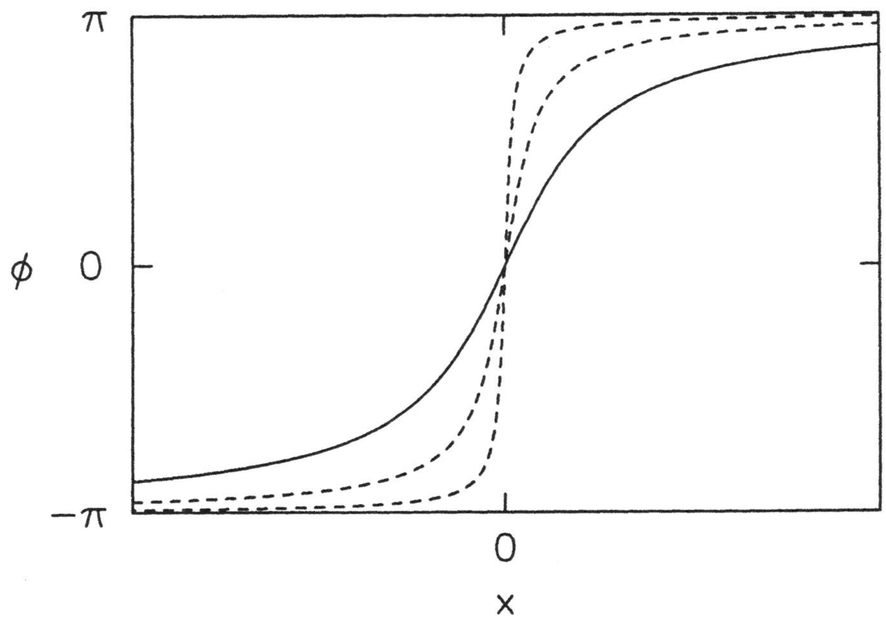

Figure 2

The solid line indicates a spherically symmetric texture configuration (x-axis scale is the radial distance). The dashed lines indicate the texture collapsing, with the gradient energy becoming concentrated towards the origin. This is representative of what happens in two and three dimensions. Remember $-\pi$ and π are to be identified.

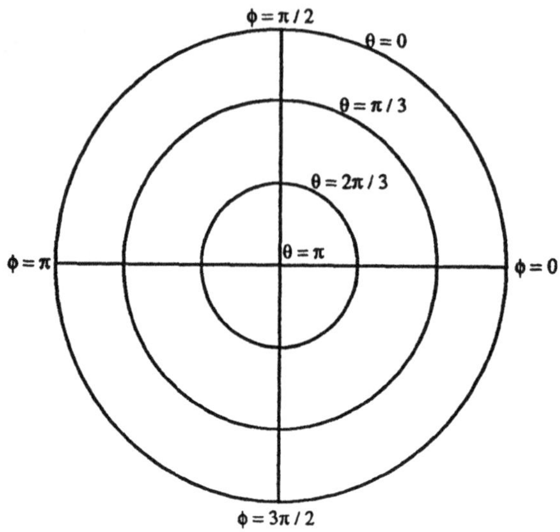

Figure 3

The spherically symmetric two dimensional texture. θ is the polar angle and ϕ the azimuthal angle on the 2-sphere. The outermost circle is at spatial infinity.

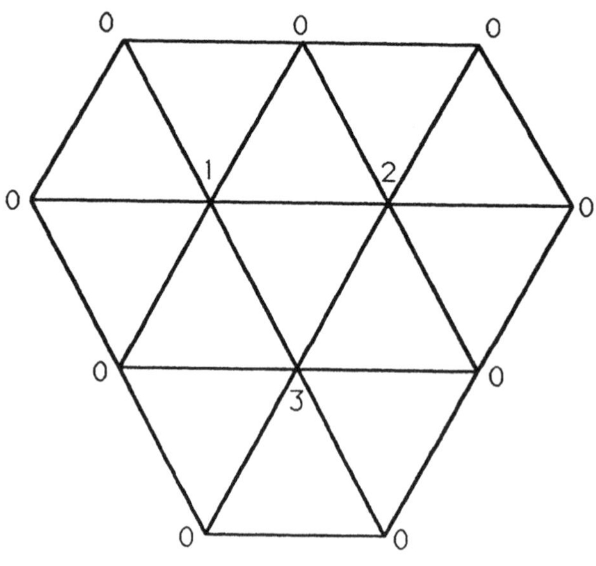

Figure 4

The discretised spherically symmetric texture in two dimensions.

(a)

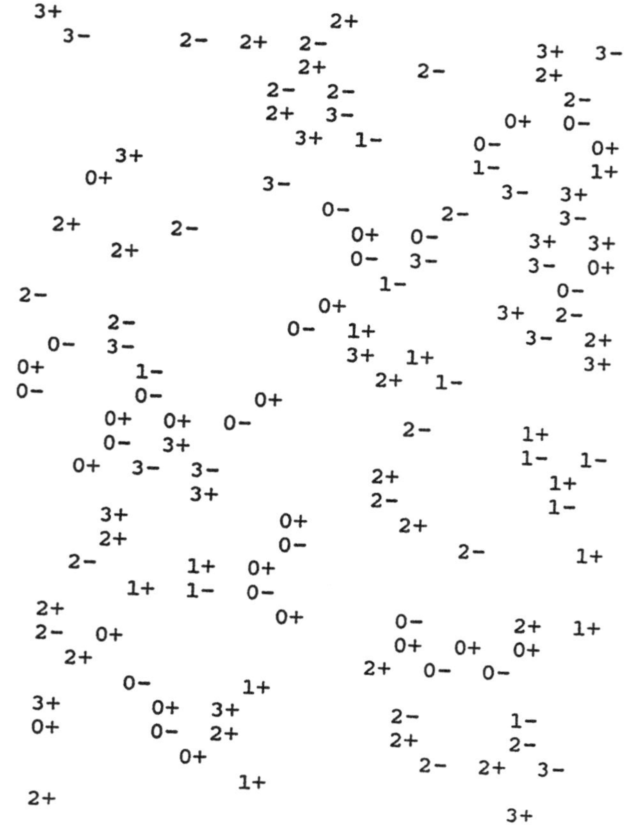

(b)

Figure 5

An initial configuration for textures in two dimensions, labelled alternately via (a) vertices and (b) partial windings. This illustrates the extremely complex nature of texture initial conditions.

The Space Telescope and the Problems of Cosmology

Philippe Crane
European Southern Observatory
Karl-Schwarzschild-Str. 2
D-8046 Garching bei München, Germany

Abstract: The status of the Hubble Space Telescope (HST) as of 1990 October is reviewed and several photographs of astrophysically interesting objects are presented to demonstrate current capabilities. Observations relevant to cosmology which may be pursued by HST are reviewed.

Introduction

The Hubble Space Telescope is a 2.4 meter optical telescope in a near earth orbit of roughly 320 nautical miles. It was launched on 1990 April 20. The telescope carries five different scientific instruments designed for specific problems as well as a fine guidance system which can be used for astrometric measurements (see Hall, 1982 for a review).

After initial deployment and check out, it was discovered that the main mirror had spherical aberration that would severely limit the kinds of science which could be pursued. Therefore, one objective of this review is to show examples of the type and quality of the science that may be pursued with the telescope in its current state.

This review is divided in the following sections: current status of the telescope and optical system, discussion of the image quality, examples of science observations, and discussion of the impact on observations of interest to cosmology.

Current status

There are two major problems which need to be solved for the observatory to reach its pre-launch expectations, and one other problem which may limit the flexibility of the observatory. The major problems are the spherical aberration, and the oscillations introduced in the telescope by the solar arrays on passage from night to day. The other problem is the less than optimal performance of the FGS sensors.

Spherical aberration of the primary mirror basically causes different annuli to be focussed at different places. This spreads the light from a point source over a larger area than desired. However, in contrast to bad seeing experienced on ground based telescopes, the images from HST maintain a sharp diffraction limited core containing from 15–20% of the light. Figure 1 shows a stellar image. Figure 2 shows an azimuthally averaged stellar profile and compares this to the expected profile. Figure 3 shows the encircled energy plot. The sharp core of the image

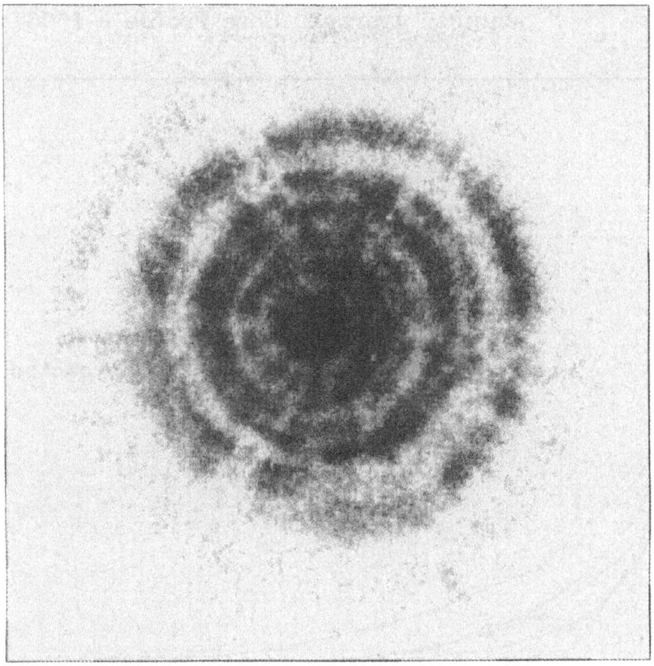

Figure 1. FOC $f/96$ stellar image at 1700 Å. Note the offset of the core from the outer diffraction rings.

can still be used to study sources at the resolution initially expected. However, the spherical aberration reduces the S/N in the image and restricts the range of observations.

The first maintenance and refurbishment mission will carry a replacement wide field camera with internal optics to correct for the spherical aberration. Corrections for some of the other instruments could take the form of optics placed just ahead of the instruments. Thus there is a strong probability that a major fraction of the performance of the first generation instruments can and will be recovered. Second generation instruments are in the process of modifying their optics to correct for the spherical aberration.

As the HST passes from night-to-day, the solar arrays heat up quickly on one side. This causes them to expand and to bend very much like a bi-metallic strip. The mechanical devices holding the arrays do not move entirely freely and the sticking friction of the mechanics introduces oscillations in the telescope pointing. Figure 4 shows a plot of the oscillations versus time. These are often large enough to cause the telescope to lose lock on the guide stars, or to have less than optimal guiding performance.

Since the oscillations are introduced at two relatively precise frequencies, modifications to the software can be made which will compensate for these oscillations to some extent. At the time of this article, it is not yet clear whether the software fixes will be able to control these oscillations to the 7 milliarcsec level which is the specification for guiding performance. Eventually, a second set of solar arrays will be installed which should not have this problem.

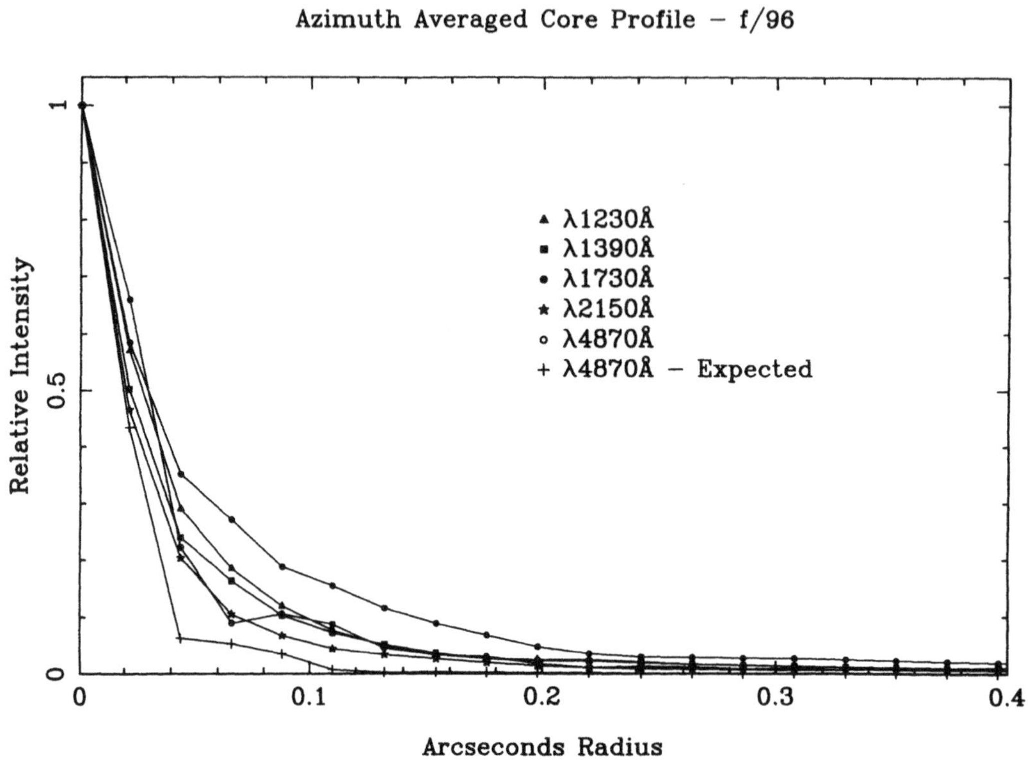

Figure 2. Azimuthally averaged profile from FOC at *f*/96. Note the relatively narrow core and the power at large distance. This latter is caused by spherical aberration.

The problems with the fine guidance sensors (FGS) are currently not well understood. Of the three sensors, one works well, one not so well and one is close to marginal performance. These problems may be related to the spherical aberration and slight decentering and tilting of the secondary, or they may be due to problems internal to the FGS's themselves. The FGS's are therefore limited in the stars which they can lock onto. The overall effect on observatory efficiency and whether or not the full performance can be achieved is not yet known.

Science observations

This section presents a few representative images obtained with HST in a program of science assessment observations. The main point is to demonstrate the kind of science which can still be done with HST images in spite of the spherical aberration. The observations are generally of high-contrast objects which can exploit the sharp unresolved core of the HST point spread function.

G2237+0305. Figure 5 is an image of a quasar which has been gravitationally lensed by a galaxy. The details of this image and its analysis are published by the FOC team (P. Crane *et al.* 1991).

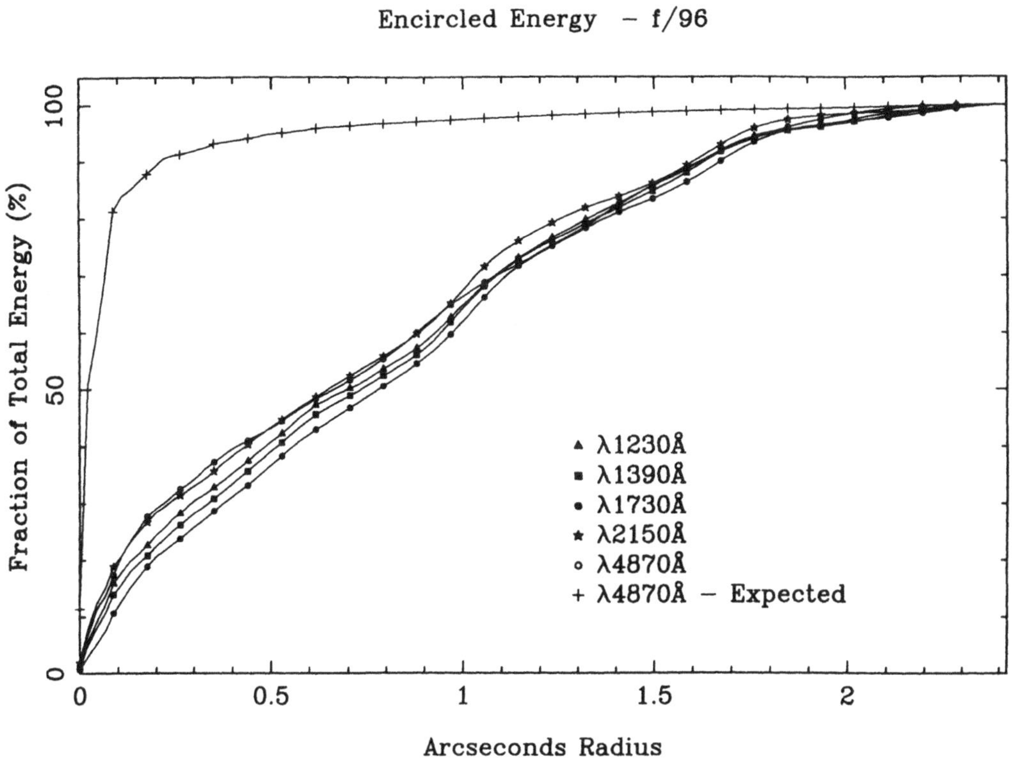

Figure 3. Plot of the encircled energy for a star imaged with the FOC *f*/96 camera. The encircled energy plots show clearly the effects of spherical aberration. The energy in the central core (~ 0."07 diameter) is about 17% of the total compared to a design criterion of 80%.

SN 1987A. These images (Figure 6) were also obtained by the FOC team (Jakobsen *et al.* 1991). Perhaps the most interesting feature of this image is the clear demonstration that the supernova is extended with a FWHM of $\simeq$ 165 milliarcsec compared to that of the inset star of 65 milliarcsec. Also seen is the ring of ionized gas surrounding the SN. Details of the analysis of this image, and the prospects for future work are discussed by Jakobsen *et al.* (1991).

Cosmological pursuits

The Space Telescope will certainly contribute to our knowledge of the origin and evolution of the Universe. However, several of the observational programs originally foreseen will have to be deferred until the second generation of instruments or until the optics can be improved on the current instruments.

Some projects where HST may contribute in the near future are mentioned below. The reader is cautioned that some of these projects are speculative or would require enormous amounts of observing time with the current optics.

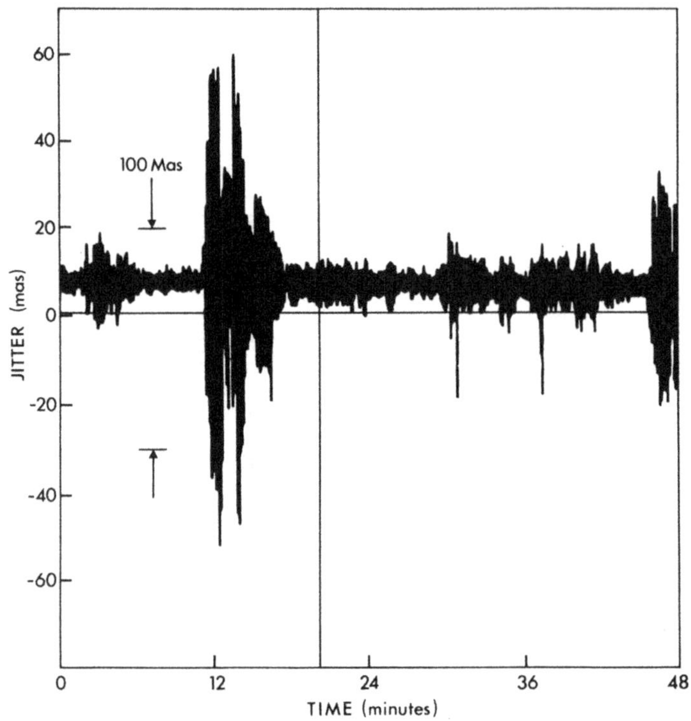

Figure 4. Plot of the oscillations in the guidance system during one part of an orbit. The large oscillations occur at the day-night crossing. The exact origin of the smaller oscillation is not clear.

Understanding of quasar absorption lines has always been hampered by the fact that we don't have any data on the absorption lines of nearby quasars. This is so since the lines are in the UV and not accessible from the ground. HST's Faint Object Spectrograph will observe nearby quasars down to Lyman α and provide a tremendous wealth of information to compare with data at higher z. Ths will finally permit studies of evolution of the clouds forming these absorption features.

Detailed studies of gravitational lens systems will eventually be used to determine the Hubble parameter H_0. HST will provide much of this data. Gravitational lenses will also be used to determine the internal dynamics of lensing systems.

Careful astrometry of the Population II binary system μ Cas will be used to determine the primordial helium abundance. Basically μ Cas is a star whose composition represents that found at the formation of our Galaxy. Astrometric measurements can be used to determine the mass of the stars which can then be used to determine the element abundances.

Morphological studies of galaxies in modestly distant ($z \approx 0.5$) clusters may be possible. This would yield data on the evolution of galaxy types in clusters. It may also shed light on the evolution of stellar content in the galaxies.

Concluding remark

In spite of grave difficulty, the HST is producing, and will continue to produce, fascinating new results in all fields of astronomy. With time and hard work, it will fulfill its goals and hopefully surpass them.

References

Crane, P., *et al.* 1991, *Ap.J. (Letters)*, 10 Mar 91

Hall, D.N.B. 1982, *The Space Telescope Observatory*, NASA CP-2244

Jakobsen, P., *et al.* 1991, *Ap.J. (Letters)*, 10 Mar 91

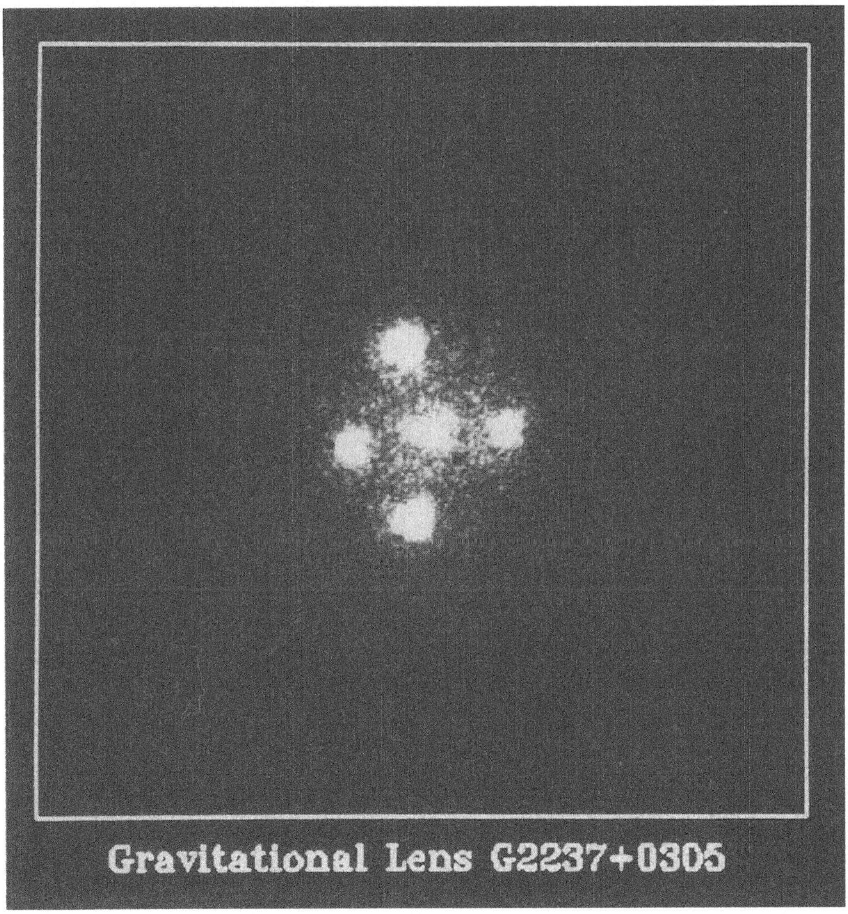

Figure 5. This shows a 15 minute FOC $f/96$ image of the gravitationally imaged quasar G2237+0305. The central object is the core of the galaxy which is doing the lensing. The four symmetrically placed images of of a single distant ($z = 1.695$) quasar behind the galaxy. A fifth image is expected near the center of the galaxy.

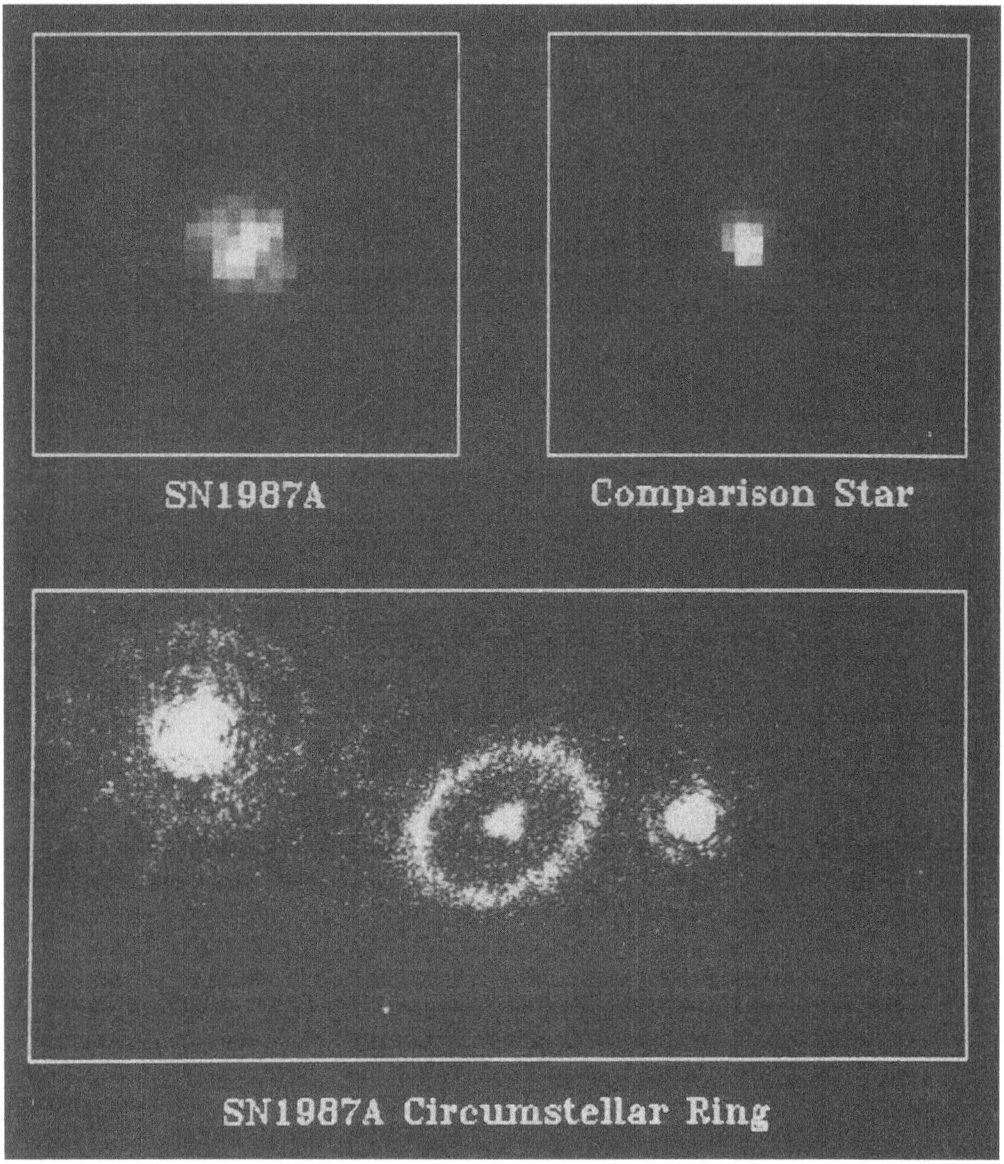

Figure 6. This shows an image of the recent supernova SN 1987a in the Magellanic Clouds. Two stars are on the edge of the field. The ring is in the light of O III and the fluctuations in the brightness are real. The central source is the supernova itself which is resolved.

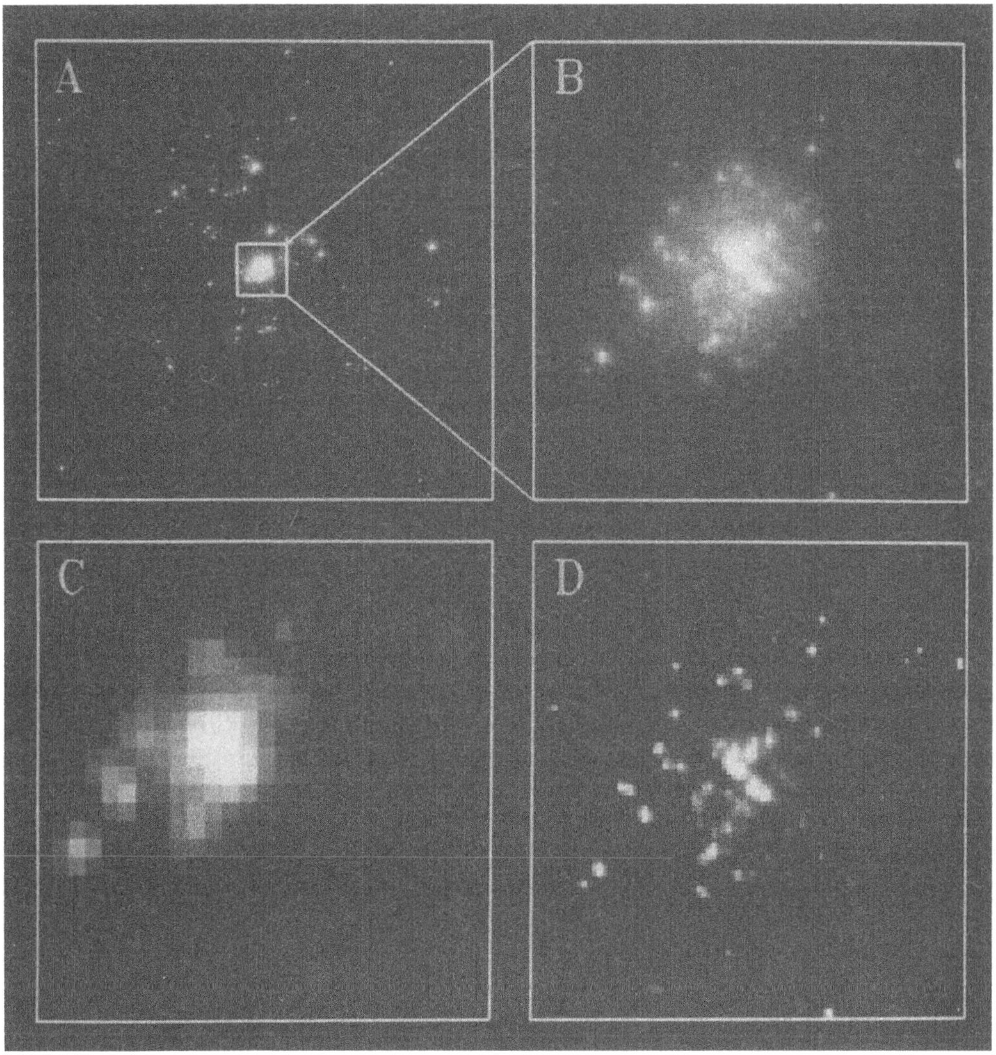

Figure 7. This is a Wide Field Camera image of the young star cluster R136a in the Large Magellanic Cloud. Panel A shows the entire frame. Panel B shows an expanded detail. Panel C shows a ground based image of stars in Panel B. Panel D shows Panel B after some image processing to remove the haloes and reduce the cores of the stars.

Self-gravitating magnetic monopoles, global monopoles and black holes

G W Gibbons

D.A.M.T.P.
University of Cambridge
Silver Street
Cambridge CB3 9EW

Introduction. The properties of the simplest magnetic monopoles in Yang-Mills-Higgs theory in flat spacetime are by now quite well understood. Moreover the fact that almost any grand unified theory will admit monopole solutions has had a profound impact on modern cosmology since it provides a very strong argument in favour of inflation as a possible solution of the monopole problem. It is therefore rather surprising that comparatively little is known about the gravitational properties of magnetic monopoles and the relation between monopoles and black holes. In the past couple of years or so this situation has begun to change and it is my purpose in these lectures to review the current situation and make some comments on it. I have included a fairly complete list of references, not all of which are referred to explicitly in the text, for the convenience of those wishing to follow up these topics.

Contents

1. Monopoles in Flat Spacetime
2. Static Solutions of Einstein's Equations without Horizons
3. Bogomolnyi Bounds for Einstein-Yang-Mills-Higgs initial data
4. Static solutions of the Einstein-Yang-Mills Equations with Horizons
5. Solutions of the Einstein-Yang-Mills-Higgs Equations with Horizons
6. Global Monopoles and Black Holes
7. Black Hole Monopole Pair Production

1. Monopoles in Flat Spacetime. The virial theorem tells us that any finite energy time-independent solution of the equations of motion must satisfy:

$$\int_{I\!R^3} T_{ik} d^3x = 0 \tag{1.1}$$

where T_{ik} are the spatial components of the energy-momentum tensor. Equation (1.1) follows from the conservation equation $T_{ij,j} = 0$ and the obvious consequence:

$$(T_{ij}x_k),j = T_{ik}$$

using the divergence theory and discarding the boundary term. The physical meaning of (1.1) is that the total stresses in an extended object must balance. In particular the components of T_{ik} cannot have a fixed sign - for example the spatial trace $\sum_i T_{ii}$ equals the sum of the "principle pressures"so there must be regions where the matter is in tension and regions where it is in compression. For a pure Higgs field (assumed through out these lectures to lie in the adjoint representation of $SU(2)$)

$$T_{ij} = +D_i\Phi.D_j\Phi - \frac{1}{2}\delta_{ij}(D_k\Phi.D_k\Phi) - \frac{1}{2}\delta_{ij}\; W(\Phi) \tag{1.2}$$

and (1.1) cannot possibly be satisfied as long as $W(\Phi) \geq 0$. The potential term gives an isotropic negative pressure. The gradient term gives a positive pressure along the gradient direct and equal tensions in the orthogonal directions. The sum of principal pressures is thus negative. In fact this result holds for an arbitrary harmonic map with non-negative potential.

By contrast for a Yang-Mills field

$$T_{ij} = -B_i.B_j + \frac{1}{2}B_k.B_k\delta_{ij} \tag{1.3}$$

where

$$B_i = \frac{1}{2}\epsilon_{ijk}F_{jk} \tag{1.4}$$

is the magnetic field strength. As Faraday taught Maxwell there is now a **tension** along the direction of B_i and an equal pressure orthogonal to the field lines. The sum of the principal pressures is now positive. Thus for pure Yang-Mills there can be no static solution either.

However as 't Hooft and Polyakov showed there **is** a solution of the combined Yang-Mills Higgs system which does satisfy (1.1) and which is stable. Infact it minimizes the "total energy":

$$\int T_{00}d^3x = \int d^3x[\frac{1}{2}B^2 + \frac{1}{2}(D\Phi)^2 + W(\Phi)] \qquad 1.5$$

where D is of course now a gauge covariant derivative. Moreover if Φ transforms by the adjoint representation of the gauge group $SU(2)$ the total energy is bounded below by

$$\frac{4\pi\eta}{e} = g\eta \tag{1.6}$$

where e is the gauge coupling constant, $g = 4\pi/e$ the magnetic charge of the monopole and $|\Phi| \to \eta$ at infinity. This Bogomolnyi bound can only be attained if $W(\Phi)$ vanishes identically (the Prasad Sommerfeld limit) and the Bogomolnyi equations:

$$\pm B_i = D_i\Phi \tag{1.7}$$

hold, moreover from (1.7), it follows that the stresses vanish pointwise i.e. $T_{ij} = 0$.

The Prasad Sommerfeld limit and the associated Bogomolnyi equations are of great mathematical interest and have many geometrical consequences. They are also related to supersymmetry: the system with $W(\Phi) = 0$ admits an $N = 4$ supersymmetric extension. It has, as a consequence, received a great deal of attention. Of greater physical relevance is the case when

$$W(\Phi) = \lambda(\Phi^2 - \eta^2)^2 \tag{1.8}$$

and the total energy is given by

$$\frac{4\pi\eta}{e} f(\lambda/e^2) \tag{1.9}$$

where $f(\lambda/e^2)$ is a dimensionless function of the dimensionless ratio λ/e^2 with $f(0) = 1$.

In addition to the finite energy solutions there is a static solution of the pure Higgs field equations (with Φ again a triplet of $SU(2)$) Φ satisfies a Hedgehog Ansatz:

$$\Phi^a = (x^a/r)F(r) \tag{1.10}$$

with $F(0) = 0$ and $F(\infty) = \eta$. Since

$$T_{00} \to \frac{\eta^2}{r^2} \tag{1.11}$$

this solution has infinite energy. It is called a **global monopole**.

It is a non-singular solution of the Higgs equations of motion. It is however well approximated by a singular solution of the non-linear σ-model obtained by enforcing the constraint that

$$|\Phi|^2 = \eta^2 \tag{1.12}$$

everywhere. Such singular solutions arise in the theory of nematic liquid crystals. They are known to be unstable in that a lower energy configuration is available with the energy concentrated along lines ("strings ") with energy per unit length equal to $4\pi\eta^2$. Global monopoles have recently been considered in connection with "cosmic textures". I will discuss them further in a later section. Let us first see to what extent these basic facts are modified when we consider self-gravitating monopoles.

2. Static Solutions of Einstein's Equations Without Horizons

Globally static metrics (i.e. time independent, time reversal invariant and without event horizons) may be cast in the form:

$$ds^2 = -V^2(x)dt^2 + g_{ij}(x)dx^i dx^j \tag{2.1}$$

The field equations are then:

$$\nabla_g^2 V = 4\pi G V(T_{\hat{0}\hat{0}} + \sum T_{\hat{j}\hat{j}}) \tag{2.2}$$

$$R_{ij}[g] = V^{-1}\nabla_i\nabla_j V + 4\pi G g_{ij}(T_{\hat{0}\hat{0}} - \sum T_{\hat{j}\hat{j}}) + 8\pi G T_{ij} \tag{2.3}$$

where ∇_g^2 is the Laplacian of g_{ij} and ∇_i its covariant derivative. $T_{\hat{0}\hat{0}}$ and $T_{\hat{j}\hat{k}}$ are the components of the energy momentum tensor in an orthonormal frame with $e_0 = V^{-1}\frac{\partial}{\partial t}$. If the metric is asymptotically flat then

$$V \sim 1 - 2Gm/r + O(\frac{1}{r^2}) \text{ and } g_{ij} \sim (1 + \frac{2Gm}{r})\delta_{ij} + 0(\frac{1}{r^2}) \tag{2.4}$$

where m is the A.D.M. mass of the spacetime. From (2.2) we have

$$m = \int_\Sigma V(T_{\hat{0}\hat{0}} + \sum T_{\hat{i}\hat{i}})\sqrt{g}d^3x \tag{2.5}$$

where Σ is a surface of constant time (assumed complete).

For some purposes it is convenient to rescale the 3-metric g_{ij} and re-write (2.1) as

$$ds^2 = -e^{2U}dt^2 + e^{-2U}\gamma_{ij}dx^i dx^j \tag{2.6}$$

The quantity U may be called the Newtonian potential. The field equations now become:

$$\nabla_\gamma^2 U = 4\pi G e^{-2U}(T_{\hat{0}\hat{0}} + \sum_{\hat{i}} T_{\hat{i}\hat{i}}) \tag{2.7}$$

and

$$\tilde{R}_{ij}[\gamma] = 2\partial_i U\partial_j U + +8\pi G(T_{ij} - \gamma_{ij}e^{-2U}\sum T_{\hat{i}\hat{i}}) \tag{2.8}$$

where $\tilde{R}_{ij}$ is the Ricci tensor of γ_{ij}. Now note:

(1) from (2.4) it follows that γ_{ij} is a complete asymptotically flat 3-metric with zero ADM mass

(2) from (2.8) the Ricci scalar $\tilde{R}$ of γ_{ij} is given by

$$\tilde{R} = 2\gamma^{ij}\partial_i U\partial_j U - 16\pi G e^{-2U}(\sum T_{\hat{i}\hat{i}}) \tag{2.9}$$

Using the positive mass theorem we now deduce the following

Theorem 1 There are no globally static asymptotically flat solutions of Einstein's with $\sum_i T_{\hat{i}\hat{i}} \leq 0$. In other words since gravity is attractive we need some pressure to resist collapse inwards. Note that to prove theorem 1 we do **not** need to assume that, the matter has positive energy.

If $T_{\alpha\beta} = 0$ theorem 1 is just Lichnerowicz's theorem If the matter is a scalar field however we obtain a new result:

Cor 1. There are no globally static asymptotically solutions of Einstein's equations with a minimally coupled scalar field source with a non-negative potential.

It is interesting to note that the solutions recently derived by Vilenkin and Bariola giving the gravitational fields of global monopoles escape cor. 1 by virtue of **not** being asymptotically flat, as I shall describe later. It is also important to point out that there **do** exist solutions in which a complex scalar field varies harmonically with time in such a way that $T_{\mu\nu}$ and the metric $g_{\mu\nu}$ are static. Such scalar fields are said to consist of Q-matter.

On the other hand for pure Einstein-Yang-Mills we cannot deduce from Theorem 1 that there are no static solutions without horizons, since $\sum T_{ii} \geq 0$. In fact for pure Einstein-Maxwell theory there are in fact no static solutions without horizons (for a proof see Breitenlohner, Gibbons and Maison (1988)). It came as a surprise therefore when Bartnik and McKinnon (1988) announced the existence of an integer's worth of static spherically symmetric solutions. Their metric ansatz was

$$ds^2 = -e^{2U(r)}dt^2 + \frac{dr^2}{1 - \frac{2Gm(r)}{r}} + r^2(d\theta^2 + \sin^2\theta d\phi^2) \tag{2.10}$$

with

$$eA = a(r)\tau_3 dt + w(r)(\tau_1 d\theta + \tau_2 \sin\theta d\phi) + \tau_3 \cos\theta d\phi \tag{2.11}$$

where τ_1, τ_2 τ_3 is the usual basis for the Lie algebra of $SU(2)$. Equation (2.11) gives spherically symmetric $SU(2)$ connection over S^2 if we set $r =$ constant, $t =$ constant. Purely magnetic solutions have $a(r) = 0$ and $eF = w'dr \wedge d\theta\tau_1 + w'\sin\theta dr \wedge d\phi\tau_2 - (1 - w^2)\sin\theta d\theta \wedge d\phi\tau_3$.

According to Bartnik and McKinnon the assumption of suitable asymptotics and of finite energy implies that the electric potential $a(r)$ must vanish. There results a system of radial ordinary differential equations for $U(r)$, $w(r)$, and $m(r)$. If $'$ denotes differentiation with respect to r we have:

$$\frac{m'}{4\pi G} = (1 - \frac{2Gm}{r})(w')^2 + \frac{1}{2}\frac{1}{r^2}(1 - w^2)$$

$$r^2(1 - \frac{2Gm}{r})w'' + (2Gm - \frac{(1-w^2)^2}{r^2})w' + (1 - w^2)w = 0$$

and

$$U = \frac{1}{2}\ln(1 - \frac{2Gm}{r}) - 4\pi G \int_r^\infty \frac{1}{r}(w\prime)^2 dr$$

These equations may be combined into a single 3rd order differential equation as shown by Ray (1978) who may be said to have anticipated some aspects of the results of Bartnik and Mckinnon. In any event Bartnik and Mckinnon presented numerical evidence that there exist solutions with

$$m(r) = O(r^3); \quad w(r) = 1 + O(r^2) \text{ at } r = 0$$

and

$$m(r) \sim m(\infty) - c^2/r^3; \quad w(r) \sim \pm(1 - c/r) \text{ at } r = \infty$$

for some constant c.

The solutions are indexed by the number k of zeros of $w(r)$; $k = 1, 1, 2\ldots$. The solutions have 3 regions. Region I is an inner core. Region III is the asymptotic region where $F = 0(\frac{1}{r^3})$ and the metric becomes Schwarzschildean. The middle region has $w \approx 0$ and the solution behaves rather like an abelian $U(1)$ Dirac monopole and the geometry resembles the throat region of an extreme Reissner-Nordstrom solution.

The results of Bartnik and McKinnon have been confirmed by Kunzle and Masood-ul-Alam (1990) and by Maison (private communication).

The question immediately arises: are these solutions stable? Bartnik and McKinnon themselves felt that the cases $k \geq 3$ were unstable. A stability analysis was carried out by Straumann and Zhou (1990)and also (private communications) by Maison. These analyses show that these solutions are unstable for **all** values of k. Presumably a small perturbation would cause them either to collapse to form a black hole or (rather less likely) to explode and dissipate. Since they have no magnetic moment the expected hole will be Schwarzschild like. A noteworthy feature of the analysis of Straumann and Zhou was that the configurations they considered were spherically symmetric and yet time-dependent. In other words Birkhoff's theorem is not valid for Einstein-Yang-Mills unlike Einstein-Maxwell.

Thus Einstein-Yang-Mills admits unstable finite energy static non-singular solutions. We have seen above that the Einstein-Higgs equations do not. What about Einstein-Yang-Mills-Higgs? It is physically clear that for small values of Gm/R where R is a typical radius and m a typical total energy the effects of gravity on a 't Hooft-Polyakov monopole will be negligible. Since typically

$$m \sim \frac{4\pi\eta}{e}$$

$$R \sim \frac{1}{e\eta}$$

gravity will be negligible so long as:

$$4\pi G\eta^2 \ll 1.$$

If, on the other hand we consider a one parameter family of static solutions labelled by the dimensionless number $4\pi G\eta^2$ (keeping λ/e^2 fixed) we might expect to encounter a critical value beyond which no static solutions are possible because they will undergo gravitational collapse. It is also likely that the static family will already have become unstable at some smaller value of $4\pi G\eta^2$. The intuition one is drawing on here is of course an analogy with the theory of white dwarf stars, the critical value of $4\pi G\eta^2$ corresponding to the famous Chandrasehkar limit.

As far as I know a detailed analysis of this situation has not been carried out until recently. Miguel Ortiz in his Ph.D thesis has begun a numerical study and his results confirm that for fixed λ/e^2 there is a maximum value of $4\pi G\eta^2$ beyond which regular solutions without event horizons cease to exist. This maximum value is about 2.5 in the Prasad-Sommerfeld limit and decreases as the quartic coupling constant λ increases.

The metric at large distances appears to approach the Reissner-Nordstrom form with an approximately minimal mass for a magnetic charge g, that is the monopole appears to collapse as soon as the Cosmic Censorship allows. The exterior gauge field and Higgs field appear to approach a Wu-Yang like configuration with the Higgs field being covariantly constant. This will be described in sections 4 and five in more detail.

The basic equations for self-gravitating 't Hooft Polyakov monopoles in the spherically symmetric case were in fact written down some time ago by Perry, Van Nieuwenhuizen and Wilkinson (1976). A variational principle was established but the equations were not analysed in detail. A qualitative physical discussion along the lines indicated in the previous paragraph has been given by Frieman and Hill (1987). Some more exact information can possibly be obtained by considering the generalizations of the Bogomolnyi bound in the gravitational setting so we now turn to that topic.

3. Bogomolnyi Bounds for Einstein-Yang-Mills-Higgs Initial Data

We shall consider time-symmetries initial data for simplicity, that is the second fundamental form K_{ij} of the initial surface is assumed to vanish. In addition we assume that the non abelian electric field vanishes, as well as the time component of the Higgs field's covariant derivative. Thus the Ricci scalar R of the 3-metric g_{ij} satisfies

$$R = 16\pi G T_{\hat{0}\hat{0}} \tag{3.1}$$

where

$$T_{\hat{0}\hat{0}} = \frac{1}{2}B^2 + \frac{1}{2}(D\Phi)^2 + W(\Phi). \tag{3.2}$$

Let us define the "total amount of matter" on the initial surface Σ (assumed to be complete) by:

$$M = \int_\Sigma T_{\hat{0}\hat{0}} \sqrt{(\ ^3g)}\, d^3x \tag{3.3}$$

Note that M does not, in general, equal the ADM mass m of the 3-metric g_{ij}. Even if it were the case that the data were such as to evolve to a static solution a comparison of (3.3) and (2.5) shows that M and m cannot be expected to coincide. Another measure of the total energy of the matter in a static spacetime would be the "Killing Energy" E defined by

$$E = \int_\Sigma \sqrt{(-g_{00})}\ T_{\hat{0}\hat{0}} \sqrt{g} d^3x \tag{3.4}$$

In general we have (when they are defined)

$$M \neq E \neq M \neq m$$

Now Bogomolnyi's original argument may trivially be "covariantised" with respect to spatial diffeomorphisms using the covariantly constant alternating tensor ϵ_{ijk} of the 3-metric g_{ij} (I prefer **not** to use tensor densities and I am of course assuming that the initial surface is oriented). Thus we have

Theorem 2. The total amount of matter M of a time-symmetric initial dates set for the $SU(2)$ Einstein-Yang-Mills-Higgs equations with non negative potential $W(\Phi)$ is bounded below by

$$M \geq g\eta \tag{3.5}$$

where g is the asymptotic magnetic monopole moment .For a single monopole $g = 4\pi/e$. Moreover equality in (3.5) implies that the covariant Bogomolnyi equations hold.

$$D_i\Phi = \pm\frac{1}{2}\epsilon_i^{\ jk}\ F_{jk} \tag{3.6}$$

hold.

The existence of solution of (3.6) on a curved metric has been studied by Floer (1987)). Although of some mathematical interest the following result shows that these solutions are never of relevance if the monopole self-gravitates.

Theorem 3. Static solutions of the Einstein-Yang-Mills-Higgs equations satisfying the Bogomolnyi equations (3.6) or equivalently saturating the Bogomolnyi bound (3.5) do not exist.

Proof The Bogomolnyi equation (3.6) imply that the spatial components of the stress tensor $T_{ij} = 0$. We can thus invoke our previous theorem 1.

It is known that to form an abelian black hole of ADM mass m and magnetic charge g we must have

$$m \geq \frac{g}{\sqrt{4\pi G}} \tag{3.7}$$

(recall that we are using rationalized units for electromagnetic or Yang-Mills fields). Equality in corresponds to the Papapetrou-Majumdar metrics describing the equipoise of an arbitrary number of extreme Reissner-Nordstrom magnetic black holes. As mentioned in section 2 equation (3.7) shows that for fixed magnetic charge g a 't Hooft Polyakov monopole cannot collapse until its ADM mass m satisfies $\sqrt{4\pi G}m \geq g$. Since $m \sim M \sim \eta g$ we need $4\pi G\eta^2 \geq 1$ which agrees approximately with what has been found by Ortiz. It thus seems very reasonable to expect that the configuration to which it gives rise is similar, if not identical to, an extreme Reissner-Nordstrom solution. We shall consider static solutions with horizons with horizons. That we in the next section. Before doing so we remark that some information about initial data for Einstein-Yang-Mills-Higgs has been obtained by Malec and and Koc (1990) and Chmaj and Malec (1989).

If one is merely interested in the Yang-Mills equations in a gravitational background it is possible to find a modified set of Bogomolnyi equations:

$$D_i\Phi = B_i - \Phi\nabla_i U \tag{3.8}$$

where U is the Newtonian potential as defined by (2.6). If the background metric satisfies

$$\nabla^2{}_g U = 0, \tag{3.9}$$

then (3.8) implies the second order Yang-Mills equations in the background (see Comtet (1980) and Comtet Forgacs and Horvathy (1984). In general (3.9) will be incompatible with the Einstein-Yang-Mills equations. An interesting case for which (3.9) is compatible with the Einstein-Yang-Mills equations is when γ_{ij} is flat. This gives the Papapetrou-Majumdar metrics for which the left and right hand sides of (3.8) are separately zero and B and Φ point in a constant direction in internal space. For more detail about these equations see Horvathy ((1987)

4. Static solutions of the Einstein-Yang-Mills equations with Horizons

It has been known for many years that one has the abelian black hole solutions with

$$A = \tau_3(\frac{q}{r}dt + g\cos\theta d\phi) \tag{4.1}$$

$$ds^2 = -\Delta dt^2 + \frac{dr^2}{\Delta} + r^2(d\theta^2 + \sin^2\theta d\phi^2) \tag{4.2}$$

$$\Delta = 1 - \frac{2Gm}{r} + G\frac{(q^2+g^2)}{4\pi r^2} \tag{4.3}$$

where m is an arbitrary constant satisfying

$$m \geq \frac{1}{\sqrt{4\pi G}}(q^2+g^2)^{\frac{1}{2}} \tag{4.4}$$

If τ_3 has a normalization such that

$$\exp(4\pi i\tau_3) = 1 \tag{4.5}$$

we must demand that

$$\frac{eg}{2\pi} = 0,\ 1,\ 3,\ 5,\ldots \tag{4.6}$$

if A is an $SO(3)$ connection and

$$\frac{eg}{4\pi} = 0,\ 1,\ 2,\ 3 \tag{4.7}$$

if A is an $SU(2)$ connection. Note that the $SO(3)$ case is only possible because the presence of the horizon means that the singularity which would otherwise result at $r = 0$ is hidden inside the horizon. It cannot occur if there are no horizons.

Note that (4.1) will always yield a spherically symmetric energy-momentum tensor although it is not spherically symmetric as an $SU(2)$ connection unless

$$\frac{eg}{4\pi} = 1 \tag{4.8}$$

This latter case corresponds to $w = 1$ in (2.11). Some authors prefer to use a different gauge. Let

$$x = r(\sin\theta\cos\phi, \sin\theta\sin\phi, \cos\theta)$$

then the connection

$$e\tilde{A}^1 = \frac{dx^2 \wedge dx^3}{r^2}(w-1) \text{ etc} \tag{4.9}$$

is gauge equivalent to (2.11) with $a(r) = 0$. Thus the Reissner-Nordstrom metric with $q = 0$ and $\tilde{A}$ given by (4.9) with $w = 0$, whence (4.8) represents the simplest purely magnetic $SU(2)$ solution. This solution will extend trivially to a solution of the Einstein-Yang-Mills-Higgs equations if one appends the covariantly constant Higgs field

$$\Phi^i = \eta\frac{x^i}{r} \tag{4.10}$$

The resulting solution is said to satisfy the Wu-Yang ansatz. Of course in the Abelian gauge it reduces to

$$eA = 4\pi\tau_3 \ \cos\theta d\phi \tag{4.11}$$

$$\Phi = \eta\tau_3$$

$$ds^2 = -(1 - \frac{2Gm}{r} + \frac{4\pi G}{e^2 r^2})dt^2 + (1 - \frac{2Gm}{r} + \frac{4\pi G}{e^2 r^2})^{-1} dr^2 \tag{4.12}$$

$$+r^2(d\theta^2 + \sin^2\theta d\phi^2) \tag{4.13}$$

Until the work of Bartnik and McKinnon it had long been felt that these abelian solutions would be the only solutions - rather by analogy with the No-Hair Theorems for the Einstein-Maxwell equations. However recent work has clearly indicated this conjecture to be false. Kunzle and Masood-ul-Alam (1990), Bizon (1990) , and Volkov and Gal'tsov (1989) have shown that there exist analogues of the Bartnik and McKinnon solutions with horizons. As one might expect these are unstable, as shown by Straumann and Zhou (1990). The no hair conjecture in its naive form thus fails. In the spherically symmetric case however Gal'tsov and Ershov (1990) have argued that as long as there is a net Yang-Mills charge measurable at infinity, i.e. that $w^2 \neq 1$ at infinity then the abelian solutions are unique.

What about the stability of the abelian solutions? This was studied some time ago by Lohiya (1982). Following the analogous analysis of singular monopoles in flat spacetime. The stability is determined by the large distance behaviour of the fields in a manner described by Coleman (19839 and Brandt and Neri (1979). The analysis shows that the purely magnetic solutions are unstable stable if the connection is topologically trivial restricted to a 2-sphere at infinity. This means that all $SU(2)$ connections i.e.

$$\frac{eg}{4\pi} \in Z$$

are unstable. Of the remaining $SO(3)$ connections only the lowest one with

$$\frac{eg}{2\pi} = 1$$

is stable.

A sufficient condition for instability of the electric solutions is that the electric charge q exceeds $3e/2$. Again this is consistent with the flat space results.

The conclusion would seem to be that in the absence of Higgs fields regardless of uniqueness the non abelian-Einstein-Yang-Mills solutions are not of very much physical interest. It is therefore appropriate to turn to the case when Higgs fields are included.

5. Solutions of the Einstein-Yang-Mills Higgs Equations with Horizons

The obvious first remark is to recall that solutions always exist if the Higgs field is covariantly constant. Choosing a gauge in which the direction of Φ is everywhere the same in internal space we see that Yang-Mills potentials associated with rotations about that direction satisfy the abelian equations and thus must belong to the Reissner-Nordstrom family described earlier. Of course the stability results of Lohiya do not necessarily apply now because the Higgs mechanism might well stabilize these Dirac-type monopoles (for sufficiently large Higgs mass) against instabilities in the non-abelian directions. To my knowledge this has not been looked at by anybody in detail.

In the case that the electric charge vanishes the results of Ortiz suggest that gravitational collapse of a 't Hooft-Polyakov monopole will result in an exterior field in which the Higgs field is covariantly constant and given by (4.10), and the gauge field by (4.9) and $w = 0$. This solution was originally written down by Cho and Freund (1975) and Bais and Russell (1975). As mentioned earlier these correspond to the Reissner Nordstrom solutions with $q = 0$ and $g = \frac{4\pi}{e}$. My conjecture is that these are indeed classically stable. Moreover quantum mechanically they should evolve by Hawking evaporaation to the extreme, zero temperature state. Such objects should behave like stable solitons and have been studied extensively by Hajicek and his collaborators from that point of view. Thus if $4\pi G\eta^2$ is large enough the monopole problem of cosmology is in fact a primordial black-hole monopole problem. In fact it it seems rather likely that 'tHooft-Polyakov monopoles will be unstable for values of $4\pi G\eta^2$ which are somewhat smaller than the maximum allowed value.

An important question now arises: are there any other solutions? For eaxmple are there any electrically charged solutions in which the electric charges are associated with the broken $SU(2)$ generators for example ? Experience and intuition based on both the physical ideas behind the Higgs mechanism (charge should be screened) and the non-hair properties of black holes would have suggested until very recently that the answer is no. At present however the answer is less clear because of two developments. The first is the Bartnik-McKinnon-Bizon-Kunzle-Masood-ul-Alam - Vokkov-Gal'tsov solutions. The second is the issue of fractional charges raised by Krauss and Wilczek (1990), see Preskill and Krauss (1990) and Preskill (1990). Even in the simpler Abelian-Higgs model the situation is not entirely clear. For that reason I will discuss what is known in that case.

The simplest question to ask is are there static solutions of the Einstein-Higgs equations with horizons? By static I mean that not only is the metric static but that the complex Higgs field which I shall now call ϕ is strictly independent of time. If one doesn't make that assumption one might expect to find shells of "Q-matter" surrounding a black hole. It is generally expected that as long as $W(\phi)$ is positive with an absolute minimum at $|\phi| = \eta$ then the only solution must have $\phi =$ constant, with the constant real with no loss of generality. If $W(\phi)$ vanishes it is easy to establish this result. If $W(\phi)$ is **convex** it is also possible to establish this result using a "Bochner Identity". Suppose, more generally, that a field $\phi^A(x)$ takes its values in some riemannian target manifold N with metric $G_{AB}((\phi)$ and potential function $W(\phi)$. The Bochner identity tells us that:

$$(\frac{1}{2}G_{AB}\frac{\partial\phi^A}{\partial x^\alpha}\frac{\partial\phi^B}{\partial x^\beta}g^{\alpha\beta})_{;\mu}^{;\mu} = \phi^{A;\alpha;\beta}\phi^B{}_{;\alpha;\beta}G_{AB}$$

$$+\phi^{A;\alpha}(G_{AB}R_{\alpha\beta} - g_{\alpha\beta}R_{ACBD}\phi^{C}_{;\mu}\ \phi^{D}_{;\nu}\ g^{\mu\nu})\phi^{B;\beta}$$

$$+\phi^{A;\alpha}(\phi^{B;\beta}\ {}_{;\beta})_{;\alpha}G_{AB} \tag{5.1}$$

where all covariant derivatives are covariant with respect to the spacetime metric $g_{\alpha\beta}$ and the target-space metric G_{AB} in the manner described by Misner (1978). The field equations are:

$$\phi^{A;\beta}\ {}_{;\beta} = G^{AB}\nabla_B W \tag{5.2}$$

and

$$R_{\alpha\beta} = 8\pi G[G_{AB}\phi^{A}_{;\alpha}\phi^{B}_{;\beta} + g_{\alpha\beta}W(\phi)] \tag{5.2}$$

If one integrates (5.1) over the region exterior to the black hole and confined within 2 spacelike surfaces $\sum_1$ and $\sum_2$ such that $\sum_1$ is the time translation of $\sum_2$ and uses the field equations and the boundary conditions:

$$\phi^{A}_{;\alpha}\ \ l^{\alpha} = 0$$

on the horizon and $\phi^A \to$ constant at ∞ and where l^α is the null generator of the horizon one obtains the following

Theorem 4 (Scalar No Hair theorem) There are no non-trivial static scalar fields on a static black hole solution of the Einstein-Higgs equations for which $W(\phi) \geq 0$, and the sectional curvature of the target manifold is non positive and $W(\phi)_{;A;B}$ is non negative

Note that unlike Cor. 1 of Theorem 1 we need a stronger assumption on $W(\phi)$ and G_{AB}. If G_{AB} is flat and ϕ^A takes its value in a vector space we could have used the simpler identity:

$$(\phi^{A}G_{AB}\phi^{B}_{;\alpha})^{;\alpha} = \phi^{A}_{;\alpha}G_{AB}\phi^{B}_{;\beta}\ \ g^{\alpha\beta} \tag{5.4}$$

and the field equation (5.2) one obtains

Theorem 5: There are no non-trivial static scalar fields on a static black hole solution of the Einstein-Higgs equations for which $\phi^A W(\phi)_{;B} \geq 0$

Remark: The proof of theorems 4 and 5 also applies to the case where infinity is replaced by a cosmological event horizon.

Neither theorem 4 nor theorem 5 applies to even the simplest case of a single real scalar field ϕ with potential

$$W(\phi) = \lambda(\phi^2 - \eta^2)^2 \tag{5.5}$$

$\lambda > 0$. Thus for non-linear field equations of this type the no-hair conjecture remains - to use a standard term in Scots law -"Not Proven "although there is some suggestive work by Sawyer (1977) and Brumbaugh (1978) .

Let us turn to the work of Adler and Pearson (1978). They assume spherical symmetry and the Einstein-Maxwell-Higgs equations, with a complex scalar ϕ. They assume that that there is only an electric field present and a gauge exists in which:

(1) ϕ is independent of time and real

(2) $A = A_0 dt$ with A_0 everywhere bounded

Actually their assumptions are unnecessarily restrictive and their arguments both incomplete and in part wrong. We shall assume to begin with that

$$A_\mu = A_0 dt \tag{5.6}$$

with

(1) ϕ, A_0 independent of time

(2) $A_0 \rightarrow 0$ at ∞

(3) the one form $A_0 dt = A_\mu dx^\mu$ has bounded "length'.' That is $g^{\mu\nu} A_\mu A_\nu < \infty$ on the horizon.

The equation for A_0 is

$$-\nabla_j(V^{-1}\nabla^j A_0) + e^2|\phi|^2 A_0 = 0 \tag{5.7}$$

where ∇ is taken with respect to the 3-metric g_{ij}. We have dropped the assumption that the metric is spherically symmetric and that ϕ is real.

If one multiplies (5.7) by A_0 and integrates over a surface of constant time one obtains:

$$\int_\Sigma \sqrt{g}(V^{-1}e^2|\phi|^2(A_0)^2 + V^{-1}(\nabla^i A_0 \nabla^j A_0)g_{ij})$$

$$= \int_{\partial\Sigma} (V^{-1} A_0 \nabla^j A_0) d\sigma_j \tag{5.8}$$

Now if $|\phi| \rightarrow \eta$ and $V \rightarrow 1$ at infinity solutions of (5.7) at infinity to like $\frac{1}{r} exp \pm e\eta r$. Thus if A_0 is to be bounded it must fall to zero exponentially and the boundary term at infinity (5.9) will vanish. On the other hand if the field strength $F_{i0} = \partial_i A_0$ is to have bounded scalar invariant on the horizon we require that $V^{-2}(\nabla_j A_0)(\nabla^j A_0)$ should be bounded near the horizon. Now if in addition $A_\mu A_\nu g^{\mu\nu}$ is to be bounded we require that A_0 vanishes at least as fast as V at the horizon and so the boundary term at the horizon in (5.9) must vanish.

We have thus established the following:

Lemma 1 There are no regular time independent electrostatic fields with time independent vector potentials and Higgs field which are bounded with bounded length $A_\mu g^{\mu\nu} A_\nu$ around a static black hole.

Unfortunately lemma 1 is not sufficient to establish that there can be no time independent electrostatic fields around a black hole because it is not clear that there should exist a global gauge in which the vector potentials and Higgs fields are both time independent and bounded. In the usual electromagetic case without symmetry breaking the potential A_μ cannot in fact be cast in a such a gauge. Thus it is necessary to investigate the case when either the gauge variant fields A_μ and ϕ vary with time or do not fall off at infinity. To my knowledge this has not been done.

Even if one assumes that the electromagnetic field vanishes and that the Higgs field is time independent and bounded and even if one assumes further that it is real I know of no rigorous proof that it must be constant in the case that the potential W has the (non-convex) form (5.5).. The argument given be Adler and Pearson for example appears to be incorrect. Although the no-hair property seems very plausible physically it is clear that much remains to be done to establish it rigorously even in the abelian case with symmetry breaking let alone in the non-abelian case.

6. Global Monopoles and Black Holes

Barriola and Vilenkin(1989) have pointed out that the gravitational field of a global monopole has some interesting properties. Far from the core one has

$$\Phi^i \simeq \eta X^i/r \tag{6.1}$$

$$T^o_o \simeq \eta^2 1/r^2 \tag{6.2}$$

$$T^r_r \simeq \eta^2 1/r^2 \tag{6.3}$$

$$T^\theta_\theta = T^\phi_\phi \simeq -\eta^2/r^2 \tag{6.4}$$

with asymptotic metric

$$ds^2 \simeq -dt^2 + \frac{dr^2}{1-8\pi G\eta^2} + r^2(d\theta^2 + \sin^2\theta d\phi^2) \tag{6.5}$$

This metric is not asymptotically flat but rather asymptotically the product of a flat time direction (i.e. the Newtonian potential tends to zero) with a 3-dimensional cone over S^2 with solid angular deficit $32\pi^2 G\eta^2$. For a non singular (but infinity total energy) Φ^i must vanish at $r = 0$, and the metric acquires some corrections. nevertheless in the "σ-model approximation" in which Φ^i always remains in the global minimum of $W(\Phi)$ one may replace the $\simeq$ in (6.1) - (6.5) by $=$ signs.

They are exact solutions of the Einstein equations with σ-model source. Moreover one may consider in addition a black hole. Then (6.1) - (6.4) continue to hold as equalities and (6.5) is replaced by

$$ds^2 = -(1 - \frac{2Gm}{r})dt^2 + \frac{dr^2}{(1-8\pi G\eta^2)(1-\frac{2Gm}{r})} + r^2(d\theta^2 + \sin^2\theta d\phi^2) \tag{6.6}$$

The metric (6.6) (which was worked out by myself and Fernando Ruiz-Ruiz) thus represents a global monopole inside a black hole. Unfortunately however, as pointed out by Goldhaber (1989), the global monopole is likely to be unstable against a sort of angular collapse in which all the Φ field energy becomes concentrated along a line defect or string leaving the point defect at $r = 0$. This string has an energy per unit length of $4\pi\eta^2$. The analysis of Goldhaber is consistent with the work of a number of people on defects in liquid crystals which are modelled using the a free-energy functional [the "Frank Oseen free energy in the one constant approximation] which particle physicists would refer to as a σ-model action and mathematicians as an harmonic map energy functional. Point defects have strings emerging from them which tend to the zero thickness limit of the σ-model cosmic strings introduced by Comtet and myself (1989). Despite this instability there continue to appear preprints analyzing there properties and those of similar objects. An interesting feature is that under some circumstances there can be repulsive gravitational effects. In particular Harari and Lousto (1990) have drawn attention to a repulsive region near the core. A similar feature was found by Ortiz near the core of a 't Hooft-Polyakov monopole. An interesting question to ask is whether for large enough $4\pi\eta^2$ gravitational collapse is inevitable and what is the critical value ? In effect this is a limiting case of the Ortiz problem when λ/e^2 is large.

The gravitational field of an infinite straight σ-model strings was given by myself and Comtet (1989). What about that due to a string emerging from a black hole? Such a string would cause the black hole to accelerate and so the appropriate solution (in the thin string approximation) is the C-metric:

$$ds^2 = \frac{1}{A^2(x+y)^2}[\frac{dy^2}{F(y)} + \frac{dx^2}{G(x)} + G(x)d\alpha^2 - F(y)dt^2]$$

where

$$G(x) = -F(-x)$$
$$= 1 - x^2 - 2GmAx^3; \qquad 0 < GmA < 1/\sqrt{27}$$
$$= 2GmA(x - x_2)(x - x_3)(x - x_4)$$

I have labelled the 3 real roots of $G(x)$ x_2, x_3, x_4 in ascending magnitude (x_2 and x_3 are both negative and x_4 is positive).

The range of the "radial" variable y is

$$-x_3 \leq y \leq -x_2$$

with $y = |x_3|$ being an acceleration horizon and $y = |x_2|$ a black hole horizon. The range of the "angular" variable x is $x_3 \leq \alpha \leq x_4$. The 2-surfaces $x = x_3$ and $x = x_4$ are axes of symmetry for the angular Killing vector $\frac{\partial}{\partial \alpha}$.

In order to understand what the coordinates used it is helpful to consider the case when the the mass parameter m vanishes. Then the metric is flat and one may transform to flat inertial coordinates using the formulae:

$$X^1 \pm iX^2 = \frac{(1-x^2)^{\frac{1}{2}}}{A(x+y)} \exp(\pm i\alpha)$$

$$X^3 \pm X^0 = \frac{(y^2-1)^{\frac{1}{2}}}{A(x+y)} \exp(\pm t)$$

Evidently the coordinate singularity at $x = \pm 1$ is a rotation axis while the coordinate singularity at $y = \pm 1$ corresponds to a pair of intersecting null hyperplanes forming the past and future event horizons for a family of uniformly accelerating worldlines. A similar interpretation may be given in the case that $m \neq 0$ but there is in addition a Black Hole horizon. A detailed description was given by Kinnersley and Walker (1970)

If $0 \leq \alpha \leq \Delta\alpha$ there will be angular deficits:

$$\frac{\delta_4}{2\pi} = \frac{\Delta\alpha - \Delta\alpha_4}{\Delta\alpha_4} \quad ; \quad \frac{\delta_3}{2\pi} = \frac{\Delta\alpha - \Delta\alpha_3}{\Delta\alpha_3}$$

where

$$\Delta\alpha_4 = \frac{4\pi}{|G'(x_4)|} \quad ; \quad \Delta\alpha_3 = \frac{4\pi}{|G'(x_3)|}$$

Since (unless $mA = 0$) $\Delta\alpha_4 \neq \Delta\alpha_3$ it is not possible to eliminate both of these by choosing $\Delta\alpha$. One can eliminate δ_3 in which case the black hole is pulled along by a string, or δ_4 in which case it is pushed along by a rod. In general the net "force" on the hole is

$$F = \frac{\delta_4}{8\pi G} - \frac{\delta_3}{8\pi G} = \frac{\Delta\alpha}{4G}(\frac{1}{\Delta\alpha_4} - \frac{1}{\Delta\alpha_3}) = \frac{\Delta\alpha m A}{8\pi}(x_4 - x_3)^2$$

The black hole event horizon area $\mathcal{A}$ is given by

$$\mathcal{A} = \frac{\Delta\alpha}{A^2}\frac{x_4 - x_3}{(x_4 - x_2)(x_3 - x_2)}$$

The black hole horizon surface gravity κ_{BH} and the acceleration horizon surface gravity κ_R are given by:

$$\kappa_{BH} = GmA^2(x_3 - x_2)(x_4 - x_2)$$

$$\kappa_R = GmA^2(x_3 - x_2)(x_4 - x_3)$$

If $GmA << 1$ one obtains:

$$\kappa_{BH} \sim \frac{1}{4Gm}$$

$$\kappa_R \sim A$$

$$\mathcal{A} \sim 8\Delta\alpha G^2 m^2$$

$$F \sim \Delta\alpha\frac{mA}{2\pi}$$

whence

$$F \simeq \frac{\mathcal{A}\kappa_{BH}}{8\pi G} \;.\; \kappa_R$$

which is equivalent to Newton's second law of motion. However for finite mA one does not obtain such a simple expression. This is perhaps not surprising since if mA is not small the Schwarzschild radius of the black hole is comparable with the radius of curvature of its world line. Nevertheless it would be nice to understand the relation between mass, acceleration and force in this non-linear situation. Some attempts in this direction, which also relate to black hole thermodynamics were made by Aryal, Ford and Vilenkin (1986),(see also Martinez and York (1990). Note that Aryal te al. use the representation of accelerating black hole metrics in terms of Weyl-metrics using the "rod representation" of Schwarzschild (Israel and Khan (1964). The relationship between this picture and the C-metric including the co-ordinate transformation between the finite rod plus semi-infinite rod (each of mass per unit length $\frac{1}{2}$) and the C-metric form quoted above may be found in (Godfrey (1972) see also (Bonnor (1983,1990).More about strings and black holes may be found in Chandraskhar and Xanthopouls (1989).

7. Black Hole Monopole Pair-Production

In the quantum theory we know that charged particle anti-particle pairs may be created by a sufficiently strong electric field - a process sometimes called the Schwinger Process. It is plausible that magnetic monopoles should similarly be created by strong magnetic fields. This process was investigated in Yang-Mills-Higgs theory by Affleck and Manton (1982) using instanton methods. The use of instanton methods to calculate the rate of production by the Schwinger process is discussed in (Affleck, Alvarez and Manton).

Some time ago I suggested that the same process should occur in quantized Einstein-Maxwell theory (Gibbons 1986). The idea has recently been taken up again by Strominger and Garfinkle (1990). Since **pure** Einstein-Maxwell theory has invariance under the duality transformation

$$F_{\mu\nu} \to (\exp\ \alpha\star)F_{\mu\nu}$$

where $\star$ is the Hodge star operation on 2-forms. There is no invariant distinction between electric and magnetic, so let us concentrate on the purely magnetic case. In any event it is this case which is physically most interesting in more realistic models.

To begin we need to model a strong magnetic field coupled to gravity. The natural choice is the Melvin solution which represents an infinitely long straight self-gravitating Faraday flux tube in equilibrium, the gravitational attraction being in equipoise with the transverse magnetic pressure (Melvin (1964). The metric is:

$$ds^2 = (1+\pi G B^2\rho^2)^2(-dt^2+dz^2+d\rho^2)+\rho^2 d\phi^2(1+\pi G B^2\rho^2)^{-2}$$

The magnetic field is given by:

$$F = \frac{B\rho d\rho \wedge d\phi}{(1+\pi G B^2\rho^2)^2}$$

The Melvin solution posesses a degress of uniqueness. For example Hiscock (1981) has shown

Theorem: The only axisymmetric, static solution of the Einstein-Maxwell field equations without an horizon which is is asymptotically Melvin is in fact the Melvin Solution.

In fact Hiscock also allows for a neutral or electrically charged black hole as well.

I myself can show:

Theorem: The only translationally invariant, static solution of the Einstein-Maxwell field equations without horizon which is asymptotically Melvin is in fact the Melvin solution.

Proof: assume the metric is static and has reflection invariance with respect to the $z-$direction. These two assumptions may easily be justified. The metric takes the form

$$ds^2 = -V^2dt^2 + Y^2dz^2 + g_{AB}dx^A dx^B$$

with $A = 1,2$. The field equations are:

$$\nabla^A(VY\nabla_A \ln(V/Y)) = VY8\pi G(T_{\hat{z}\hat{z}} + T_{\hat{0}\hat{0}})$$

$$\nabla^A(VY\nabla_A \ln(VY)) = VY8\pi G T^A_A$$

$$V^{-1}\nabla_A\nabla_B V + Y^{-1}\nabla_A\nabla_B Y = Kg_{AB} - 8\pi(T_{AB} - \frac{1}{2}g_{AB}(T^A_A + T_{\hat{z}\hat{z}} + T_{\hat{0}\hat{0}}))$$

where K is the Gauss curvature of the 2-metric g_{AB}. The electromagnetic field is assumed to be of the form:

$$F = \frac{1}{2}F_{AB}dx^A dx^B.$$

It follows that $T_{\hat{0}\hat{0}} + T_{\hat{z}\hat{z}} = 0$ and hence:

$$\nabla^A(VY\nabla_A(V/Y)) = 0.$$

Now V/Y tends to one at infinity (asymptotic boost invariance) and so we may invoke tha Maximum Principle to show that $V = Y$ everywhere. Thus the metric must be boost invariant.It now follows that

$$\nabla_A\nabla_B V = fg_{AB}$$

for some scalar f. Thus

$$K^A = \epsilon^{AB}\nabla_B V$$

is a Killing vector field of the 2-metric g_{AB} and since $K^A\partial_A V = 0$ it is also a Killing vector field of the entire 4-metric. It is not difficult to see that this Killing vector field corresponds to rotational symmetry of the solution.

The argument just given will generalise in various ways to cover some other stress tensors and as mentioned above the staticity asummption and the assumption that $g_{\alpha z} = g_{zz}\delta_{\alpha z}$ is not difficult to justify using standard methods on the global theory of balck holes. Interestingly however it does not seem to be possible to show using this method that the metric of a local cosmic string must be axisymmetric. Even in flat spacetime this seems to be a very difficult problem, i.e. to show that all time independent Nielsen-Olesen vortex solutions of the abelian Higgs model (in the non-supersymmetric case) must have axial symmetry. Having established the credentials of the Melvin solution as uniquely suitable model of a static magnetic field in general relativity we turn to looking for instanton solutions representing the creation of a black hole monopole anti-monopole pair. If there were no external magnetic field the obvious candidate instantons would be the magnetically charged C-metric for which

$$G(x) = 1 - x^2 - 2GmAx^3 - G(g^2/4\pi)A^2x^4.$$

However this has nodal singularities. In fact since the metric is boost invariant it has zero ADM mass and thus it cannot be regular by the positive mass theorem generalised to include apparent horizons. However it was pointed out by Ernst ((1976) that the nodal singularity may be eliminated if one appends a suitable magnetic field. The resulting metric is of the same form as (6.7) but the first three terms are mutiplied by and the last term divided by the factor:

$$(1 + GBgx/2)^4.$$

If $m = 0 = g = A$ we get the Melvin solution but the limit must be taken carefully. The nodal singularity may be eliminated if B is chosen so that

$$G'(x_3)/(1+GBgx_3/2)^4 \quad + G'(x_4)/(1+Ggx_4/2)^4 = 0.$$

Where x_3 and x_4 are two larger roots of $G(x)$ and we assume now that there are 4 roots. The smallest root x_1 is thus inside the acceleration horizon. This equation may be regarded as an equation for B the magnetic field necessary to provide the force to accelerate the magnetically charged black hole. It is difficult to find an explicit solution in terms of g, m and A except when GmA is small in which case one finds the physically sensible result:

$$gB \approx mA.$$

In order to obtain an instanton which is regular on the Riemannian section obtained by allowing the time coordinate t to be pure imaginary it is necessary that the $\tau = it$ is perodic with period given by the surface gravity. This leads to the condition that

$$G'(x_2) + G'(x_3) = 0.$$

It appears that the the only way to satisfy this condition is to set:

$$m = |g|/\sqrt(4\pi G)$$

Note that this equation **does not** mean that the horizons have vanishing surface gravity as I mistakenly asserted in (1986).It is not difficult to see that the topology of the Riemann section is $S^2 \times S^2$ with a point removed. In fact topologically one can obtain this manifold from R^4, which is the topology of the Melvin solution, by surgery along an S^1. That is by cutting out a neighbourhood of a circle which has topology $D^3 \times S^1$ with boundary $S^2 \times S^1$ and replacing by $S^2 \times D^2$ which has the same boundary. This surgery is also what is needed to convert $R^3 \times S^1$ to $R^2 \times S^2$ i.e. to convert a manifold with the topology of "Hot Flat Space" to that with the topology of the Riemannian section of the Schwarzschild solution.

The existence of this instanton would seem to be rather important. It seems to imply for example that it would be **inconsistent** not to consider the effects of black hole monopoles since given strong enough magnetic fields they will be spontaneously created. Once they are created they should evolve by thermal evaporation to the extreme zero temperature soliton state. Another reason why I believe that this process is so important is that it seems to show that while one may have one's doubts about the effects of wormholes because of the absence of suitable solutions of the classical equations of motion with positive definite signature, the solutions described here do indicate that some sort of topological fluctuations in the structure of spacetime **must** be taken into account in a satisfactory theory of gravity coupled to Maxwell or Yang-Mills theory.

References

S L Adler and R P Pearson, No Hair theorem for the Abelian Higgs and Goldstone models. Phys Rev **D18**, 2798-2803 (1978)

I K Affleck, O Alvarez and N S Manton, Pair production at strong coupling in weak external fields. Nucl. Phys **B197** 509-519 (1982)

I K Affleck and N S Manton Monopole pair production in a magnetic field. Nucl Phys **B194** 38-64 (1982)

F J Almgren and E H Lieb, Counting Singularities in liquid crystals. Proc. IXth International Congress on Mathematical Physics, eds B Simon, A Truman and I M Davies, Adam Hilger 1989

F Almgren and E Lieb, Singularities of energy minimizing maps form the ball to the sphere: examples counter examples and bounds, Ann of Math. **128** 483-430 (1988)

M Aryal, L H Ford and A Vilenkin, Cosmic Strings and Black Holes, Phys Rev **D34** 2263-2266 (1986)

A Ashtekar and T Dray, On the Existence of solutions to Einstein's Equation with Non-Zero Bondi News. Commun. Math. Phys **79** 581-589 (1981)

F A Bais and R J Russell, Magnetic-monopole solution of the non-Abelian gauge theory in curved spacetime. Phys Rev **D11** 2692-2695 (1975)

R Bartnik and J McKinnon, Particle like solutions of the Einstein-Yang-Mills equations. Phys Rev Lett **61** 141-144 (1988)

M Barriola and A Vilenkin, Gravitational field of a global monopole. Phys. Rev. Lett **63** 341-343 (1989)

J Bicak, The motion of a charged black hole in an electromagnetic field. Proc. Roy. Soc. **A371** 429-438 (1980)

W B Bonnor, The sources of the vacuum C-metric. Gen. Rel. Grav. **15** 535-551 (1983)

W B Bonnor, The C-metric in Bondi's coordinates. Class. Quant. Grav. **7** L229-L230 (1990)

P Bizon, Colored Black Holes. Phys. Rev. Letts **64** 2844-2847 (1990)

P J Braam, A Kaluza-Klein approach to hyperbolic three-manifolds. Enseign. Math **34** 275-311 (1985)

R A Brandt and F Neri, Stability Analysis for Singular Non-Abelian Magnetic monopoles Nucl. Phys. **B161** 253-282 (1979)

P B Breitenlohner, G W Gibbons and D Maison, 4-dimensional Black Holes from Kaluza-Klein Theory. Commun. Math. Phys. **120** 295-334 (1988)

H Brezis, J M Coron and E Lieb, Harmonic Maps with defects. Commun. Math. Phys **107** 649-705 (1986)

B E Brumbaugh, Nonlinear scalar field dynamics in Schwarzschild geometry. Phys. Rev. **D18** 1335-1338 (1978)

S Chandrasehkar and B C Xanthopoulos two Black Holes attached to strings. Proc. Roy. Soc. **A423** 387-400 (1989)

T Chmaj and E Malec, Magnetic monopoles and gravitational collapse. Class and Quantum Grav. **6** 1687-1696 (1989)

Y M Cho and P G O Freund, Gravitating 't Hooft monopoles. Phys. Rev. **D12** 1588-1589, (1975)

S Coleman in "The Unity of Fundamental interactions"ed. A Zichichi (Plenum, New York) (1983)

A Comtet, Magnetic Monopoles in curved spacetimes. Ann. Inst. H Poincare **23** 283-293 (1980)

A Comtet, P Forgacs and P A Horvathy, Bogomolnyi-type equations in curved spacetime. Phys. Rev **D30** 468-471 (1984)

A Comtet and G W Gibbons, Bogomol'nyi Bounds for Cosmic Strings. Nucl. Phys. **B299** 719-733 (1989)

A D Dolgov, Gravitational Dipole. JETP Lett. **51** 393-396 (1990)

T Dray, On the Asymptotic Flatness of the C Metrics at Spatial Infinity. Gen.Rel.Grav. **14** 109-112 (1982)

T Dray and M Walker, On the regularity of Ernst's generalized C-metric. L.I.M.P. **4** 15-18 (1980)

F J Ernst, Black holes in a magnetic universe. J.M.P. **17** 54-56 (1976)

F J Ernst, Removal of the nodal singularity of the C-metric. J.M.P. **17** 515-516 (1976)

F J Ernst, Generalized C-metric. J.M.P. **19** 1986-1987 (1978)

F J Ernst and W J Wild, Kerr black holes in a magnetic universe. J.M.P. **17** 182-184 (1976)

A A Ershov and D V Gal'tsov, Non Existence of regular monoples and dyons in the SU(2) Einstein-Yang-Mills theory. Phys. Letts. **150A** 159-162 (1990)

J A Frieman and C T Hill, Imploding Monopoles. SLAC-PUB-4283 Oct. 1987 T/AS

A Floer, Monopoles on asymptotically Euclidean manifolds. Bull. AMS **16** 125-127 (1987)

D V Gal'tsov and A A Ershov, Non-abelian baldness of coloured black holes. Phys. Letts **A138** 160-164 (1989)

D Garfinkle and A Strominger, Semi-classical Wheeler Wormhole Production. UCSB-TH-90-17

G W Gibbons Non-existence of Equilibrium Configurations of Charged Black holes. Proc. Roy. Soc. **A372** 535-538 (1980)

G W Gibbons, Quantised Flux-Tubes in Einstein-Maxwell theory and non-compact internal spaces, in Fields and Geometry Proc. of XII Karpac Winter School of Theoretical Physics 1986, ed A Jadczyk, World Scientific

B B Godfrey, Horizons in Weyl metrics exhibiting extra symmetries. G.R.G. **3** 3-15 (1972)

A Goldhaber, Collapse of a Global Monopole. Phys. Rev. Letts **63** 2158(c) (1989)

Gu Chao-hao, On Classical Yang-Mills Fields. Phys. Rep. **80** 251-337 (1981)

P Hajicek, Wormhole solutions in the Einstein-Yang-Mills-Higgs system. I General theory of zero-order structure. Proc. Roy. Soc A **386** 223-240 (1983)

P Hajicek, Wormhole solutions in Einstein-Yang-Mills-Higgs system II Zeroth-order structure for $G = SU(2)$. J. Phys. **A16** 1191-1205 (1983)

P Hajicek, Classical Action Functional for the system of fields and wormholes. Phys. Rev. **D26** 3384-2295 (1982)

P Hajicek, Generating functional Z_0 for the one-wormhole sector. Phys. Rev. **D26** 3396-3411 (1982)

P Thomi, B Isaak and P Hajicek, Spherically Symmetric Systems of Fields and Black Hole. I Definition and properties of Apparent Horison. Phys. Rev. **D30** 1168-1171 (1984)

P Hajicek, Spherically symmetric systems of fields and black holes. II Apparent horizon in canonical formalism. Phys. Rev. **D30** 1178-1184 (1984)

P Hajicek, Spherically symmetric systems of fields and black holes. III Positively of enemy and a new type of Euclidean action. Phys. Rev. **D30** 1185-1193 (1984)

P Hajicek, Spherically symmetric systems of fields and black holes. IV No room for black hole evaporation in the reduced configuration space? Phys. Rev. **D31** 785-795 (1985)

P Hajicek, Spherically symmetric systems of fields and black holes. V Predynamical properties of causal structure. Phys. Rev. **D31** 2452-2458 (1985)

P Hajicek, Quantum theory of wormholes. Phys. Letts **106B** 77-80 (1981)

P Hajicek, Quantum wormholes (I). Choice of the classical solutions. Nuc. Phys. **B185** 254-268 (1981)

P Hajicek, Duality in Klein-Kaluza Theories. BUTP-9/82

P Hajicek, Exact Models of Charged Black Holes. I: Geometry of totally geodesic null hypersurface. Commun. M. Phys. **34** 37-52 (1973)

P Hajicek, Exact Models of Charged Black Holes II: Axisymmetries Sationary Horizons. Commun. M. Phys **34** 53-76 (1973)

P Hajicek, Can outside fields destroy black holes. J. Math. Phys. **15** 1554 (1974)

D Harari and C Lousto, Repulsive gravitaional effects of global monopoles. Buenos Aires preprint GTCRG-90-4

W A Hiscock, Magnetic Monopoles and evaporating black holes. Phys. Rev. Letts **50** 1734-1737 (1982)

W A Hiscock, On black holes in magnetic universes. J. Math. Phys. **22** 1828-1833 (1981)

W A Hiscock, Magnetic monopoles and evaporating black holes. Phys. Rev. Lett **50** 1734-1737 (1983)

W A Hiscock, Astrophysical bounds on global monopoles.

W A Hiscock, Gravitational particle production in the formation of global monopoles.

P A Horvathy, Bogomolny-type equations in curved space. Proc. 2nd Hungarian Relativity Workshop (Budapest 1987) ed. Z Peres, World Scientific

H S Hu, Non existence theorems for Yang-Mills fields and harmonic maps in the Schwarzschild spacetime (I). Lett. Math. Phys. **14** 253-262 (1987)

H S Hu and S Y Wu, Non existence theorems for Yang-Mills fields and harmonic maps in the Schwarzschild spacetime (II). Lett Math. Phys. **14** 343-351 (1987)

W Israel and K A Khan Collinear paricles and Bondi dipoles in general relativy. Nuovo Cimento **33** 331 (1964)

W Kinnersley and M Walker, Uniformly accelerating charged mass in general relativity. Phys. Rev. **D2** 1359-1370 (1970)

L M Krauss and F Wilczek, Discrete gauge symmetry in continuum theories. Phys Rev Letts **62** 1221-1223 (1989)

H P Kunzle and A K M Masood-ul-Alam, Spherically symmetric static $SU(2)$ Einstein-Yang-Mills fields. J. Math. Phys. **31** 928-935 (1990)

A S Lapedes and M J Perry, Type D Gravitational instantons. Phys. Rev. **D24** 1478-1483 (1983)

D Lohiya, Stability of Einstein-Yang-Mills Monopoles and Dyons. Ann. Phys. **14** 104-115 (1982)

M Magg, Simple proof for Yang-Mills instability. Phys. Letts **74B** 246-248 (1978)

E Malec and P Koc, Trapped surfaces in monopole-like Cauchy data of Einstein-Yang-Mills-Higgs equations. J. Math. Phys. **31** 1791-1795 (1990)

J E Mandula, Classical Yang-Mills potentials. Phys. Rev. **D14** 3497-3507 (1976)

J E Mandula, Color screening by a Yang-Mills instability. Phys. Letts **67B** 175-178 (1977)

J E Mandula, Total Charge Screening. Phys. Lett. **69B** 495-498 (1977)

E A Martinez and J W York, Thermodynamics of black holes and cosmic strings. IFP-342 UNC: May 1990

A K M Masood-ul-Alam and Pan Yanglian (Y L Pan) Non Existence theorems for Yang-Mills fields outside a black hole of the Schwarschild spacetime. Lett in Math. Phys. **17** 129-139 (1989)

M A Melvin, Pure magnetic and electric geons. Phys. Lett **8** 65-67 (1964)

C W Misner, Harmonic maps as models for physical theories. Phys. Rev **D18** 4510-4524 (1978)

M J Perry, Black holes are coloured. Phys. Letts **71B** 234 (1977)

J Preskill and L M Krauss, Local discrete symmetry and quantum-mechanical hair. Nucl. Phys. **B341**50-100 (1990)

J Preskill Quantum Hair. Caltech preprint CALT-68-1671 (1990)

D Ray, Solutions of coupled Einstein-$SO(3)$ gauge field equations. Phys. Rev. **D18** 1329-1331 (1978)

P Ruback A New Uniqueness Theorem for Charged Black Holes. Class. and Quant. Grav. **5** L155-L159 (1988)

R F Sawyer, The possibility of a static scalar field in the Schwarzschild geometry. Phys. Rev. **D15**, 1427-1434 (1977). Erratum Phys. Rev. **D16** (1977) 1979

P Sikivie and N Weiss, Screening Solutions to Classical Yang-Mills theory. Phys. Rev. Letts **40** 1411-1413 (1978)

N Straumann and Z-H Zhou, Instability of colored black hole solution. Phys. Letts **141B** 33-35 (1990)

N Straumann and S-H Shou, Instability of the Bartnik-McKinnon solutions of the Einstein-Yang-Mills equations. Phys. Letts **237** 353-356 (1990)

P van Nieuwenhuizen, D Wilkinson and M J Perry, Regular solution of 't Hooft's magnetic monopole in curved space. Phys. Rev. **D13** 778-784 (1976)

M Y Wang, A solution of coupled Einstein-SO(3) gauge field equations. Phys. Rev. **D12** 3069-3071 (1975)

M S Volkov and D V Gal'tsov, Non-Abelian Einstein-Yang-Mills black holes. JETP. Letts **50** 346-350 (1989)

D V Galt'sov and A A Ershov. Yad. Fiz. **47** 560 (1988)

Understanding Large-Scale Cosmic Structure *

Bernard J.T. Jones

Astronomy Centre, Sussex University, UK

and

Rien van de Weygaert

Sterrewacht, Leiden, NL

'Inner constitution of things' ...
'Inner meaning of the Universe' ...
That is all false, it all means nothing.
It is incredible that people can think about those things.

Fernando Pessoa as Alberto Caiero

1: INTRODUCTION

During the past decade, research into the large scale structure of the universe has played a key role in our emerging understanding of cosmology. There are several reasons why this particular direction, rather than say galaxy evolution, should have proved so productive.

We believe in the first place that the large scale structure has been assembled into its present configuration by the action of the force of gravity alone. Furthermore, the present day deviations from homogeneity and isotropy on scales in excess of a few tens of Megaparsecs are relatively small and were, according to our hypotheses, even smaller in the past. This means that there is a well defined and simple theoretical framework within which to understand the large scale structure of the universe. That is provided by the weakly perturbed Friedman-Lemaitre solutions of the Einstein Field Equations. Given this framework, it turns out that carefully considered Newtonian limits of these solutions

* Lecture presented to the XII Autumn School "The Physical Universe", Lisbon, October 1 - 5, 1990

are generally more than adequate for handling many problems of large scale structure formation.

Secondly, there has been an explosion of the amount and variety of data with which to confront theoretical notions. These data are generally part of large systematic surveys, and often the observational selection effects are well under control. We now have a situation where theories are suggested by observations, to be supported or later discredited by further data gathering. The observations are constraining the theories.

Of course the methods of data analysis have to be more sophisticated. The correct procedure is to interpret data in the light of a given model, and to assess the validity of the model by doing some kind of maximum likelihood reconstruction. The existence of the simple gravitational framework simplifies this interpretive procedure.

The purpose of this lecture is mainly to discuss new techniques that have been developed in order to best interpret various kinds of galaxy catalogues. With these techniques we leave, for the first time, the idea that we are forever confined to knowing only about the universe as projected onto the sky. We are able to construct three dimensional models of the universe. There are indeed some crucial assumptions that go into this, but they are clearly understood and appear to be entirely reasonable. It remains an important problem for the future to validate these assumptions.

A summary of the necessary cosmological equations are given in the Appendix. We will be continually referring to "peculiar" velocities of galaxies. By this we mean the following. If the galaxies in the universe participated in simple homogeneous and isotropic expansion at a rate given by the Hubble parameter H_0, the radial velocity of a galaxy at distance r from the observer would be $v_r = H\ r$. This is the ideal Hubble Expansion Law. Because of the influence of other forces acting on galaxies, they will have an additional, "peculiar", component of radial velocity, v_{pec}, so that the total observed velocity is $cz = v_r + v_{pec}$. z is called the observed *redshift* of the galaxy. Observers measure cz and it has become one of the major goals of cosmology today to try to determine its components v_r and v_{pec}. The existence of nonzero v_{pec} is thought to be due to the existence of inhomogeneities in the cosmic mass distribution, and theories of structure evolution provide the link between v_{pec} and the density inhomogeneity $\delta\rho$. The link does depend on particular hypotheses for how structure grows in the Universe. The methods we shall be discussing are systematic ways of deriving these two quantities.

2: METHODS

2.1: Types of Survey

The Large Scale Structure of the Universe is perceived in many different kinds of astronomical observations. The earliest studies were simple observations of the projection of the galaxy distribution on the sky. Then came the redshift surveys allowing us to plot the spatial locations of galaxies as estimated from their redshift (assuming the Hubble expansion law). And finally we have the surveys where we have redshift independent distance estimates for each galaxy, allowing us to plot a map of the "real" spatial distribution of galaxies, albeit with a substantial error.

Maps of the distribution of galaxies projected on the sky show large scale features such as galaxy clusters and voids, and there are even suggestions of filamentary structures. It was with such projected surveys that the two-point correlation function for galaxies, $w(\theta)$ was measured, and it is arguable that this still provides the best estimate of the three-dimensional correlation function $\xi(r)$. Such maps now exist showing several million galaxies in the optical wavebands (Maddox et al., 1990). There are also substantial catalogues of galaxies in the Infrared wavebands culled from the IRAS survey. These latter surveys have the advantage of complete sky coverage, the Milky Way does not interfere.

Radial velocity surveys, even though they contain far fewer galaxies (less than around 10,000), substantially enhance the impression gained from the projected data. The famous CfA slice of de Lapparent et al. (1988) probably did more to paint an impression of the large scale distribution of galaxies than any other observation. Subsequent slices and surveys have strengthened this impression, revealing still larger structures ("Great Walls", Geller and Huchra, 1990).

The concern has always been that the redshift surveys are plotting the galaxy positions in velocity space, under the assumption that the Hubble expansion law holds precisely. This might artificially enhance features such as filaments and voids. It would obviously be better to use the true distances to galaxies if these were known.

The surveys of galaxies containing both redshifts and redshift independent distance estimates, while not quite as extensive as the simple redshift surveys, have yielded dramatic results. Comparing the radial velocity estimated from the distance and the Hubble Law with the observed radial velocity gives the radial component of the velocity of the galaxy relative to the Hubble flow. At present there are substantial uncertainties in the velocity independent distance estimates. Nonetheless, it appears that these "peculiar velocities" are many hundreds of kilometres per second, and, what is most surprising, they appear to be correlated over very large volumes of the space.

2.2: Analysis of redshift survey data

Vast amounts of literature has been devoted to the analysis of galaxy catalogues of various kinds. We could do no more than present a few salient points here, though later we shall delve deeper into the analysis of the redshift surveys using iteration techniques to reconstruct the three dimensional density and velocity fields.

The analysis of the projected galaxy samples has yielded two-point, three- point and higher order angular correlation functions, from which one infers the three dimensional correlation functions using the assumption of statistical isotropy. While there are interesting alternative measurements of the two point correlation function from redshift surveys, there are as yet no measurements of the higher order functions from these redshift surveys.

The redshift surveys do provide a first look at the true three dimensional galaxy distribution, and so have been the subject of extensive analysis. What is remarkable about the redshift surveys is that the striking structure they reveal (the voids and filaments) is all on very large scales where the galaxy position correlation functions are very weak or undetectable. Characterising that structure has been a major goal requiring more powerful techniques than correlation analysis and several of these techniques have been developed, notably the Void Probability Function, the Genus of the distribution and Multi-Fractal analysis. These techniques, however, merely provide descriptive information about the statistical nature of the clustering. They leave open some important dynamical questions such as "how big are the non-Hubble velocities and how much is our impression of the clustering influenced by these peculiar velocities?".

These dynamical questions can been answered in two ways, one from all sky redshift surveys and the other from redshift surveys in which there are velocity independent distance estimators for each galaxy in the sample.

A redshift survey of IRAS galaxies has been undertaken by Strauss, Davis, Yahil and Huchra (1990) ("SDYH") who go on to shown how to reconstruct the entire density and three-dimensional peculiar velocity fields from this survey. While this does not answer the question of the nature of the clustering, it does provide direct quantitative information about the distribution and dynamics of matter in the universe. There is a complementary study of IRAS galaxies by a conglomerate from Queen Mary College, Durham, Oxford and Toronto ("QDOT", Efstathiou et al. 1990) who are obtaining redshift data for a sparse sample of galaxies (1 in 6) down to a fainter flux level than SDYH. Both these IRAS-based surveys have 2000 - 3000 galaxies.

There are several smaller surveys with velocity-independent distance estimates for the galaxies in the sample. However, taken together, these surveys form a rather heterogeneous sample of about a thousand galaxies probing the peculiar cosmic velocity field out to a redshift of some 6000 $km.s.^{-1}$. In a very important paper, Bertschinger and Dekel (1989) showed how to reconstruct from this data a map of the three-dimensional peculiar velocity field in the sample volume. Since the method has the virtue of treating galaxies like sensors of the local gravitational field, it can handle this kind of heterogeneous data.

The possibility of reconstructing the full three dimensional cosmic flow is as fundamental a step forward in cosmology as the introduction by Oort of the Galactic A and B constants was to understanding the dynamical state of our Galaxy.

2.3: Two Approaches

There are essentially two approaches to reconstructing the three dimensional galaxy distribution.

2.3.1: IRAS based surveys

If we have no independent distance estimators to individual galaxies we have to use the observed density distribution together with a dynamical model to estimate the peculiar velocities that are consistent with the inferred density. The peculiar velocities can then be subtracted from the observed radial velocities to give a better impression of the density field. The process can then be cycled until convergence is obtained. We can start the process off with a redshift survey to make a first estimate of the local density fluctuations on the assumption the redshift reflects true distance.

As we shall see below, the peculiar velocity field generated by a density fluctuation field $\delta(\mathbf{x}) = \delta\rho/\rho$ is in linear perturbation theory:

$$\mathbf{v} = \frac{2}{3H\Omega^{0.4}}\mathbf{g}$$

where, by virtue of the perturbed Poisson equation:

$$\mathbf{g}(\mathbf{x}) = Ga(t)\rho_b \int \frac{\mathbf{x}' - \mathbf{x}}{|\mathbf{x}' - \mathbf{x}|^3} \; \delta(\mathbf{x}', t) \; d^3\mathbf{x}'$$

The integral should involve the entire universe, but in practise only the survey volume can be used. There is also a problem that arises when the density field becomes nonlinear, the velocity field can then become triple valued (due to shell crossing). Hence the velocity field prediction will not be as good as the density field prediction in such a method.

Another complication is also evident here. We cannot observe $\delta\rho/\rho$, but only the fluctuation in the distribution of the luminous component. In order to cope with this, we have to introduce another parameter, the bias parameter, b:

$$\frac{\delta n}{n} = b\frac{\delta\rho}{\rho}$$

Obviously b could be a function of the local density, $b = b(n)$, or even worse it might not be a universal function. At this first exploratory stage the simplest assumption to make is that $b = constant$, and to try and determine this constant as yet another one of the constants of cosmology. The iteration method clearly allows b to be one of the parameters involved in making the fit and so b (in fact $b\Omega^{-0.6}$) can be determined self- consistently. If it should turn out that the data analysis gives rise to inconsistencies with prior expectations based on models, then this is obviously the first place to look for a resolution of the problems.

The two surveys based on IRAS use slightly different selection criteria for finding galaxies in the IRAS catalogue. The Strauss et al. survey uses all nonstellar objects having a $60\mu m$ flux greater than 1.9 Jansky, yielding a sample of over 2500 objects out to a distance corresponding to a redshift of $\simeq 3000\ \mathrm{km.s^{-1}}$. The QDOT survey selects 1 galaxy randomly out of 6 down to a flux limit of 0.6 Jansky, giving a sample of more than 2000 galaxies out to a distance corresponding to a redshift of $\simeq 7000\ \mathrm{km.s^{-1}}$.

Because of the far infrared nature of the IRAS survey, elliptical and early- type galaxies are absent from the surveys, and hence the surveys do not contain most of the galaxies in rich galaxy clusters like Coma. Nonetheless, they do appear to trace the known features in the universe. Accordingly, a correction has to be applied to take account of the missing galaxies when reconstructing the galaxy density field. This correction comes over and above any correction for the bias factor.

2.3.2: Surveys with independent distance estimates

If we have redshift independent distance estimates, then we can avoid the step of deducing the peculiar velocity field from the apparent density inhomogeneities. This approach is called the "POTENT" method and has been used taken by Bertschinger and Dekel and their collaborators. We shall describe this at some length below.

Redshift independent distance estimates can come either from Fisher-Tully type relationships relating the peak rotational velocities of disk galaxies to their absolute magnitude in some waveband, or from the so-called $D_n - \sigma$ relationships relating the central velocity dispersions of elliptical galaxies to the actual diameter of a particular isophote in the galaxy light distribution. The root mean square errors of these techniques are 15% and 21% respectively, and this gives rise to quite large distance errors. The analysis therefore has to be done rather carefully.

The redshift independent distances provide a direct first order measure of the peculiar velocities. These can then be used to produce a consistent map of the density field via a relationship like

$$\nabla \cdot \mathbf{v} = -\Omega^{0.6}\ H\ \delta.$$

Of course, that map can be iterated in just the same way as before, only this time we go the other way around and use the peculiar velocity field to deduce the density field.

This process has a number of advantages. The velocity field divergences is a direct measure of the mass distribution, and so what lies outside the data volume is not relevant. The data sample does not even have to be complete, where we have data we can rebuild the density field. However, there is the serious disadvantage that redshift independent distances are not available for large numbers of galaxies and so the sampling of the universe is necessarily sparser than we would like. Currently, a sample of around 1000 galaxies has been used.

Comparison between the results from these two approaches is very interesting. If the assumptions underlying the methods are correct, there should be substantial agreement between the reconstructed maps of the universe. In particular, we shall be interested in comparing the bias parameters b (or the bias functions $b(n)$) derived from the two methods.

3: THE UNIVERSE IN THREE DIMENSIONS

3.1: The problem and the POTENT solution

The galaxy redshift surveys for which there are velocity independent distance estimates for the sample galaxies have revealed substantial deviations from uniform Hubble flow. There might be, because of the uncertainty in the distance estimation procedure, some room for scepticism. However, the fact that we already observe a "large" velocity for the Local Group relative to the microwave background radiation and that different samples yield consistent results encourages us to go on and ask the question "what is the cosmological implication of these deviations?".

The first studies of the deviations showed large scale coherence in the peculiar (ie: non-Hubble) component of the velocity field. In particular they showed evidence for the so-called "Great Attractor" in the direction of the Hydra-Centaurus clusters of galaxies. These early studies relied either on the use of the radial component of the peculiar velocity, or on the fitting of a specific model for the Great Attractor and its environment (Lynden-Bell et al. 1988). While such models give an indication of what the Great Attractor is, one is left with a very large parameter space of possible models none of which has an a priori dynamical justification.

This model-fitting situation has been dramatically improved by the recent discovery of Bertschinger and Dekel (1989) that one could, on the basis of a few reasonable assumptions, reconstruct the entire three dimensional velocity field given only the radial peculiar velocity data for a sample of galaxies. Moreover, the sample does not have to be a complete sample (though where there are most galaxies the reconstruction of the cosmic flow field is obviously most reliable). Bertschinger, Dekel, Dressler and Faber, in a recent series of papers, have applied the technique to a compendium of redshift samples that allow the universe to be mapped out to a distance of 6000 km.s.$^{-1}$.

This discovery has given new impetus to radial velocity surveys and to getting velocity independent distance estimates for individual galaxies.

3.2: The Basic Premises

We shall suppose, as is commonly done, that galaxies evolve from initially small amplitude density inhomogeneities that are amplified by the action of gravitational forces alone. We shall also suppose that the universe was initially (ie. near to the singularity) close to an Einstein de Sitter universe. With these assumptions, the evolution of small amplitude density perturbations is very simple. There are several descriptions of this, the simplest one probably being the famous Zel'dovich (1970) approximation, which is claimed to also give a reasonable approximation to the not-so-linear phases of evolution.

One of the main problems in getting to grips with this subject is the diversity of mathematical notation used by different authors. We shall write several of the more important formulae in different ways so that the reader can make comparisons with the original literature. The Appendix gives a brief review of the underlying cosmology, focusing in particular on the various coordinate systems which appear in the following discussion.

3.3: Zel'dovich's Quasi-Linear Theory

3.3.1: The Zel'dovich approximation

We can label the *Ath* particle in the universe by the position, $\mathbf{q}^A$, it would have had, had the universe been uniform, and then create the nonuniformity by displacing the particles from their initial positions. Suppose that at some fiducial time t_0 the Ath particle is at comoving coordinate

$$\mathbf{x}^A(t_0) = \mathbf{q}^A + \delta\mathbf{q}^A.$$

Suppose the displacement $\delta \mathbf{q}^A$ is written in terms of a position dependent displacement potential $S(\mathbf{q}^A, t_0)$:

$$\delta \mathbf{q}^A = \frac{\partial S}{\partial \mathbf{q}^A} \equiv \psi(\mathbf{q}^\mathbf{A})$$

Writing the displacement in this way guarantees that the associated velocity field is initially (ie. at t_0) irrotational. If certain conditions are satisfied the flow will remain irrotational thereafter (see, for example, the discussion of the Kelvin Circulation Theorem in Jones, 1976).

In what follows, we shall suppress the superscript numbering the particles and use subscripts to denote components of vectors. Thus $\mathbf{q} \equiv \{q_i\}$.

In the linear theory the displacement δq_i grows as a simple function of the cosmic time, t, and so we can write the position of the particle in *comoving coordinates* as any of

$$\mathbf{x} = \mathbf{q} + b(t) \frac{\partial S}{\partial \mathbf{q}}\bigg|_{t_0}$$

$$\begin{aligned} x_i(t) &= q_i + b(t) \frac{\partial S}{\partial q_i}\bigg|_{t_0} \\ &= q_i + D_1(t)\psi_i(q_j). \end{aligned} \tag{1}$$

where Zel'dovich's b is just Peebles' D_1:

$$b(t) \equiv D_1(t)$$

Note that it is the velocity potential $S(\mathbf{q}, t_0)$ at one fiducial instant of time that determines the entire evolution in this approximation. It is the fact that the "velocity" term is separable as a function of the time and space coordinates that ultimately limits the applicability of the Zel'dovich approximation to the quasi-linear regime.

$D_1(t)$ ($b(t)$) is referred to as the *linear growth factor* and is determined from the theory of the growth of perturbations (We shall come to that in the next subsection). The form of the function $b(t)$ (or equivalently $D_1(t)$) depends on the cosmological model. Peebles

(LSSU equ. 11.16) gives an exact formula for $D_1(t)$, and the approximation (LSSU p65)

$$D_1(t) = H_0 a^2 f(\Omega_0), \qquad f(\Omega_0) \simeq \Omega_0^{0.6}$$

Zel'dovich gave a simple approximation for b expressed as a function of redshift, $b(z)$:

$$b(z) = \frac{1}{1+\tilde{\Omega} z}, \qquad \tilde{\Omega} \equiv \frac{2.5\Omega_0}{1+1.5\Omega_0}$$

In the case $\Omega_0 = 1$ it is simply

$$b(t) = (t/t_0)^{2/3} \qquad \Omega_0 = 1.$$

Note that as $t \to 0$, the particles are at $x_i(0) = q_i$.

The equation of continuity relates the field S to the fluctuation in mass density relative to the mean:

$$\frac{\delta\rho}{\rho} = -b\frac{\partial^2 S_0}{\partial \mathbf{q}^2} \tag{2}$$

and that is why b is referred to as the "linear growth factor". This equation only makes sense while $b(t)$ is such that $\delta\rho/\rho \ll 1$.

3.3.2: Gravitational Potential Fluctuations

The fluctuating part of the gravitational potential is the solution of Poisson's equation, which in comoving coordinates is expressed as

$$\frac{\partial^2 \phi}{\partial \mathbf{x}^2} = 4\pi G a^2 \delta\rho$$

In a sense, ϕ, being defined as a function of the comoving $\mathbf{x}$ coordinates, should be referred to as the "comoving potential".

There is a general solution of this equation, giving ϕ in terms of an integral over the distribution of density fluctuations $\delta\rho/\rho$. We shall exploit this approximation in the next section, when we follow the standard "LSSU" development. In the case of the Zel'dovich approximation, however, this simplifies considerably to

$$\phi = 3a\ddot{a}bS_0(\mathbf{q})$$

There are two important points to note. Firstly, in the case $a(t) \propto t^{2/3}$ and $b(t) \propto t^{2/3}$ the function $\phi(q)$ is constant in time. Thus even as $t \to 0$ there are perturbations in the metric, despite the fact that the density fluctuations appear to tend to zero in this limit. Secondly, the approximation is self-consistent in the sense that the initial particle displacements δq_i are consistent with having been driven by this constant fluctuating potential acting on the initially uniform (from the point of view of $\delta\rho/\rho$) state at $t = 0$.

Peebles (LSSU equ. 8.2) introduces the peculiar acceleration vector $\mathbf{g}$, derived from ϕ by

$$\begin{aligned} \mathbf{g}(\mathbf{x}) &= -\frac{\nabla_{\mathbf{x}}\phi}{a(t)} \\ &= Ga(t)\rho_b \int \frac{\mathbf{x}' - \mathbf{x}}{|\mathbf{x}' - \mathbf{x}|^3} \frac{\rho(\mathbf{x}') - \rho_b}{\rho_b} d^3\mathbf{x}' \end{aligned}$$

(see also LSSU eq. 14.2). With the dynamical equation of motion for small perturbations the physical peculiar velocity can be written in terms of the acceleration $\mathbf{g}$ as

$$\mathbf{V} = \frac{2}{3}\frac{1}{\Omega H^2}\frac{\dot{b}}{b}\mathbf{g}.$$

Peebles (LSSU equ. 14.8) chooses to write this as

$$\mathbf{V} = \frac{2f}{3H\Omega}\mathbf{g}, \qquad f = \frac{d\log b}{d\log a}.$$

It turns out that f depends only on Ω and is well approximated by

$$f(\Omega) \simeq \Omega^{0.6}.$$

3.3.3: The peculiar Velocity

The actual trajectory (ie. in physical coordinates) of the particle is thus

$$r_i(\mathbf{q},t) = a(t)q_i + a(t)b(t)\frac{\partial S}{\partial q_i}\bigg|_{t_0}$$

$$\equiv a(t)(q_i + D_1(t)\psi_i(\mathbf{q}))$$

depending on which notation you prefer. The peculiar velocity (again in real coordinates) of a particle is thus

$$\mathbf{V} = \mathbf{u} - \frac{\dot{a}}{a}\mathbf{r} = a\frac{d\mathbf{x}}{dt}$$

This can be variously expressed as

$$\mathbf{V} = a(t)\dot{b}(t)\frac{\partial S_0(\mathbf{q})}{\partial \mathbf{q}} \tag{3}$$

$$= a(t)\dot{D}_1(t)\psi(\mathbf{q}).$$

Combining equations (2)(3) we get

$$\nabla_{\mathbf{q}} \cdot \mathbf{V} = -a\frac{\dot{b}}{b}\left(\frac{\delta\rho}{\rho}\right) \tag{4}$$

which enables us to get from the q-space peculiar velocity field to the density fluctuation field.

3.4: Reconstructing the 3-d Flow

3.4.1: The comoving coordinate q*-space*

In the above approximation (equation (4)), the peculiar velocity is the gradient of a velocity potential Φ, and so we can write

$$\mathbf{V} = -\nabla_{\mathbf{q}}\Phi(\mathbf{q}).$$

$$\Phi(\mathbf{q}, t) = -a\dot{b}S(\mathbf{q}, t_0).$$

There is a formal solution to this equation giving Φ in terms of a line integral of $\mathbf{V}$:

$$\Phi(\mathbf{q}) - \Phi(\mathbf{O}) = \int_{\mathbf{O}}^{\mathbf{q}} \mathbf{V} \cdot d\mathbf{l}$$

The integral can be taken over any path from $\mathbf{O}$ to $\mathbf{q}$, and in particular a radial path in the comoving coordinate $\mathbf{q}$-space. This particular choice of path involves only the *radial* component of the velocity. In $\mathbf{q}$-space spherical polar coordinates (q, θ, ϕ):

$$\Phi(\mathbf{q}) = \int_0^q V_r(q', \theta, \phi) dq'.$$

We have set the potential equal to zero at the origin since we don't need its value, only its derivatives. Having got Φ at all points we can then determine the *three-dimensional* velocity field from it by doing

$$\mathbf{V} = -\nabla_{\mathbf{q}}\Phi(\mathbf{q}). \tag{5}$$

The projection of this velocity along the line of sight is the contribution of the peculiar velocity to the observed recession velocity. Thus we can improve our estimate of the true distance to the galaxy.

We seem to have got something for nothing! In fact it was not for free. The price we had to pay was the assumption that the velocity field was derivable from a potential.

3.4.2: Getting to grips with physical space

Given a galaxy with radial velocity cz and velocity independent distance estimate r, the peculiar radial velocity is

$$V_r = cz - H_0 r.$$

In principle, one could just plug this into the integral on the right hand side of equation for Φ, get Φ and differentiate to get $\mathbf{V}$. This is unfortunately not so easy because this last equation refers to quantities measured in the present physical space, not the comoving coordinate space demanded by the equations (5).

It could of course be argued that we walked into a trap by starting off with the Zel'dovich approximation, which is a Lagrangian description of what goes on, rather than an Eulerian description.

The preceding discussion is that it all takes place in the comoving coordinate **q**-space. We don't know what this looks like until we know the relation between the present location and velocities of galaxies and the initial conditions. It is just this relationship that is expressed by the Zel'dovich approximation (1). So the situation is somewhat circular, we have to guess what the **q**-space looks like, calculate where the galaxies ought to be today, and then correct our guess.

3.4.3: Real Data: sparseness and noise problems

So far, everything has been theoretical, dealing with continuous fields. These fields are, however, sampled at discrete points where observed galaxies happen to lie. The data is also very noisy in the sense that the error bars on the distance estimators are relatively large.

The galaxy sample comes from a number of quite different catalogues. Within a sphere of radius 8000 km.s.$^{-1}$ there are around 500 E/S0 galaxies taken from the "S7" survey, from surveys of individual southern hemisphere clusters and from a survey of the great attractor region. In addition to that there are some 200 S galaxies coming from galaxy cluster surveys. Within a sphere of radius 3000 km.s.$^{-1}$ there are around 200 nearby field spirals distributed over the whole sky.

The distance estimators are different for the various subsamples, but generally have an accuracy of around 20%. This can give rise to errors in the peculiar radial velocity estimate for a single galaxy of 1000 km.s.$^{-1}$ or more.

The way this problem is handled is to smooth the data over large scales, and to use maximum likelihood to estimate the bulk flow within spherical Gaussian windows.

3.4.4: Windows and Biases

BDF in their various papers discuss at length how to best do the smoothing, taking particular care with regard to the biases that may be introduced by the smoothing process. Suppose the catalog consists of galaxies at locations $\{\mathbf{r}_i\}$ having velocities u_i. The smoothed velocity field at any point $\mathbf{r}$ can be determined from a convolution of the form

$$u_s(\mathbf{r}) = \sum_i \mathcal{W}(\mathbf{r}, \mathbf{r}_i) u_i$$

The choice of the smoothing function $\mathcal{W}$ is one of the most important parts of the reconstruction procedure, because it must not produce biases in the data. The issue is discussed at great length in the papers of BDF so it is sufficient to discuss some salient points here.

In the first place, the function $\mathcal{W}$ cannot be a simple function of the scalar $|\mathbf{r} - \mathbf{r}_i|$. It has to be a tensor function of $\mathbf{r}$ and $\mathbf{r}_i$ in order to account for the fact that the objects in the sample are observed along different radial directions (see also Szalay, 1988 for a different angle on this). If $\hat{\mathbf{r}}$ denotes the unit vector in the direction of $\mathbf{r}$, then $\mathcal{W}$ has the form

$$\mathcal{W}(\mathbf{r}, \mathbf{r}_i) = [\hat{\mathbf{r}} \cdot \mathbf{A}^{-1} \cdot \hat{\mathbf{r}}_i] \;\; W(\mathbf{r}, \mathbf{r}_i)$$

$$\mathbf{A} = \sum_i W(\mathbf{r}, \mathbf{r}_i) \;\; \hat{\mathbf{r}}_i \; \hat{\mathbf{r}}_i$$

W being a standard (ie. scalar) smoothing window which in general would be a function of $|\mathbf{r} - \mathbf{r}_i|$.

A simple choice for the smoothing window W is:

$$W_B(\mathbf{r}, \mathbf{r}_i) = \frac{1}{\sigma_i^2} \exp\left[-\frac{|\mathbf{r} - \mathbf{r}_i|^2}{2R_w^2} \right]$$

where R_w is a suitable window radius and σ_i is a normalizing factor, both constants being independent of position. This introduces a "sampling gradient" bias in which the velocity field from regions of high density pollutes region of low density within the same effective window volume. This can be of importance if there are steep velocity gradients.

The filter can be adjusted to cure this by using the fixed-radius window function

$$W_F(\mathbf{r}, \mathbf{r}_i) = \frac{V_i}{\sigma_i^2} \exp\left[-\frac{|\mathbf{r} - \mathbf{r}_i|^2}{2R_w^2}\right]$$

where V_i is a measure of the specific volume occupied by each object in the sample. In this way, objects in dense areas carry somewhat less weight. Experimentally, the authors chose $V_i = R_4^3(\mathbf{r}_i)$ where $R_4(\mathbf{r}_i)$ is the distance from the object i to its fourth nearest neighbour.

The final touch is to allow the window radius R_w to vary with the local conditions:

$$W_V(\mathbf{r}, \mathbf{r}_i) = \frac{V_i}{\sigma_i^2} \exp\left[-\frac{|\mathbf{r} - \mathbf{r}_i|^2}{2R_5^2}\right]$$

where R_5 is the fifth nearest neighbour to the ith object. This makes it possible to handle almost empty regions since the smoothing volume is always bound to contain several objects. Note however that it is not straightforward to interpret maps that have been generated using variable width smoothing functions.

3.4.5: Reconstructed flows

The original papers give a lot of detail about the various reconstructions of the peculiar velocity and density fields from numerical simulations and from the presently available data. The density field in the supergalactic plane slice is illustrative and is reproduced here. The reconstruction used the variable window radius filter W_V.

The centre of the picture is the Local Group and the large peak is the Great Attractor with the Virgo Cluster Southern extension a hillock nestled in to the right of it. The lesser peak below the Great Attractor is the Pavo-Indus- Telescopium cluster complex. These are features that we would expect to find in this slice. Seeing them through this kind of analysis is encouraging since the method does not assume ab initio that mass traces light.

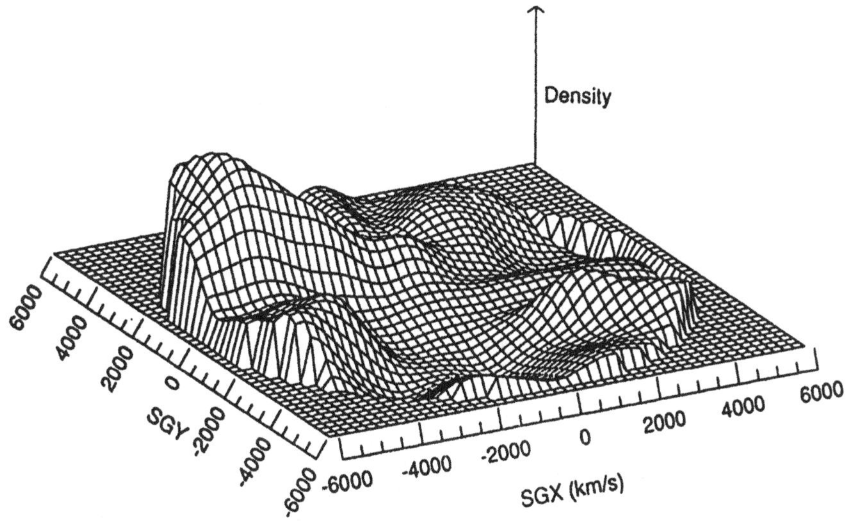

The mass density distribution in the supergalactic plane as reconstructed by POTENT. The Local Group lies at the center of the plot and the Great Atractor is the large hump on the left. The density contrast of the Great Attractor is $\simeq 0.7$. *(From Bertschinger and Dekel, 1990)*

In the original analysis of BDF, the Perseus-Pisces super-cluster was far weaker than might have been expected and there is a very strong density peak in the reconstruction (in the vicinity of the cluster A400) which has no obvious luminous counterpart. Since then, it has been reported (Dekel, 1990) that inclusion of the latest data on the Perseus-Pisces region (Willick, 1990) does reveal the Perseus-Pisces complex with a density enhancement of $\simeq 1.0 \pm 0.4$.

3.5: Ω_0 and biasing

Taking the reconstructions literally, one can compare the divergence of the reconstructed peculiar velocity field with the reconstructed galaxy number count map obtained from, say, the IRAS catalog. The results provide evidence that mass and light distributions are related. The simplest suggestion ("linear biasing") is that the fluctuations in galaxy number density might be proportional to the fluctuations in the mass density:

$$\frac{\delta n_{gal}}{n_{gal}} = b\frac{\delta \rho}{\rho}$$

b is the "linear bias parameter". Just how they are related will be one of the long term outcomes of this kind of study. Doing this gives a value of the free parameter

$$\frac{f(\Omega_0)}{b} = 1.00^{+.26}_{-.38}, \qquad 93\% \text{ confidence}$$

(Bertschinger, 1990). Since we have only poor knowledge of b, we cannot fix Ω_0, but the result is extremely interesting.

4: Conclusions

Cosmology has entered a new phase where large data samples are leading to detailed three-dimensional maps of the distribution of luminous matter in our universe. The science of Cosmography has been born and now it will be the job of numerical models to provide an understanding of the birth processes that gave rise to the observed structure.

Already, the new data has revealed apparent conflicts with various model predictions. There appears, for example, to be manifest lack of large scale power in the standard cold dark matter models. It is early days yet to jump to the conclusion that this indicates some fundamental lack in these models. It could be, for example, that our biasing models are too naive to allow a proper prediction of the luminous matter distribution in what is otherwise a purely gravitational simulation. In any case there is abundant room for "fixing" the numerical models and cosmology is not yet being thrown into any crisis. We could invoke initially non-Gaussian fluctuations - these are a natural consequence of models of the early universe dominated by strings or textures. What is important is that the data is already hinting at such things and that, as these surveys grow, we can look forward to a deep understanding of the early universe in the not too distant future.

Acknowledgements

Bernard Jones would like to thank Alfredo Henriques for the opportunity to attend this superbly organised meeting and Orfeu Bertolami and Jose Mimoso who provided the education in Trouxas, Vinho Verde and Fernando Pessoa and other fine things Portuguese.

References

Bertschinger, E. talk presented at *2nd. Rencontres de Blois*, Physical Cosmology, Aug. 28 - Sept. 1st. (1990).

Bertschinger, E. and Dekel, A. *Astrophys. J. Lett* **336**, L5 (1989).

Bertschinger, E. and Dekel, A. in *Large Scale Structures and Peculiar Motions in the Universe*, ASP Conference Series, ed. D.W. Latham and L.N. Da Costa (1990).

Bertschinger, E., Dekel, A., and Faber, S.M., preprint (1990).

Bertschinger, E., Dekel, A., Faber, S.M. and Dressler, A., preprint (1990).

Dekel, A. talk presented at *2nd. Rencontres de Blois*, Physical Cosmology, Aug. 28 - Sept. 1st. (1990).

de Lapparent, V., Geller, M.J. and Huchra, J. *Astrophys. J.* **332**, 44 (1988).

Efstathiou, G., Kaiser, N., Saunders, W., Rowan-Robinson, M., Lawrence, A., Ellis, R.S. and Frenk, C.S. *Mon. Not. R. astr. Soc*, (1991).

Geller, M.J. and Huchra, J.P. *Science* **246**, 897 (1989).

Jones, B.J.T. *Rev. Mod. Phys.* **48**, 107 (1976).

Lynden-Bell, D., Faber, S.M., Burstein, D. Davies, R.L., Dressler, A. Terlevich, R.J. and Wegner, G. *Astrophys. J.* **326**, 19 (1988).

Maddox, S.J., Efstathiou, G. and Sutherland, W.J. *Mon. Not. R. astr. Soc.* **246**, 433 (1990).

Martinez, V., Jones, B.J.T., Dominguez-Tenreiro, R. and van de Weygaert, R. *Astrophys. J.* **357**, 50 (1990).
Peebles, P.J.E. *Large Scale Structure of the Universe* , Princeton (1980).
Strauss, M.A., Davis, M., Yahil, A. and Huchra, J.P. *Astrophys. J.* **361**, 49 (1990).
Weinberg, S *Gravitation and Cosmology* , John Wiley (1972).
Zeldovich, Ya.B. *Astron. Astrophys.* **5**, 84 (1970).

APPENDIX: The background cosmological model

We present here, for the sake of completeness, a brief description of the cosmological formalism that we use. This is to be found in almost any standard textbook on cosmology, but we wish to focus here on the issue of the coordinate systems that are used in describing inhomogeneous universes.

The Friedman-Lemaitre Equations

The dynamics of the expansion are calculated by considering the motion under gravity of a sphere of cosmic material containing a fixed mass M:

$$\frac{d^2 l}{dl^2} = -\frac{GM}{l^2}$$

If the pressure is zero the equation can be written as

$$\frac{1}{l}\frac{d^2 l}{dl^2} = -\frac{4\pi G}{3}\rho$$

There is also an equation for the conservation of mass which, again in the zero-pressure case, takes the form

$$\frac{d\rho}{dt} + 3\frac{\dot{l}}{l}\rho = 0$$

The first integral (energy integral) of these dynamical equation is

$$\left(\frac{\dot{l}}{l}\right)^2 = \frac{8}{3}\pi G\rho(t) - \frac{k}{l^2}$$

and this involves the infamous integration constant k which we were told in ancient times could (for unspecified reasons usually) take on the values $-1, 0, +1$. Since we abhor non-physical quantities we today express this constant in terms of directly observable quantities, the density parameter Ω and the expansion rate parameter h.

Ω and h

The present value of the Hubble constant is defined by

$$H_0 = \frac{\dot{l}_0}{l_0}$$

and its current value, being somewhat uncertain, is written as

$$H_0 = 100\ h\ \text{km.s}^{-1}\text{Mpc}^{-1}$$

There is a relationship between the density and expansion rate of a universe with zero energy, the "Einstein de Sitter model":

$$\rho_c = \frac{3{H_0}^2}{8\pi G}$$

This may not be the actual density, so we introduce the density parameter Ω telling us how far from the critical density we are:

$$\Omega = \frac{\rho}{\rho_c}$$

In ancient days this was called $2q_0$, and q_0 was referred to as the deceleration parameter. Underlying that name was the idea that we might be able to measure it directly from the Hubble diagram. We have long since given up. Note that in general the density parameter Ω is a function of time, unless $\Omega = 1$.

We should use the notation that Ω_0 denotes the present value of Ω. In terms of real numbers, the cosmic mass density is

$$\rho_0 = \Omega\rho_c = 1.8 \times 10^{-29}\Omega h^2\ \text{gr.cm}^{-3}$$

In terms of these observable parameters the equation for the expansion becomes

$$\left(\frac{\dot{l}}{l}\right)^2 = \frac{8}{3}\pi G\rho(t) - H_0^2(\Omega_0 - 1)\left(\frac{l_0}{l}\right)^2$$

To solve this equation we need to have an equation telling us how ρ behaves with the size of the sphere whose current radius is l_0. The solution then tells us the evolution of the size of a sphere whose current size is l_0. (Alternatively, we can regard l_0 as the initial size of a sphere at some fiducial initial epoch).

The Scale Factor

Traditionally the amount of cosmic expansion is measured by the so-called "cosmic scale factor". Again this is a rather loose term used by theorists who never have any intention of measuring anything. We have introduced above the notion of the size of a box, $l(t)$ that partakes in the cosmic expansion. We can define the scale factor of the universe at a time t to be the ratio of the value of the size of the box $l(t)$ at that time, to its present time l_0:

$$a(t) = \frac{l(t)}{l_0}$$

(With this definition, the present value of the scale factor is 1). Since the universe is by hypothesis homogeneous and isotropic, this number will be the same at a given time t whatever box size l_0 is chosen for reference.

It is easily shown that for a dust universe with $\Omega_0 = 1$

$$a(t) = \left(\frac{t}{t_{now}}\right)^{\frac{2}{3}}$$

where t_{now} is the current epoch. For a radiation universe with equation of state $p = \frac{1}{3}\rho c^2$, we have the behaviour

$$a(t) = \left(\frac{t}{t_{now}}\right)^{\frac{1}{2}}$$

For other values of Ω_0, see for example the book of Weinberg (1971).

Background Comoving Coordinates

For many applications in cosmology it is useful to normalize all length scales relative to the current value of the scale factor $a(t)$ appropriate to the background universe. Thus a galaxy which at time t is at $\mathbf{r}(t)$ is assigned the *comoving coordinate* $\mathbf{x}$:

$$\mathbf{x} = \frac{\mathbf{r}(t)}{a(t)}$$

In a homogeneous and isotropic universe, the trajectory of a particle is then simply $\mathbf{x} =$ *constant*. When looking at a model for the evolution of a homogeneous model universe in such a coordinate system, all particle appear stationary relative to this coordinate system.

In an evolutionary model of an inhomogeneous universe we can still normalise all particle positions with the *background* scale factor $a(t)$. This removes the overall expansion, but the particles still move relative to the background comoving coordinate frame. (That is why we emphasise that these are coordinates that are comoving in the background, they are not in fact comoving coordinates). This motion is described in the linear and quasi-linear regimes by Zel'dovich's equation (1).

A particle can be labelled by the value of its position in background comoving coordinates at some initial time. It keeps this labelling throughout its motion (it is the particle that was at position $\mathbf{q}$ at time t_i). This labelling corresponds to a choice of Lagrangian (ie. comoving) coordinates. The difficulty in general is to relate the Lagrangian and Eulerian descriptions of a fluid flow. The Zel'dovich approximation provides that relationship for the cosmic flow, at least in the linear and quasi-linear regimes.

The Zel'dovich equation, written as

$$\mathbf{x}(t) = \mathbf{q} + b(t)\left.\frac{\partial S}{\partial \mathbf{q}}\right|_{t_0}$$

$$\mathbf{r}(t) = a(t)\mathbf{x}.$$

illustrates the use of three coordinate systems:

$\mathbf{q}$	Lagrangian coordinate
$\mathbf{x}$	Background comoving coordinate
$\mathbf{r}$	Physical coordinate

$\mathbf{q}$ is the value of $\mathbf{x}$ at $t = 0$ and is fixed for a given particle. $\mathbf{x}$ is where the particle is relative to a coordinate system that co-expands with the background universe, and $\mathbf{r}$ tells us where it is in physical space.

We observe galaxies in $\mathbf{r}$-space, but the calculations are to be done in $\mathbf{q}$-space.

Metric Space as a Model of Spacetime: Classical Theory and Quantization

C.J. Isham
Blackett Laboratory, Imperial College, London SW7 2BZ, UK

Yu.A. Kubyshin*
CFN, Universidade de Lisboa, Av. Prof. Gama Pinto, 2
P-1699 Lisboa Codex, Portugal

P.Renteln
Joint Science Department, The Claremont Colleges
11th and Dartmouth Avenues, Claremont, CA 91711, USA

10 October 1990

Abstract

An approach to describing the dynamics of spacetime, in which the basic variable is the distance function (metric) on a metric space, is considered. We study the set $\mathcal{M}(X)$ of all metrics on a set X and show that, for a finite set X, almost all such metrics can be obtained by embedding X into a vector space V and then varying the norm on V. A quantum theory of norms on V is constructed. A simple illustrative model which can produce a change in metric topology is presented.

1. Most of approaches to the quantization of gravity are based on a number of standard assumptions (see, for example, [1], [2]). In the present paper we

*Permanent address: Nuclear Physics Institute, Moscow State University, 117899 Moscow, USSR

will concentrate our attention on the following two which lie in the foundation of the entire ediface of quantum gravity:

QG1) Spacetime is a pseudo-Riemannian differentiable manifold M, so the differential geometry is deemed to be the appropriate mathematical tool for describing spacetime.

QG2) The object which is to be quantized is the metric tensor, and the quantization scheme is an adaption of one of those used in standard quantum field theory (e.g. canonical quantization, quantization by means of a functional integral, etc.)

The first assumption seems to be in very good agreement with physical reality at distances larger than the electroweak scale $L_{ew} \simeq 10^{-16}cm$. But there is no a priori reason to believe that it will be valid at small distances of the order of the Planck length, $L_P \simeq 10^{-33}cm$, where quantum gravity effects are presumably essential. We will present an approach to the description of quantum phenomena of spacetime dynamics that is based on more general structures than smooth manifolds. Namely we will substitute the assumption (QG1) by the following one:

QMT1) Physical space is modelled by a metric space X, i.e. a set of points equipped with a distance function, and the appropriate framework for its description is general metric topology.

In accordance with (QMT1) the assumption (QG2) will be modified in the following way:

QMT2) The object which is to be quantized is the distance function, and the scheme employed is to be one of those used in conventional quantum field theory (e.g. canonical quantization, quantization by means of a functional integral, etc.).

Obviously one should have good reasons for giving up (QG1) and the whole realm of differential geometry. So we will start by presenting some arguments supporting our approach before embarking on a detailed technical discussion.

Extrapolating the predictions of both quantum mechanics and general relativity to the Planck length scales, we see that the metrical structure of spacetime is determined by vacuum fluctuations whose energies are of the order of the Planck mass $M_P = 10^{19}GeV$. These fluctuations are sufficiently energetic to form black holes with radii of order L_P. Hence one expects the very fabric of spacetime to be highly chaotic and therefore not to be described by a smooth manifold. Such a situation was called "spacetime foam" by Wheeler in his paper [3] and was discussed by Hawking [4] in the context of his Euclidean quantum gravity programme.

In the latter approach, the basic entity is the transition probability amplitude $K(g_1, \Sigma_1; g_2, \Sigma_2)$ for going from a metric g_1 on a 3-manifold Σ_1 to a metric g_2 on a 3-manifold Σ_2 given formally by the functional integral

$$K(g_1, \Sigma_1; g_2, \Sigma_2) = \sum_M \mathcal{K}_M(g_1, g_2), \tag{1}$$

$$\mathcal{K}_M(g_1, g_2) = \int \exp^{-S(g)} dg, \tag{2}$$

where $S(g)$ is the classical action, the sum is over all 4-manifolds M interpolating between Σ_1 and Σ_2, and the integral is over all Riemannian metrics on M which induce the given 3-metrics g_1 and g_2 on Σ_1 and Σ_2 respectively. There is no known non-perturbative way of evaluating this functional integral, but general experience from quantum field theory and numerical simulations of random surfaces in two dimensional quantum gravity and string theory (see for example [5] and references therein) suggests that the configurations (M, g) which contribute essentially to the integral are likely to be considerably more singular than their classical counterparts. In particular the sum in (1) should include contributions not only from smooth manifolds but also from more general spaces.

Now, the topological structure of a manifold can be completely recovered from the distance function

$$d(x, y) := \inf_\gamma \int \left(g_{ab}(\gamma(t)) \frac{d\gamma^a}{dt} \frac{d\gamma^b}{dt} \right)^{1/2} dt$$

where the infinum is over all piecewise smooth curves connecting the pair of points x, y. It is known that the topologies induced by different choices of g_{ab} are equal. However, if the metric tensor becomes a quantum field then $d(x, y)$ will itself become an operator-valued distribution and, as such, could induce a topology that is inequivalent to that derived from any classical metric. In particular, the "quantized" topology may not be a smooth manifold at all.

It is considerations such as this which have led us to consider as the fundamental mathematical model for spacetime spaces which are more general than smooth manifolds. Other attempts of this kind are discussed in [6] - [10].

2. Although in the present paper we depart from the traditional framework of smooth differential geometry, we wish to retain some of its underlying structure. In particular, we suppose that, while spacetime is no longer a smooth manifold modelled on Euclidean space, it is still meaningful to speak of points as being 'near' each other, and with the 'nearness' being measured by a distance function. A distance function or a *metric* on a set X is defined as usual to be a map $d : X \times X \to R_+$ (R_+ denotes the non-negative real numbers) which satisfies the three axioms

(M1)	(positive definiteness)	$d(x,y) \geq 0$ and $= 0 \Leftrightarrow x = y$
(M2)	(symmetry)	$d(x,y) = d(y,x)$
(M3)	(triangle inequality)	$d(x,z) \leq d(x,y) + d(y,z)$.

If (M1) is relaxed, so that there exist distinct points with zero distance between them, then d is said to be a *pseudo-metric*. For convenience, we will use the word 'metric' for both metrics and pseudo-metrics. A set X equipped with a metric is called a metric space. According to the supposition (QMT1) we describe physical space by a metric space X and focus on metric topologies induced by open sets which are arbitrary unions of the open balls $B_\varepsilon(y)$ defined for any $\varepsilon > 0$ and $y \in X$ by

$$B_\varepsilon(y) := \{x \in X | d(x,y) < \varepsilon\}. \tag{3}$$

Note that different metrics on a set X may induce topologies which are homeomorphic to each other or even equal. Every metric induces a topology via (3), but there are many topologies which do not arise from any metric. Recently, one of us has suggested an approach to quantizing on the set $\tau(X)$ of *all* topologies on X ([8]). Here we pursue the idea that the topology is necessarily defined via a metric and concentrate on quantum metric topology. This can be regarded as an intermediate step between conventional quantum gravity on a fixed manifold and a full quantization of $\tau(X)$. Thus real numbers enter our approach at a fundamental level and this commits us to a certain position regarding the nature of space. We think that it is also easier to connect conventional general relativity with a theory of quantized topology when the latter is restricted to the metric case.

Thus the space $\mathcal{M}(X)$ of all metrics on X plays the role of the configuration space of our system. Its structure for a set X with a finite number of points $|X|$ will be discussed in the next section. To quantize such a system means to construct 'two-point' operator fields $\hat{d}(x,y)$, which satisfy the conditions (M1)-(M3) in some appropriate operator sense, and a Hilbert space $\mathcal{H}$ of quantum states $|\Psi>$ in which these operators act. The expectation value of the metric on X in the state $|\Psi>$ is given by

$$d_\Psi(x,y) = <\Psi|\hat{d}(x,y)|\Psi> . \tag{4}$$

We consider $d(x,y)$ as an observable and assume that each classical metric $d(x,y)$ in the configuration space $\mathcal{M}(X)$ can be recovered as a value of the observable in some state $|\Psi> \in \mathcal{H}$. This means that the spectrum of the metric operator spec $\hat{d}(x,y)$ contains $\mathcal{M}(X)$. Quantum transitions $|\Psi_1> \longrightarrow |\Psi_2>$ can be viewed as metric changing transitions or even topology changing transitions if the expectation values d_1 and d_2, corresponding to the initial and final states respectively, induce different topologies.

Our quantization procedure will be described in Sect. 4 and resembles canonical quantization to some extent although we are unable to construct the variable which is conjugate to d. Table illustrates the main features of our approach and shows the analogy between conventional quantum mechanics and quantum metric topology. An interesting attempt, based on ideas of the paper ([9]), to quantize metric topology via functional integral was developed recently in ([10]).

We would like now to make a few remarks concerning the possible connections of our approach to conventional general relativity. Since the metric satisfying the conditions (M1)-(M3) is non-negative it is not clear how the pseudo-Riemannian structure of spacetime of general relativity and the notion of time could emerge in quantum metric topology. Thus the approach can be considered either as a generalization of the Euclidean version of gravity, or as a description of the spatial part of physical spacetime. In the latter case time remains an additional parameter labelling the metric in the quantum metric topology approach.

For a general state $|\Psi>$ our system is not a differentiable manifold, and the dimension of X can not necessarily be defined even locally. But it is possible that for certain states the corresponding observable d is such that a local dimension *can* be determined. Thus if A is a compact subspace of X

	Quantum mechanics	Quantum metric topology
....	....	....
1.main object	particle in R^3	metric space X
2.configuration space	$R^3 = \{q_1, q_2, q_3\}$	$\mathcal{M}(X) = \{d(x,y)\}$
3.phase space	$\mathcal{F} = \{q_i, p_j\}$	?
4.quantization	$(q_i, p_j) \to (\hat{q}_i, \hat{p}_j)$	$d(x,y) \to \hat{d}(x,y)$
	$[\hat{q}_i, \hat{q}_j] = 0$	$[\hat{d}(x_1,x_2), \hat{d}(x_3,x_4)] = 0$
	$[\hat{p}_i, \hat{q}_j] = -i\delta_{ij}$	?
5.observables	$q_i, f(q)$	$d(x,y)$
	spec $\hat{q} = R^3$	$\mathcal{N}(x) \subset$ spec $\hat{d}$
6.Hilbert space of	$\lvert\Psi\rangle \in \mathcal{H}$	$\lvert\Psi\rangle \in \mathcal{H}$
quantum states	$f_\Psi(q) = \langle\Psi\lvert\hat{q}_i\rvert\Psi\rangle$	$d_\Psi(x,y) = \langle\Psi\lvert\hat{d}(x,y)\rvert\Psi\rangle$

one can count the minimal number $N(\varepsilon, A)$ of open balls $B_\varepsilon(y)$ defined by (3) needed to cover A. Then the dimension of A is given by the Kolmogorov capacity [11]

$$\dim_K A := \limsup_{\varepsilon\to 0} \frac{\log(N(\varepsilon, A))}{\log(1/\varepsilon)}. \tag{5}$$

If for some particular states of the system the dimension is the same for all points of X, then one can speak about the dimension of X as a single number. We also expect that for certain conditions (for example, when the energy is smaller than the Planck energy) the system spends most of its time in the semiclassical regime with X representing a smooth manifold of some dimension n. In this case we can cover X by open neighbourhoods U and construct homeomorphisms from U to R^n ,hence introducing local coordinates. Then the Riemannian metric tensor g_{ab} is given by

$$g_{ab}(x) = \frac{1}{2} \lim_{x\to y} \frac{\partial}{\partial x^a} \frac{\partial}{\partial y^b} [d(x,y)]^2,$$

where x^a, y^b are local coordinates of the points x, y. This recovers the relation with standard general relativity. At this point an interesting problem arises: which requirements should the metric $d(x,y)$ satisfy in order to be a metric for an underlying smooth manifold structure on X? Some issues of this kind are discussed in [12].

3. In this section we address the problem of constructing a concrete representation of the set $\mathcal{M}(X)$ of all metrics on X which plays the role of the classical configuration space. This representation will be used later for quantization. In the present paper only the case of X with a finite number of points $|X| < \infty$ will be examined. We hope to consider in a future paper the more realistic case with infinite X.

To get an idea of the structure of this space let us consider the very simple case where the set X consists of just three points x_1, x_2, x_3, $|X| = 3$. Each metric $d \in \mathcal{M}(X)$ is characterized by three non-negative numbers $s_1 = d(x_1, x_2), s_2 = d(x_2, x_3), s_3 = d(x_3, x_1)$ so that the metric d can be viewed as a point in R^3. Because of the triangle inequality (M3) the coordinates $s_i, i = 1, 2, 3$ of the point satisfy the following constraints

$$s_1 \leq s_2 + s_3, s_2 \leq s_3 + s_1, s_3 \leq s_1 + s_2. \tag{6}$$

Thus the configuration space $\mathcal{M}(X) \cong T \subset R^3$, where T is the subspace in R^3 restricted by three planes (6). It consists of the interior and the boundary of the conical domain bounded by these planes with the apex positioned at the origin of R^3 (see Fig. 1). It seems to be surprisingly hard to carry out the quantization of the system with such a onfiguration space to say nothing about the cases with $|X| > 3$. Some ideas concerning this case are discussed in [13].

Therefore we will approach the problem in another way and apply an embedding trick which is a usual device in metrisation theorems. We embed X isometrically into some vector space V on which one can define a seminorm, and from which X then inherits its metric. Recall that a siminorm is a map $N : V \longrightarrow R_+$ which satisfies the axioms

(N1)	(positive homogeneity)	$N(\lambda v) = \lvert\lambda\rvert N(v) \forall \lambda \in R, \forall v \in V$
(N2)	(subadditivity)	$N(u + v) \leq N(u) + N(v) \forall u, v \in V.$

If, in addition, N satisfies the requirement

$$\text{(N3) (non-degeneracy) } N(v) = 0 \Rightarrow v = 0$$

then N is said to be a norm. In what follows we will use the word 'norm' both for norms and seminorms.

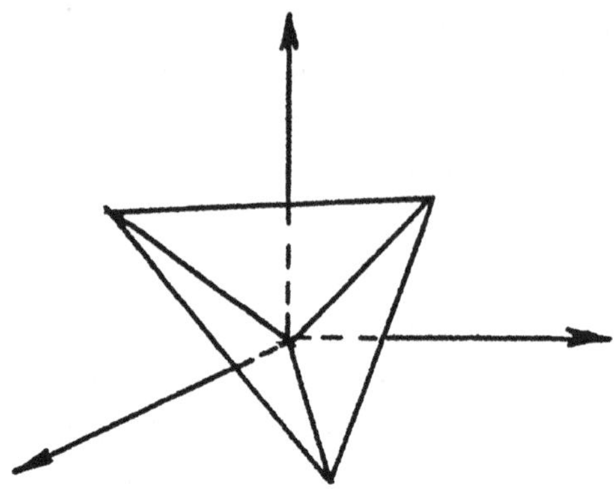

Figure 1: The space $\mathcal{M}(X)$ for $|X| = 3$.

Now we take an embedding $i : X \longrightarrow V$. Then given a norm N on V one can define a metric on X by the following construction

$$d(x, y) = N(i(x) - i(y)). \tag{7}$$

A priori, both the norm and the embedding are variable. The approach to quantizing $\mathcal{M}(X)$ in [13] was based on the idea of fixing the norm and varying the embedding. In the present paper we will adopt the alternative strategy of varying the norm on V and fixing the embedding i. For a finite set X with n points $x_1, \ldots, x_n$ we choose the canonical embedding in the following way:

$$i(x_k) = e_k \text{ for } k = 1, \ldots, n-1; i(x_n) = 0,$$

where $\{e_1, \ldots, e_{n-1}\}$ is an ordered basis set in R^{n-1}. Hence, in this case, $V = R^{n-1}$. It can be shown [14] that for every metric d on X one can find a norm N on V such that the metric will be recovered by the formula (7) for the embedding i described above. Moreover, there is 'gauge' degeneracy in which different norms on V may yield the same metric on X using the canonical embedding.

Now we are going to find a suitable representation for the space $\mathcal{N}(V)$ of all norms on V. To be exact we will describe a set of basic norms such that all norms on V can be approximated in some sense arbitrarily closely

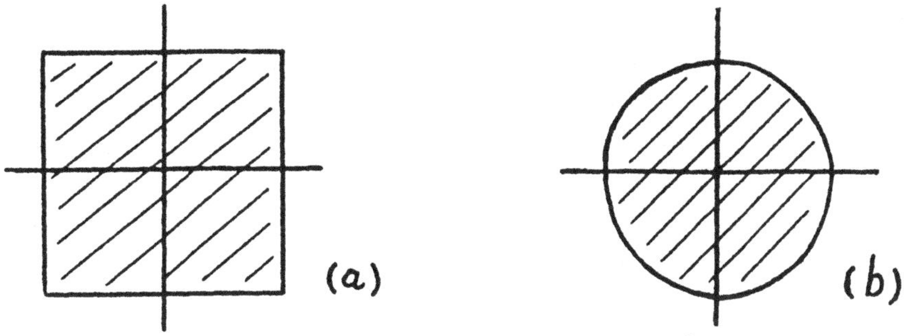

Figure 2: Examples of the unit semiballs of norms on R^2. (a) The sup norm $N(v) = \max_i |v_i|$. (b) The Euclidean norm $N_0(v) = \{v_1^2 + v_2^2\}^{1/2}$.

by norms from this set. The details of this construction are presented in [14], only a sketch of the derivation is given here. The main idea is based on the fact that there is a 1 - 1 correspondence between norms on V and subsets $C \subset V$ called barrels (absorbing closed convex sets which include the origin of V and are symmetric about it). Given a norm N we associate with it the barrel C_N which coincides with its closed unit semiball [15]

$$C_N := \{v | N(v) \leq 1\}.$$

Examples of the unit semiballs for some norms are presented in Fig. 2. The most elementary barrels are closed, symmetric polytopes [16]. In view of the 1 - 1 correspondence described above every polytope can be obtained as the unit semiball of one of the norms of the following type [17]:

$$N_{\infty,\Theta}(v) := \sup_{\xi_i \in \Theta} | < \xi_i, v > |, \tag{8}$$

where $< \cdot, \cdot >$ is the scalar product in V. Denote the set of such norms by $\mathcal{N}_0(V)$. These norms are characterized by sets $\Theta = \{\xi_1, \ldots, \xi_m\}$ of vectors from V. If Θ spans V then (8) is a proper norm (a norm characterized by m vectors corresponds to a polytope with $2m$ vertices).

The set C of all convex sets in V can be metrised using the Hausdorff metric, and it can be shown that the polytopes are dense in C equipped with this topology (see [18] for example). So, one can construct a sequence of polytopes that converges to a given convex set (one can take a sequence of inscribed polytopes for example). But this means that any norm on V can be approximated arbitrarily closely by a norm from $\mathcal{N}_0(V)$. This provides us with a useful representation of the space $\mathcal{N}(V)$ (and thus of the classical configuration space $\mathcal{M}(X)$) which will be employed in the next section.

4. Now we are going to construct a "quantum norm theory" associated with a target space V which is a finite dimensional vector space R^{n-1} for our case $|X| = n$. By this we mean a family of Hermitian operators $\hat{N}(v), v \in V$ defined on some Hilbert space $\mathcal{H}$ which commute

$$[\hat{N}(u), \hat{N}(v)] = 0 \text{ for all } u, v \in V, \tag{9}$$

and satisfy the operator analogous of the conditions (N1) - (N2) for a norm. The operator inequalities are to be understood in the usual way as referring to the properties of the diagonal matrix elements.

Being unable to construct variables conjugate to $N(v)$ and carry out the canonical quantization programme (see [19] for example) directly, we present a partial solution to the problem under consideration. We make two additional suppositions at this point, some considerations justifying them are given in [14].

1) There are basic annihilation $\hat{a}(\xi)$ and creation $\hat{a}\dagger(\xi)$ operators ($\xi \in V$) satisfying the standard bosonic commutation relations

$$\begin{aligned} [\hat{a}(\xi), \hat{a}\dagger(\xi')] &= \delta(\xi - \xi'), \\ [\hat{a}(\xi), \hat{a}(\xi')] &= [\hat{a}\dagger(\xi), \hat{a}\dagger(\xi')] = 0, \end{aligned} \tag{10}$$

and all other operators of the system are functions of them.

2) The norm operators $\hat{N}(v)$ are labelled by an integer label p and have the following commutation relations with the basic operators

$$\begin{aligned} [\hat{N}_{(p)}(v)^p, \hat{a}(\xi)] &= -| < \xi, v > |^p \hat{a}(\xi) \\ [\hat{N}_{(p)}(v)^p, \hat{a}\dagger(\xi)] &= | < \xi, v > |^p \hat{a}\dagger(\xi). \end{aligned} \tag{11}$$

The algebra of operators defined by the formulae (9) - (11) satisfies the Jacobi identities. Validity of the operator version of the relations (N1) and (N2) is ensured by the form of the commutators (11). To satisfy the positivity property of a norm we choose the norm operator $\hat{N}_{(p)}(v)$ to be a positive function of $\hat{a}(\xi), \hat{a}\dagger(\xi)$. That $\hat{N}(v)$ is *some* function of $\hat{a}(\xi)$ and $\hat{a}\dagger(\xi)$ can be ensured by requiring the representation to be irreducible when restricted to the subalgebra (10).

One natural choice for $\hat{N}_{(p)}(v)$ is the following

$$\hat{N}_{(p)}(v) = \{\int | < \xi, v > |^p \hat{a}\dagger(\xi)\hat{a}(\xi)d\xi\}^{1/p}. \tag{12}$$

Now it is natural to take the Fock representation of the subalgebra (10) and to realize the Hilbert space $\mathcal{H}$ as the bosonic Fock space. As usual, the vacuum vector $|0>$ is defined by the condition $\hat{a}(\xi)|0>= 0$ and the $m-$vector states are given by

$$|\xi_1, \xi_2, \ldots, \xi_m > := \hat{a}\dagger(\xi_1)\hat{a}\dagger(\xi_2)\ldots\hat{a}\dagger(\xi_m)|0> . \tag{13}$$

Then one easily gets

$$\hat{N}_{(p)}(v)|\xi_1, \xi_2, \ldots, \xi_m >= \left(\sum_{i=1}^{m} | < \xi_i, v > |^p\right)^{1/p}$$

Thus, in a normalized Fock space state $|\psi >= |\xi_1, \xi_2, \ldots, \xi_m >$ the expectation value of $\hat{N}_{(p)}(v)$ is

$$< \psi|\hat{N}_{(p)}(v)|\psi >= \left(\sum_{i=1}^{m} | < \xi_i, v > |^p\right)^{1/p}.$$

If we take the formal limit $p \longrightarrow \infty$ of $\hat{N}_{(p)}(v)$, we get the well-defined operator $\hat{N}(v) \equiv \hat{N}_{(\infty)}(v)$ on the Fock space whose expectation value in the state $|\psi>$ is

$$< \psi|\hat{N}(v)|\psi >= \sup_{\xi_i \in \Theta} | < \xi_i, v > |. \tag{14}$$

But these are the norms $N_{\infty,\Theta}(v)$ with $\Theta = \{\xi_1, \xi_2, \ldots, \xi_m\}$ which approximate every norm on V. We have shown that the Fock representation gives us a full quantum norm theory.

5. In this section we demonstrate the existence of a simple mechanism for metric topology change and provide a simple model of some of the effects which might take place in our approach. We suppose that the dynamics of our system is governed by some "Hamiltonian" and that there is a metric d_0 on X which corresponds to the ground state of the system (i. e. which minimizes the energy). We will choose the Euclidean norm on V as the norm associated with the ground state metric.

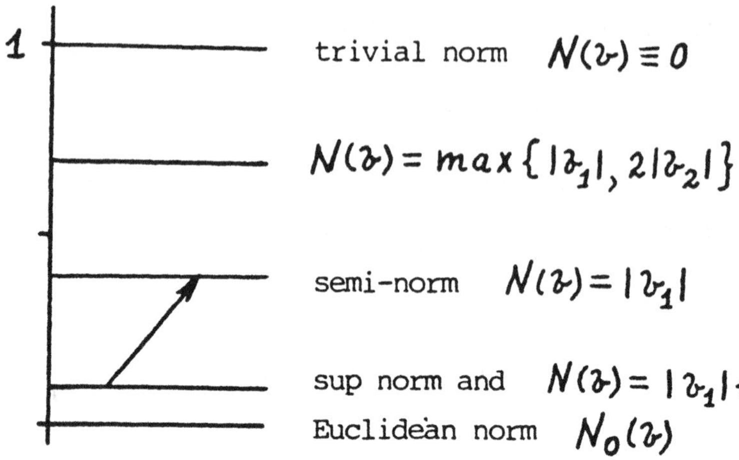

Figure 3: The energy spectrum in the toy model and a "topological" transition induced by $\hat{H}_{int}$.

In our simple model the ground state represents the quantum gravitational vacuum while the spacetime foam is represented by the excited states. If we assign zero energy to the ground state with the topology induced by the Euclidean norm via (7), we would expect that it is the highly energetic fluctuations which give topology change. As we will see later, the states, for which the induced topology is different from the ground state, correspond to seminorms and have very few (spanning) vectors ξ in them. So we will assign a very high energy to states with only a few vectors ξ, while assigning a low energy to states with many vectors.

A simple "free" Hamiltonian which commutes with the norm operator $\hat{N}(v)$ (so that the norm eigenstates are energy eigenstates) and which possesses the desired spectral properties is

$$\hat{H}_0 := k \int_{S^{n-2} \subset V} |N_0(v) - \hat{N}(v)| dv, \tag{15}$$

where $k = \text{const}$ and $N_0(v) := \{\sum_{i=1}^{n} v_i^2\}^{1/2}$ is the Euclidean norm. Clearly this "phenomenological" Hamiltonian measures the deviation of a norm eigenvalue from the Euclidean norm. The spectrum of the model for the case $|X| = 3$ is shown in Fig. 3 (generalization to the case $|X| = n < \infty$ is straightforward). We choose the coefficient k in (15) in such a way that the state $|0>$ corresponding to the trivial norm $N(v) \equiv 0$ has energy one. The state corresponding to the Euclidean norm obviously has energy equal to zero. All other states (with the vectors ξ_i being properly normalized) have energies in between these two extremes, with the seminorms having energies at the high end of the scale.

In order to induce transitions between different levels, we need an operator which does not commute with the norm operator. By analogy with the usual phenomenological Hamiltonians arising from the interaction of a field with some classical source, we propose an interaction Hamiltonian of the form

$$\hat{H}_{int} = \int f(\xi)[\hat{a}\dagger(\xi) + \hat{a}(\xi)]d\xi, \tag{16}$$

where $f(\xi)$ is some real function. This Hamiltonian describes transitions between states which differ by one vector.

Let us consider an example of a quantum transition. We take the state $|\psi_{in}>=|e_1, e_2>$ (e_1 and e_2 are unit basis vectors in $V = R^2$) as an initial state and $|\psi_{fin}>=|e_1>$ as a final one. The initial norm corresponds to the sup norm $N_{in}(v) = \max(|v_1|, |v_2|)$, the final state corresponds to the seminorm $N_{fin}(v) = |v_1|$. Using the definition of the canonical embedding $i : X \longrightarrow V = R^2$ and the formulae (4) and (7) we calculate the metric on $X = \{x_1, x_2, x_3\}$ in the initial state

$$d_{in}(x_1, x_2) = d_{in}(x_1, x_3) = d_{in}(x_2, x_3) = 1.$$

On the other hand the pseudo-metric associated with the final state is given by

$$d_{fin}(x_1, x_2) = d_{fin}(x_1, x_3) = 1, d_{fin}(x_2, x_3) = 0.$$

We see that the distance between the points x_2 and x_3 in X is zero and so these points can be identified. Since $< \psi_{fin}|\hat{H}_{int}|\psi_{in} >\neq 0$ for the interaction Hamiltonian (16), our model describes a transition from the (discrete) topology on the set $\{x_1, x_2, x_3\}$ to the discrete topology on the set $\{x_1, x_2 = x_3\}$ (this transition is indicated in Fig. 3). This can be viewed as a topological transformation in our toy model with finite X.

Acknowledgements

We would like to thank O. Bertolami, J. Céspedes, J. Mourão and E. Verdaguer for useful discussions and insightful suggestions. One of us (Yu.K.)

would like to thank the GTAE, CFNUL and CFMC (Lisbon) for financial support and Prof. B. Henriques and the Gravitation group for their warm hospitality extended to him during his stay in Lisbon where the present paper was prepared.

References

[1] Hartle J.B., Hawking S.W. *Phys. Rev.* **D28** (1983) 2960.

[2] Kuchar K. in: *Quantum Gravity 2: A second Oxford Simposium* ed C.J. Isham, R. Penrose and D.W. Sciama (Oxford: Oxford University Press, 1981).

[3] Wheeler J.A. In: *Relativity, Groups and Topology.* ed C. DeWitt and J.A. Wheeler (New York: Benjamin, 1964)

[4] Hawking S.W. in: *General Relativity: An Einstein Centenary Survey* ed S.W. Hawking and W. Israel (Cambridge: Cambridge University Press, 1979.

[5] Kazakov V.A., Migdal A.A. *Nucl Phys* **B311** (1988/9) 171
Baig M., Espriu D., Wheater J.F. *Nucl. Phys.* **B314** (1989) 609
Ambjorn J., Durhuus B., Jonsson T. *Nucl. Phys.* **B316** (1989) 526
Harnish R.G., Wheater J.F. The crumpling transitions of crystalline random surfaces. *Preprint* OUTP - 90 - 13P. University of Oxford (1990).

[6] Penrose R. in: *Quantum Theory and Beyond* ed T. Bastin (Cambridge: Cambridge University Press, 1971)
't Hooft G. in: *Recent Developments in Gravitation, Cargese 1978* ed M. Levy and S. Deser (New York: Plenum, 1979).

[7] Bombelli L., Lee J., Sorkin R. *Phys. Rev. Lett.* 59 (1988) 521
Finkelstein D. Quantum Topology and Vacuum. *Preprint* Georgian Institute of Technology (1989)
Brightwell G., Gregory R. The Structure of Random Discrete Spacetime. *Preprint* FERMILAB - Pub - 90/141-A (1990).

[8] Isham C.J. *Class. Quant. Grav.* 6 (1989) 1509.

[9] Alvarez E. *Phys. Lett.* **210B** (1988) 73.

[10] Alvarez E., Céspedes J., Verdaguer E. Dynamical Generation of Space Time Dimensions. *Preprint* CERN - TH. 5764/90 (1990).

[11] Hurewicz W., Wallman H. *Dimension Theory* (Frinceton: University Press, 1949)
Nagami K. *Dimension Theory* (New York: Academic Press, 1970).

[12] Busemann H. *The Geometry of Geodesics* (New York: Academic, 1955).

[13] Isham C.J. *Proc. Osgood Hill Conference on Conceptual Problems in Quantum Gravity.* In press (1990).

[14] Isham C.J., Kubyshin Yu.A., Renteln P. *Class. Quant. Grav.* **7** (1990) 1053.

[15] Treves F. *Topological Vector Spaces, Distributions and Kernels* (New York: Academic, 1967).

[16] Robertson S.A. *Polytopes and Symmetry* (Cambridge: Cambridge University Press, 1984).

[17] Herz C.S. *Proc. Am. Math. Soc.* **14** (1963) 670.

[18] Gruber P.M. in: *Convexity and Its Applications* ed P.M. Gruber, J.M. Wills (Basel: Birkhäuser Verlag, 1983).

[19] Isham C.J. in: *Relativity, Groups and Topology II (Les Houches 1983)* ed B.S. DeWitt and R. Stora (Amsterdam: North Holland, 1984).

Galaxy Formation - An Update

M.S. Longair
Cavendish Laboratory, Madingley Road, Cambridge CB3 0HE

1 Congratulations and Introduction

It is a great pleasure to give the introductory series of lectures at this Autumn School on "The Physical Universe". Astronomy, astrophysics and cosmology in Portugal have received an enormous boost recently through Portugal becoming a member of European Southern Observatory (ESO) and I wish all Portuguese astronomers every success in exploiting the great opportunities which this opens up. We look forward to working with our Portuguese colleagues on many programmes which, because of their nature, have to involve significant international collaboration.

⋆ ⋆ ⋆

I have been asked by the organisers to provide a gentle introduction to the subject of Galaxy Formation. Specifically, I was asked to provide a rerun and update of the course "Galaxy Formation" which I gave at the First Astrophysics School of the European Astrophysics Doctoral Network (Longair 1989; hereafter GF). I will therefore assume that the reader has access to that review since it is unnecessary to repeat all the arguments which were discussed in considerable detail there. Here, I will discuss some of the more important advances which have occurred since that review was written. This review will in no sense be a complete picture of all aspects of galaxy formation. Springer-Verlag has asked me to expand my lecture notes into a book which will be called **Galaxy Formation** and that will provide more complete coverage of many aspects of the subject.

There have been a number of important meetings since my review of Galaxy Formation which was written in late 1988. Among these are:

1. **Large-Scale Motions in the Universe**, eds V.C. Rubin and G. Coyne (1988). Pontificia Academia Scientiarum.
2. **The Post-Recombination Universe**, eds. N. Kaiser and A. Lasenby (1988). Kluwer Academic Publishers.
3. **The Epoch of Galaxy Formation**, eds. C.S. Frenk, R.S. Ellis, T. Shanks, A.F. Heavens and J.A. Peacock. (1989). Kluwer Academic Publishers.
4. **Physics of the Early Universe**, eds. J.A. Peacock, A.F. Heavens and A.T. Davies. (1990) SUSSP Publications. There are two outstanding reviews of astrophysical cosmology in this volume, one by Simon White on "Physical Cosmology" and the other

by George Efstathiou on "Cosmological Perturbations". The other chapters provide excellent introductions to many aspects of particle physics relevant to cosmology.

These volumes provide good summaries of what are perceived to be the key issues by many of the most distinguished workers in the field. In the following sections I will concentrate upon the following topics: The Large Scale Structure of the Universe, the Standard Hot Big Bang, the Origin of Structure in the Universe and various aspects of Galaxy Evolution and the Search for Youg Galaxies. Much of this material is taken from a review presented at the COSPAR Symposium "The Infrared and Submillimetre Universe at High Redshifts" which took place in June 1990.

2 The Large Scale Structure of the Universe

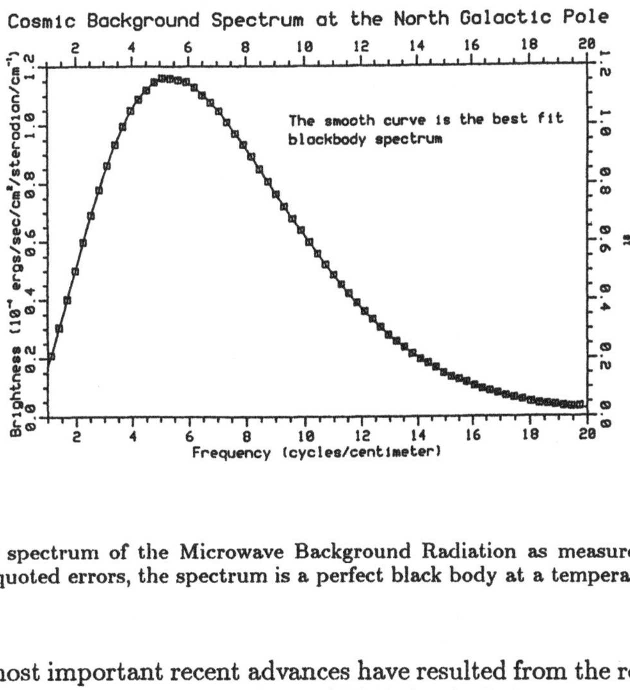

Fig. 1. The spectrum of the Microwave Background Radiation as measured by the COBE satellite. Within the quoted errors, the spectrum is a perfect black body at a temperature of 2.736 K (Mather et al 1990).

The most important recent advances have resulted from the remarkable observations of the Cosmic Background Explorer (COBE) which has measured the spectrum and isotropy of the Microwave Background Radiation as well as other background components throughout the infrared waveband with very high precision (Mather et al 1990). The spectrum of the Microwave Background Radiation is now known to be very precisely of black-body form (Fig. 1). Conservatively, the error on the radiation temperature is quoted as $T = 2.736 \pm 0.06$ K. In fact, if you look closely at the published spectrum, it can be seen that the error boxes, which are shown as 1% of the peak intensity, are overestimates since the best fitting Planck curve runs right through the centre of all the error boxes. The COBE investigators are now tackling the problem of reducing the systematic errors present in their magnificent data-set and it is expected that within a few years much refined error estimates will be available. In the meantime, it is obvious that the deviations from a black-body spectrum must be very small. Measures of the precision with which the spectrum is known to be a black body are provided by the limits to the Compton optical depth $y = \int (kT_e/m_e c^2) d\tau$ and the dimensionless chemical potential μ which measures the deviation of a Bose-Einstein distribution from a Planck function at the same temperature. The quoted limits are as follows:

$$y \leq 10^{-3} \qquad \mu \leq 10^{-2}$$

Since the Microwave Background Radiation is the remnant of the hot very early phases of the Universe, the physical meaning of these limits is that there cannot have been major releases of energy which could have distorted the black-body spectrum of the radiation in the pre- and post-recombination Universe (Sunyaev and Zeldovich 1980). This is perhaps not too unexpected if one considers what would have to have happened to produce a significant distortion - really enormous energy releases would be required. For the average cosmologist in the street, the good news is that the standard Hot Big Bang model can account extremely well for the spectrum of the Microwave Background Radiation.

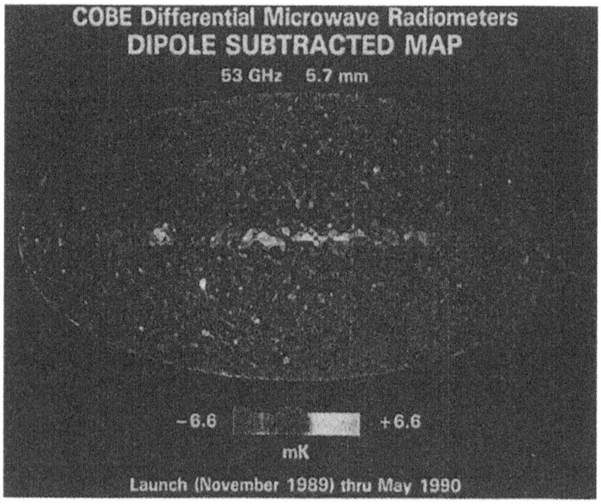

Fig. 2. The map of the whole sky as observed in the millimetre waveband at a wavelength of 5.7 mm by the COBE satellite once the dipole component associated with the motion of our Galaxy through the background radiation has been removed. The residual radiation from the plane of the Galaxy can be seen. Away from the galactic plane, there is no evidence for any anisotropy in the distribution of the radiation on the sky. (Courtesy of John Mather, the Goddard Space Flight Centre and NASA)

A second major result of the COBE project is the isotropy of the Microwave Background Radiation. Excluding the perfect dipole component which can be attributed to the motion of our Galaxy through a frame of reference in which the background radiation would be isotropic on the 360° scale, there is no evidence for any anisotropy in the distribution of the radiation on the sky on all angular scales greater than about 7° (Fig. 2). The limits to the intensity fluctuations are:

$$\frac{\Delta I}{I} \leq 4 \times 10^{-4}$$

These observations result in real problems for the simplest theories of galaxy formation as we will indicate in a moment. The COBE workers hope that eventually they will be able to push these limits below the level $\Delta T/T \leq 10^{-5}$.

Another area in which there have been very significant advances has been in the description of the large scale distribution of galaxies. The most detailed picture of the distribution of galaxies in the nearby Universe has been produced by Margaret Geller, John Huchra and their colleagues at the Harvard-Smithsonian Center for Astrophysics. Their survey of a complete sample of about 20,000 galaxies has provided the best delineation of the large scale structure so far (Fig. 3). The striking features of these maps of the spatial distribution of galaxies are the gross large scale inhomogeneities which are clearly shown in Fig. 3. On the one hand, there are filaments and "sheets" of galaxies, including the feature known as "the Great Wall" and, on the other, there are the huge voids in which the number density of galaxies is very significantly depressed relative to the mean value. The clusters and groups of galaxies are always found within the sheets and filaments, rich clusters being found at the intersection of these structures. The voids have size up to about 100 Mpc and yet the filaments are very thin. The velocity width of the sheets and filaments is less than about 50 km s^{-1}. In other words, however they were formed, they have succeeded in losing their "binding energy" very effectively, using this term in a rather loose sense. Similar results have been found in deeper surveys by Broadhurst et al (1990) but the velocity resolution is lower because their samples of galaxies are fainter. According to the CfA workers, the size of the largest structures they observe is set by the limited area of sky which they have surveyed properly - even larger structures may exist.

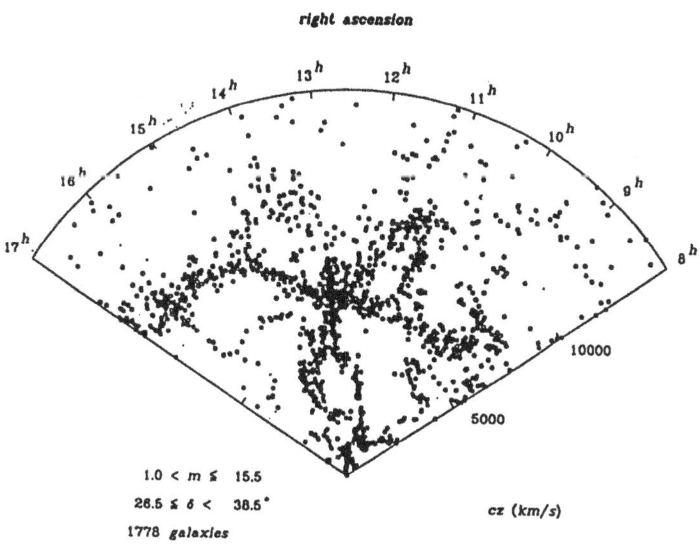

Fig. 3. A section through the large scale distribution of galaxies in the nearby Universe as determined from the Harvard-Smithsonian survey of galaxies. The figure shows the velocities of galaxies within a narrow strip of sky which is centred at declination 32.5° and is 12° wide. If the distribution of galaxies were uniform, the distribution of galaxies in this two dimensional representation would be uniform (from Geller and Huchra 1988).

It is convenient to have a mental picture of what this structure looks like. The topological studies by Gott and his colleagues (1987) show that we can think of the distribution of galaxies as being sponge-like. The material of the sponge represents the location of the galaxies and the holes in the sponge represent the large voids seen in the galaxy distribution. The intriguing thing about this topology is that the material of the sponge is all joined together and, in addition, the holes are all joined together too - this can happen in three dimensions but not in two. This form of topology has implications for the properties of the fluctuations out of which these structures must have formed.

The big cosmological problem is how to reconcile the amazing smoothness of the Microwave Background Radiation with the grossly irregular, sponge-like structure seen in the distribution of galaxies.

3 The Standard Hot Big Bang Model

The essentially perfect Planck spectrum of the Microwave Background Radiation and the two other key cosmological observations - the **Hubble flow** and the **production of the light elements** in the early stages of the Hot Big Bang - are the basis of the **Standard Hot Big Bang Model** of the Universe. The fact that these three independent observations can be reconciled in a remarkably unique way within the framework of the standard Hot Big Bang model is distinctly encouraging and is why the model is taken so seriously by cosmologists. There can be little doubt but that the Universe went through a phase when it was very dense and hot and also that it must have been remarkably isotropic on the very largest scale for essentially all that time (see GF for a discussion of the reasoning behing these assertions).

Another triumph for the standard model has come from the recent results of the LEP experiment at CERN. The LEP experiment has measured rather precisely the number of neutrino species and found that there are only three, the standard electron, muon and tau neutrinos. This is exactly what was predicted by the standard Hot Big Bang model. If there were more neutrino species present, the variation of temperature with epoch through the crucial phases when the protons and neutrons maintained in their thermal equilibrium abundances would be different from the standard model and too much helium would be produced by primordial nucleosyntheses. Thus, within the realm of known laboratory physics, the Hot Big Bang is in rather good shape.

There are, however, fundamental problems and it is a intriguing question how severe one believes these problems to be. The four great problems of the Hot Big Bang are:

1. Why is the Universe so isotropic?
2. Why is the Universe asymmetric with respect to matter and antimatter?
3. Why is the density of the Universe so close to the critical value, $\Omega = 1$?
4. What is the origin of the fluctuations from which galaxies formed?

Within the framework of the standard Hot Big Bang model of the Universe, these are all ad hoc initial conditions which have to be put in arbitrarily in order to "explain" the large scale properties of the Universe now. Fortunately, help is at hand in the form of the **inflationary model of the early Universe**. I like to think of the inflationary model in three parts. The first is inflation without physics, in the sense that, if the Universe expanded exponentially by an enormous factor in its very early phases, we are able to eliminate the first and third of the great problems. In the second part, if we put in a little bit of physics, similar to that which accounts for the asymmetry of elementary processes in the "low-temperature limit", specifically the charge asymmetry of K^0 decay, we can explain the matter-antimatter asymmetry (see the lectures by John Barrow). In the third step, if one is much bolder, one uses the best theories we have of elementary particles to identify forces which could cause the exponential expansion of the early Universe. I am far too much of an amateur in these areas to pass judgement as to whether or not this is a convincing physical picture but I can appreciate that processes are required by the theories of elementary particles which bear a close resemblance to what is needed to produce the exponential expansion of the very early Universe. Specifically, the equation of state at very high energies has to be a negative energy equation of state $p = -\rho c^2$ and this is a property of the scalar fields needed to account for the masses of elementary particles. I am therefore quietly optimistic that the inflationary picture of the early Universe can be made physically respectable although a convincing picture for these processes has yet to be found.

A problem for the astronomer is that there is not a great deal that can be done observationally to assist the theorists in understanding the first three of these problems. Obviously, it would be very nice to obtain definitive values for the density parameter Ω and the deceleration parameter q_0 from observations but these are very difficult observations indeed. I believe it will be some time before good estimates are agreed upon by most workers. The one area in which observations we make now can cast light upon some of the necessary features of the early Universe is in the fourth problem, the origin of the fluctuations from which galaxies formed. We can hope to do something about this problem observationally, although it is not easy. Proponents of the inflationary model of the early Universe claim to be able to account for the origin of the fluctuations from which galaxies formed and so it is important to provide them with as precise information as possible about the initial spectrum of fluctuations which has to be explained. I therefore consider the problems of understanding the origin of the large scale structure of the Universe and the origin of galaxies as being perhaps the most important of the cosmological problems which can be addressed by observations which can be made now or in the very near future.

4 The Origin of Structure

The challenge to the theorist is to reconcile the extraordinary smoothness of the Microwave Background Radiation with the requirement to create galaxies, clusters of galaxies and the structures observed on even larger scales by the present epoch. The standard Hot Big Bang model tells us that the photons of the Microwave Background Radiation were last scattered at a redshift of about 1000 and so the smoothness of the background tells us that it was not significantly perturbed by the collapse of large scale structures at that time. The requirement is to create the large scale structure of the Universe as we know it without violating this constraint.

The theory of galaxy formation in the context of the Hot Big Bang model has been intensively studied over the last 25 years and there is agreement about the nature of the basic problem to be solved. As is well known, gravitational perturbations grow very slowly in the expanding Universe. I have given a simple introduction to these problems (GF Sections 3 and 4) and a more comprehensive treatment is given by Efstathiou (1990). Stated most simply, the problem is that small density perturbations in the expanding Universe grow only **linearly** with redshift z or scale factor $R = 1/(1+z)$. The result, first derived by Lifshitz, can be expressed

$$\frac{\Delta\rho}{\rho} \propto (1+z)^{-1} \qquad \Omega z < 1 \tag{1}$$

where $\Delta\rho$ is the perturbation in density ρ and Ω is the density parameter of the Universe now.

This is a crucial result and is the origin of the good and bad news about galaxy formation. I have shown how this result can be derived from consideration of the dynamics of model universes (GF Section 3.3). I have developed another way of looking at this result which gives further insight into why this comes about. It is well known that gravitational collapse under gravity in a static medium results in exponential growth of any infinitesimal perturbation. A good example of this is the collapse of a pole which is balanced on one end. If it is slightly perturbed, the equation of motion for the angle to the vertical θ is most simply derived from the equation of conservation of energy:

$$\frac{1}{2}I\dot{\theta}^2 = \frac{mgl}{2}(1-\cos\theta) \tag{2}$$

where g is the acceleration due to gravity and m, l and I are respectively the pole's mass, length and moment of inertia about its base. In the case of small perturbations, $\theta \ll 1$, we can approximate $1-\cos\theta = \theta^2/2$ and then we find:

$$I\dot{\theta}^2 = \frac{mgl}{2}\theta^2 \tag{3}$$

If we write $mgl/2I = k^2$, we find

$$\theta = \theta_0 \exp(kt) \tag{4}$$

This represents exponential growth of the instability under gravity and is just the standard **Jeans' instability** in a stationary medium - in other words, if there are no pressure gradients to hold the cloud up, it collapses exponentially under gravity.

It is intriguing to contrast this behaviour with what happens in an expanding medium. We are able to understand the dynamics of the expanding Universe by asking what the decelerating force is on unit mass located at the surface of a spherically expanding volume (GF Section 2.4). Gravity slows down the expansion but now, because of the expansion of the volume, the gravitational decelerating force decreases with time. If the sphere has radius r, then the gravitational force acting on unit mass at r is just GM/r^2 where M is the mass within radius r. Then, we can write down the equation of motion for the radius of the sphere as

$$\frac{d^2r}{dt^2} = -\frac{GM}{r^2} = -\frac{4\pi r^3 G\rho}{3r^2} = -\frac{4\pi G\rho}{3}r \tag{5}$$

where ρ is the density of the sphere. This equation describes the background continuum out of which the perturbations develop. If we are dealing with a stationary medium, then we write $\rho = \rho_0 =$ constant. In the expanding medium, however, the density decreases with time as $\rho = \rho_0(r_0/r)^3$. We can therefore think of the instability in this case as being growth under gravity but with a gravitational constant which decreases as r^{-3}. This is where we have to put in the way in which the scale factor of the Universe changes with time. In the critical Einstein-de Sitter model ($\Omega = 1$), we know that $R \propto t^{\frac{2}{3}}$ and hence the effective gravitational constant decreases as $G \propto t^{-2}$. Now notice the remarkable thing which happens when our pole tries to collapse in a gravitational field which is decreasing as t^{-2}. Substituting into equation (3), we see that the equation becomes

$$\dot{\theta}^2 = \frac{A}{t^2}\theta^2 \qquad (6)$$

We can see immediately that the solutions of this equation must be a power-law of the form $\theta \propto t^{-1}$ i.e. linear growth of the perturbation rather than exponential growth. The algebraic law of growth of the perturbation (1) has exactly the same origin. We can also see how sensitive the growth of the perturbation is to the dependence of the scale factor $R(t)$ upon time. If we use a low density model with $\Omega \ll 1$, the dynamics approximate to those of the empty model for which $R \propto t$. Then we find that $g \propto t^{-3}$ and then the perturbation grows even more slowly than in the case of the critical model. In the limit of the completely empty model, there is no gravitational force and no growth of the perturbation at all. These results exactly parallel the full treatment.

The law of growth of the perturbationss (1) refers to matter-dominated Universes which is known to be a good approximation for redshifts $z < 1000$. Since we know that large scale structures exist at $z \sim 1$, it follows that there must have been fluctuations with amplitude at least $\Delta\rho/\rho \sim 10^{-3}$ at a redshift of 1000. Since the matter and radiation are still closely coupled at this epoch, we expect fluctuations in the radiation temperature of the same order in the Microwave Background Radiation. This qualitative description is confirmed by detailed analyses (see e.g. Efstathiou (1990)).

The reason these results bring us both good and bad news is as follows: it is plainly bad news that the simplest theory predicts larger amplitude fluctuations than are observed and we will find that we have to go to all sorts of contortions to try to resolve this problem. But the good news is that, because the fluctuations grow so slowly, they must contain information about the structure which was present in the very early Universe. If the fluctuations had grown exponentially, that might be good for making galaxies but the study of their properties would tell us very little indeed about the early Universe. I take the view that it really is a very good thing that the growth of perturbations is at best algebraic in expanding Universe. John Barrow describes other features of the expanding Universe in which algebraic rather than exponential functions are found because the processes take place in an expanding substratum.

This clear discrepancy between the expectations of the simplest model of galaxy formation and the lack of fluctuations in the Microwave Background Radiation is, in my view, one of the strongest motivations for taking seriously Dark Matter models for the origin of the large scale structure of the Universe. In these, the Universe is of high density $\Omega \sim 1$ and most of the matter is in some unknown form. The basic idea behind these models is well known. If the density of the Universe really corresponds to $\Omega = 1$, then the dark matter must be in some non-baryonic form or else too little deuterium is produced by primordial nucleosynthesis. It is especially helpful if the matter is in some weakly interacting form so that it decouples from the baryonic matter early in the

Universe. The fluctuations in the dark matter then grow independently of the baryonic matter so long as they are dynamically dominant. In this scenario, the fluctuations in the dark matter can be quite large by the epoch of recombination whilst the flucuations in the baryonic component are still below the level of detectability.

The two favorite forms of the dark matter are **cold** and **hot dark matter**. Both pictures are only partially successful but for quite different reasons. A popular form of the **hot dark matter** model is one in which the matter is in the form of neutrinos with mass $\sim$ 10 eV. This model is not very popular with theorists because this mass is now very close to the experimentally determined upper limit to the mass of the electron neutrino. Nonetheless, if this were the case, too much structure would be created on very large scales because the process of phase mixing wipes out small scale structure very effectively. In addition, galaxies have to form rather late in the Universe in this picture since they form out of the pre-existing large scale structures. It is, however, successful in creating narrow, large-scale structures.

The **cold dark matter** picture has the opposite problem. It has problems in producing enough structure in the form of filaments and sheets of galaxies on the very largest scales. This can be understood intuitively in the sense that the formation of structure in this model is by the aggregation and clustering of smaller mass entities into larger ones. These dynamical processes tend to wipe out small scale structure and produce roughly spheroidal or ellipsoidal rather than elongated sheet-like structures. It is difficult to account for the large voids seen in the CfA plots of the distribution of galaxies without some further ad hoc adjustment to the models. These aspects of the cold and hot dark matter pictures are nicely illustrated by the numerical simulations of Frenk (1986) and Efstathiou (1990). In a recent paper, Saunders et al (1991) have used data on a complete sample of IRAS galaxies to show that in this sample too there is too much structure in the distribution of galaxies on the very largest scales as compared with the expectations of the simplest cold dark matter theory. The remarkable thing about this paper was the fact that it represented a recantation by some of the most ardent supporters of the cold dark matter theory. It would be a mistake, however, to believe that this work excludes the theory. There are many ways in which the theory can be patched up to produce consistency with the data, for example by biassing.

A further recent problem for the cold dark matter picture is that there are no particularly obvious candidates for the cold dark matter particles. It might have been hoped that these would be the types of particles suggested by Grand Unified Theories and supersymmetric theories of elementary particles but the recent limits to the number of neutrino species from the LEP experiments suggests that there cannot be any massive neutrino-like particle with mass less than about 40 GeV in addition to the three known types of neutrino. This is a significant result because a favoured mass range for heavy neutrinos which could close the Universe is about 1 to 10 GeV. A literal interpretation of this result would be that the cold dark matter particles would have to be some as yet unknown form of particle. John Barrow has, however, pointed out a way around this problem if the particles are very massive (see his article).

Central to this whole story is the question of the fluctuations in the Microwave Background Radiation. It is worthwhile asking if any genuine fluctuations in the background radiation have been observed. There are three good to convincing pieces of evidence. First, there is no question about the dipole component. Second, Davies et al (1987) have presented evidence for fluctuations in the background radiation at a level

$$\frac{\Delta I_\nu}{I_\nu} = 3.7 \times 10^{-5}$$

on angular scales $\theta \sim 8°$. This observation needs to be confirmed independently and, in due course, COBE should be able to attain this level of sensitivity. The third good example concerns the search for decrements in the Microwave Background Radiation in the direction of clusters of galaxies. The most recent review by Birkinshaw convinces me that these decrements have now been observed at the level $\Delta I_\nu / I_\nu \approx 10^{-4}$ in the direction of rich clusters of galaxies which are known to be strong X-ray emitters (Birkinshaw 1990). One of the arguments which I find persuasive, even if it is back-to-front, is the fact that the observed decrements result in values of Hubble's constant which are in the same range as the conventional determinations. Sunyaev (1990) has emphasised that these forms of fluctuations must occur and, with a factor of 10 increase in sensitivity, the study of fluctuations in the Microwave Background Radiation will become an important observational tool for all astrophysics and cosmology.

The whole problem of understanding the origin and evolution of the large scale structure of the Universe would be much easier if we had more direct observations of the properties of the Universe at large redshifts. It is for this reason that I believe it is crucial to attempt to define directly by observation the evolutionary behaviour of all classes of object. The hope is that, from these types of observation, we can infer how large scale structures in the Universe came about. In the sections which follow, I will discuss (1) evidence for evolution in different wavebands; (2) illustrate the problems which arise by reference to the evolution of the quasar and radio galaxy populations at radio, infrared and optical wavelengths and (3) survey some approaches to the search for young galaxies.

5 The Cosmic Evolution of Discrete Sources

I provide here a catalogue of pieces of evidence for cosmological evolutionary phenomena associated with discrete sources.

Observations of Discrete Radio Sources I will use observations of discrete radio sources in the next section to provide a case study of the problems of interpretation which occur, even when there is a great deal of good data. One class of observation which I will not mention in that Section is the **angular diameter-redshift relation** (or the **angular diameter-flux density relation**). The evidence for evolutionary effects is twofold. First, carefully selected samples of double radio sources show that the physical separations of double radio sources are smaller at redshifts of 1 than they are at small redshifts (see e.g. Kapahi 1987). Second, there is convincing evidence that the radio structures of quasars which are strong double radio sources are much more distorted at redshifts greater than 1 than they are at small redshifts (e.g. Barthel 1986, Miley 1987). Presumably both of these pieces of evidence are telling us about the environments of the radio sources at large redshifts - apparently, the source components cannot penetrate so far through the intergalactic gas and also there must be much larger velocities present in the ambient intergalactic gas at large redshifts as compared with redshifts less than one.

IRAS Counts of FIR Galaxies The counts of galaxies which are intense far infrared emitters (FIR galaxies) do not yet extend to large redshifts but nonetheless there is a small excess of faint sources relative to the expectations of uniform world models. I do not believe that this result should be taken too seriously until the counts extend to significantly fainter flux densities. Because of the steepness of the source spectra, quite small and subtle effects can influence the counts. Nonetheless, this is obviously a topic of importance.

Optical Counts of Galaxies and Quasars There is now good agreement that the counts of faint galaxies display an excess relative to predictions based upon nearby samples of galaxies. The excess becomes significant only at magnitudes greater than about 22 and extends at least to about magnitude 28 (Tyson 1990, Ellis 1987). The excess is particularly marked in counts made in the blue wavelength region. It is natural to interpret these results in terms of increased star formation activity with increasing redshift. The problem with these studies is that, if the galaxies do not display emission lines, it is very difficult to obtain redshifts for galaxies fainter than about 22nd magnitude. Interestingly, Ellis and his colleagues found that the redshift distribution for their sample of faint galaxies is similar to what would have been expected if there were no evolutionary effects present. The redshift distribution did not extend beyond about $z = 0.5$ despite the presence of a significant number of faint blue galaxies. The ability to obtain large numbers of spectra of faint galaxies is one of the most important challenges for the next generation of very large telescopes.

Clusters of galaxies There has been evidence for evolutionary changes in the properties of clusters of galaxies since the pioneering efforts of Butcher and Oemler (1978). They found that there are many more blue galaxies in clusters at redshifts of about 0.5 as compared with those at small redshifts. The presence of many more emission line galaxies is not dissimilar from the excess of blue galaxies found in the counts.

Quasar surveys Some of the strongest evolutionary effects have been found in quasar surveys, by which I mean quasar samples selected at optical wavelengths. Since the early work of Schmidt (1970), it has been convincingly demonstrated that there is a large excess of faint quasars. The V/V_{max} test has been used to show that the comoving space density of quasars increases dramatically with increasing redshift out to redshifts of the order 2 to 2.5. Beyond that redshift the comoving space density of sources decreases gradually. This is a difficult observational programme because it relies upon accurately calibrated surveys of large regions of sky to discover the very few, very large redshift quasars present. The APM surveys by Warren and his colleagues and the CCD surveys of Schmidt, Gunn and Schneider have, however, provided good evidence that the strong evolution observed at redshifts less than 2 does not extend to larger redshifts. What has been remarkable has been the ability of these groups to discover significant numbers of quasars with redshifts greater than 4. At the time of writing, there are 16 known quasars with redshifts greater than 4. These are naturally very important cosmological probes. I will come back to these important observations when we discuss the results of recent radio source surveys.

X-ray Source Counts The counts of faint X-ray sources show the tantalising result that their source count is almost exactly the "Euclidean" source count $N(\geq S) \propto S^{-1.5}$. This result is derived from the comparison of the numbers of bright X-ray sources with those found in the Einstein deep survey. Since the sources are known to have redshifts which extend out about 1, this means that there must be some evolutionary effects present or else the count would be much flatter than is observed. This is therefore definite evidence for the evolution of this population of objects. This is not entirely surprising since a significant fraction of the objects consists of quasars and active galaxies. Traditionally, estimates of the counts of X-ray sources have been tied to the evolution of the population of active galaxies in general. The problem of modelling these data is the lack of large samples of X-ray sources at faint X-ray flux densities. This situation will be completely changed with the deep all-sky ROSAT survey which has just been completed at the time of writing.

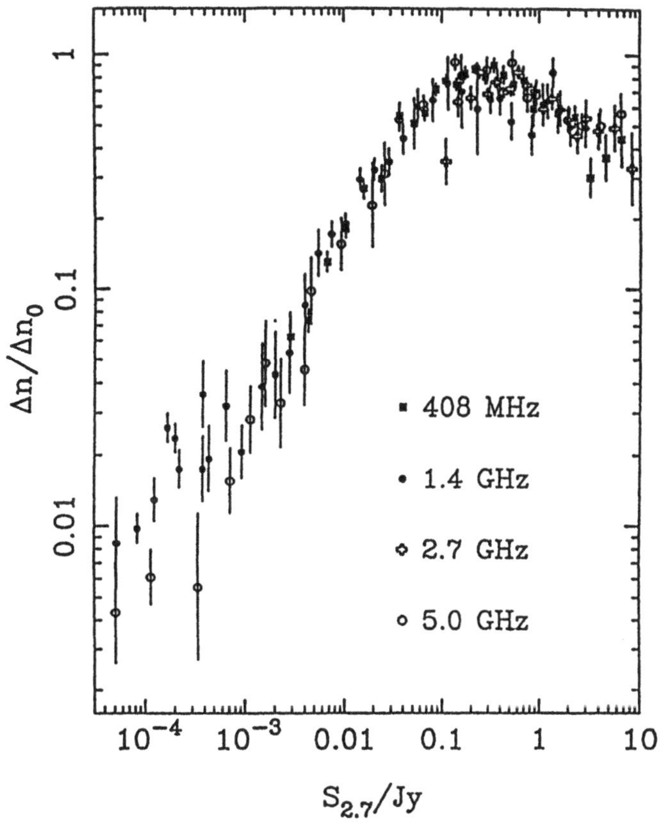

Fig. 4. The differential counts of steep spectrum radio sources presented in the standard form of $\Delta N/\Delta N_0$ where ΔN_0 is the expectation of a Euclidean model. The counts at 408 MHz, 1.4 GHz, 2.7 GHz and 5 GHz have been scaled to the counts at 2.7 GHz assuming a spectral index $\alpha = 0.85$ (Dunlop and Peacock 1990).

Quasar Absorption Lines There is now agreement that the numbers of Lyman-α absorbing systems observed in the spectra of distant quasars show evolutionary changes in the sense that their comoving number densities increase with increasing redshift (Sargent 1987). Again this must be of cosmological significance and the interpretation depends upon an understanding the nature of these absorbing systems. Other cosmological evolutionary effects may be associated with the "metal-rich" absorption systems.

6 Case Study 1 - the Evolution of the Radio Source Population

I describe briefly two case studies to illustrate the problems of studying the cosmological evolution of particular classes of source. In both cases, the objects studied are very luminous, rare objects and therefore they are not necessarily typical of galaxies in general. The catch is that only very luminous objects can be studied at redshifts greater than one. The classes of object I will describe have the additional advantage of having strong emission lines which makes their spectroscopic study feasible.

The first case study concerns the evolution of the radio source population. These studies have advanced steadily over the last 20 years. I will refer particularly to the work recently completed by my colleagues James Dunlop and John Peacock (1990). The excess of faint radio sources in the radio source counts is a long-standing problem of interpretation. Figure 4 shows a recent compilation of counts by Dunlop and Peacock. Studies of extragalactic radio sources have the enormous advantage that one automatically selects objects at large redshifts because the typical radio objects are extremely luminous.

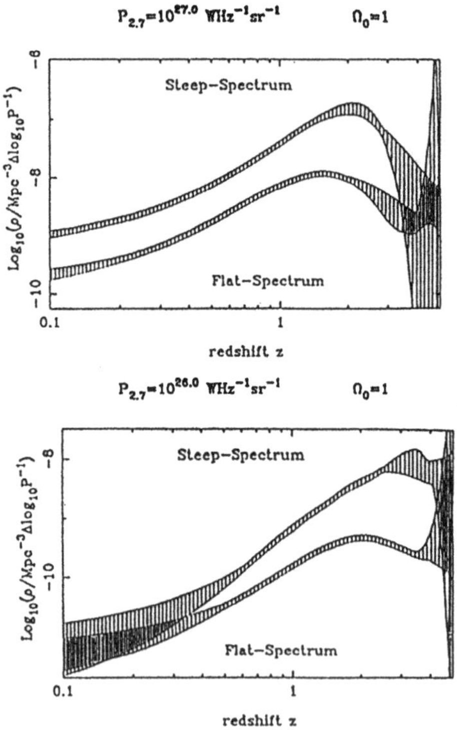

Fig. 5. The evolution of the comoving space density of radio sources with flat and steep radio spectra with redshift z. The shaded areas indicate the uncertainty with which the space densities are determined (Dunlop and Peacock 1990).

In their most recent studies, they have concentrated upon a complete sample of radio sources with flux densities corresponding to the maximum of the excess of faint sources, $S_{2.7} \geq 0.1$ Jy at 2.7 GHz (see Figure 4). The sample consists of 178 sources carefully chosen in 6 selected areas. VLA maps were obtained for all of the radio sources and this was followed by optical and infrared identifications and by optical and infrared photometry. Finally, spectra were obtained for as many objects as was feasible. The problem with this programme is that many of the identifications are very faint but, despite this, it was possible to obtain complete identifications for the sample from the infrared observations and more or less complete identifications in the optical waveband. As we will see this is a remarkable result. The spectroscopy for all the quasar candidates was straightforward. Obtaining redshifts for the radio galaxies is more difficult because it requires a great deal of observing time. However, all the radio galaxies observed so far continue to fall beautifully upon the standard infrared $K-z$ relation and so approximate redshifts can be determined from that empirical relation.

Putting together all the data available on the redshifts and distances of complete samples of radio sources, Dunlop and Peacock have been able to set tight constraints upon the evolutionary behaviour of the radio source population. Examples of the evolutionary behaviour which come out of these studies are shown in Figure 5 — the analysis is performed using the free-form techniques developed by Peacock (1985). The key results are the following:

1. Figure 5 shows the dramatic increase in the comoving space density of the radio sources with cosmic epoch out to redshifts of about 2 to 2.5. Beyond that redshift, the evolution "flattens out" and shows definite signs of decreasing at large redshifts. At a redshift of 4, the comoving space density of sources is about five times less than that at a redshift of 2.

2. This is an important result because it is entirely independent evidence for a "cut-off" in the distribution of sources at large redshifts found in optically selected samples of quasars by Warren and his colleagues (1987) and by Mitchell and Miller (1990). Notice that the evolution is similar for both steep-spectrum and flat-spectrum radio sources.

3. The source samples consist of both quasars and radio galaxies. Indeed, in the steep spectrum sample, the numbers are dominated by radio galaxies. It is important that the radio galaxies and the quasars have radio properties which evolve in the same way with cosmic epoch.

4. For both the optically and radio selected samples, the data suggest that the epoch of maximum high energy activity occurred at a redshift of about 2 to 2.5 and not at remotely large redshifts. This must be a genuinely astrophysical effect because the radio luminosity of the objects which is the primary selection criterion is not affected by dust obscuration or by scattering. In addition, all the quasars and radio galaxies in the samples have been identified and so nothing has remained hidden.

One convenient way of describing the form of evolution of the population is to use models in which the strong radio sources simply have greater luminosity at earlier epochs. The forms of luminosity evolution shown in Figure 6 can account satisfactorily for all the data — they correspond to an increase in the luminosity of the radio sources as compared with those observed at the present epoch by a factor of about 10. Because of the shifting of the radio luminosity function, this corresponds to a much greater increase in the number density of objects of a given radio luminosity.

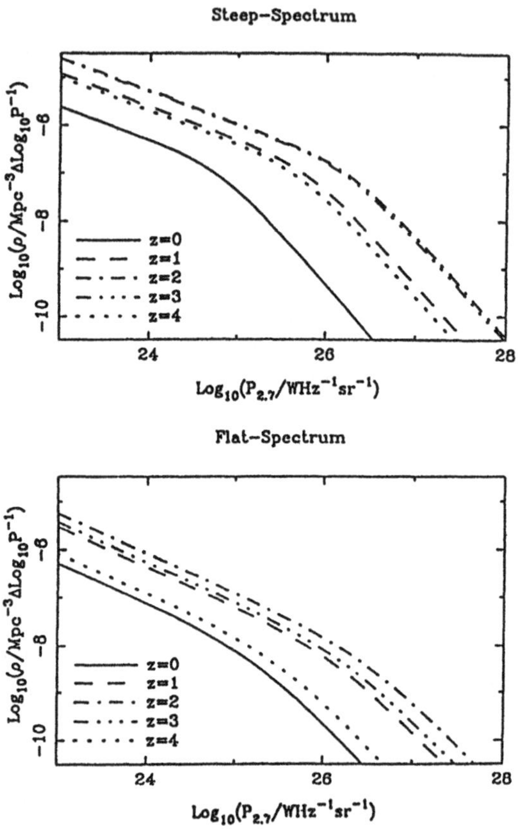

Fig. 6. Illustrating the evolution of the radio luminosity function of extragalactic radio sources in terms of luminosity evolution of the local luminosity function of powerful radio sources. The space densities refer to comoving number densities per logarithmic unit of radio luminosity (Dunlop and Peacock 1990).

7 Case Study 2 - the Stellar Populations of Distant Radio Galaxies

The effects of cosmological evolution described in Section 6 are very strong but they have the disadvantage that they are related to poorly understood pieces of high energy astrophysics, namely the physics of the origin of the quasar and radio source phenomena. The theories for these objects have not reached a state of maturity at which strong predictions can be made about the cause of the evolution and why it changes in the way it does, except in the most general terms. The advantage of studying the stellar populations of the radio galaxies is that, in principle, we should be able to use more secure astrophysical tools to interpret the optical and infrared spectra and hence understand more about the astrophysical state of the parent objects.

The importance of infrared observations of distant radio galaxies was first appreciated in about 1980 when the Riekes and our group showed that distant radio galaxies can be readily detected at 2 μm. From the beginning, this story has been one of extraordinary good luck. It has turned out that the $K - z$ relation for the radio galaxies in bright source samples is remarkably tight and that the dispersion in absolute magnitude remains narrow out to the largest redshifts observable. Second, the large redshift radio galaxies turned out to have the ideal spectra for these studies. They have very intense, narrow emission lines which enable the redshifts of the galaxies to be determined with precision but which leave the underlying continuum spectrum, which is the integrated light from the stellar populations in the galaxies, unaffected. As alluded to above, the miracle has continued as more and more galaxies have been added to the $K - z$ diagram.

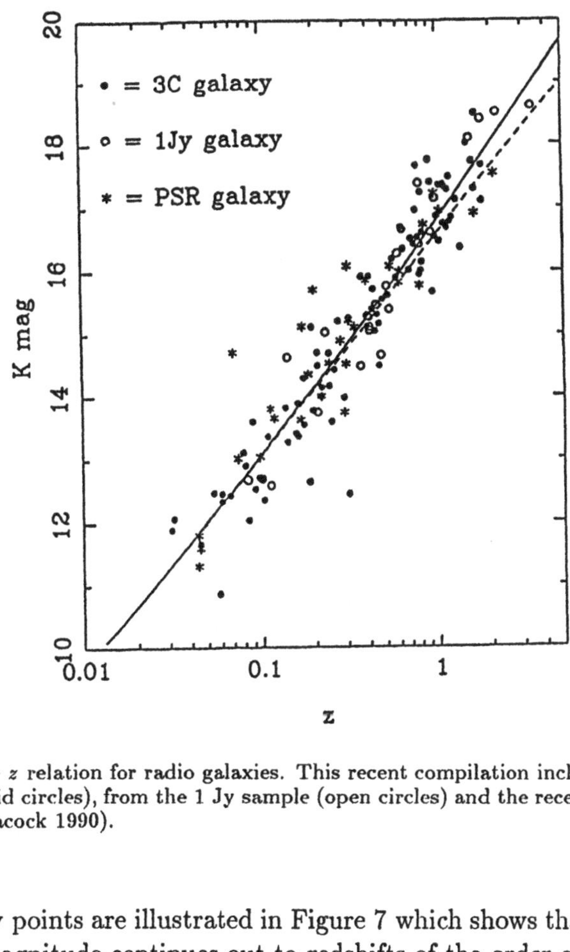

Fig. 7. The $K - z$ relation for radio galaxies. This recent compilation includes radio galaxies from the 3CR sample (solid circles), from the 1 Jy sample (open circles) and the recent 0.1 Jy samples (asterisks) (Dunlop and Peacock 1990).

These key points are illustrated in Figure 7 which shows that the narrow dispersion in apparent magnitude continues out to redshifts of the order of 2. Thus, whatever the correct interpretation of this relation, it is clear that there is a well-defined systematic trend with epoch, despite the fact that the Universe was only about one third of its present age or less when the light was emitted from the largest redshift objects. A literal interpretation of the $K - z$ relation is that the value Ω is about $7 - 8$. Our preferred interpretation is that this diagram shows the effects of the evolution of the stellar populations of these galaxies with cosmic epoch. It is a general result of models of the evolution of the stellar populations of galaxies that galaxies at a redshift of 1 are expected to be about 1 magnitude brighter than they are now (see e.g. Gunn 1978, Longair 1989). The dominant effect is the fact that the galaxy is only about half its present age at a redshift of 1 and consequently brighter stars are evolving at a greater rate off the main sequence resulting in a greater giant branch luminosity. This form of evolution of the stellar populations of galaxies is bound to occur, provided the galaxy has created most of its stars by a redshift of 1. Once we make this correction, the estimated value of the density parameter Ω lies somewhere in the range 0 to 2, which is much more reassuring. We interpret this result as evidence for the evolution of the stellar population of these large redshift galaxies with cosmic epoch.

Unlike the 2 μm infrared obsevations which are associated with the oldest populations of stars in galaxies, optical observations are associated with stars which have ages much less than the age of the Universe and therefore their optical luminosities are much more susceptible to "contamination" by subsequent bursts of star formation. Indeed, this has been found to be the case for these radio galaxies. The optical–infrared colours of the galaxies are all "bluer" than is expected for a galaxy model in which all the stars are put in at the beginning and the spectral evolution is entirely attributed to the evolution and death of this first generation of stars. This result has been confirmed by subsequent observations. Even in our first sets of observations (Lilly and Longair 1984), there was evidence that this phenomenon might be associated with the radio source events.

The most recent work we have completed on this topic concerns the optical– infrared colours of the radio galaxies in the 0.1 Jy survey. The striking result of these observations is the fact that we have been able to identify all the galaxies at infrared wavelengths and most of them at optical wavelengths. This is surprising because, if the galaxy spectra were unchanged with cosmic epoch, the galaxies should be much brighter at infrared wavelengths than at optical wavelengths, simply because of the redshifting of their spectral energy distributions, and many of the most distant galaxies would be undetectable at optical wavelengths. The fact that they can be observed optically means that they must be much brighter optically than expected which is exactly the same as saying that there must be significant star formation activity in these galaxies.

To quantify what this means in terms of star formation rates as a function of cosmic epoch, we have compared the observed optical-infrard colours as a function of redshift with the expectations of the evolutionary models of Guiderdoni and Rocca-Volmerange with whom we have been collaborating. To indicate the nature of the problem, I show in Figure 8 the raw data compared with what is expected if the energy spectra of giant elliptical galaxies are redshifted, **without including the effects of stellar evolution.** Two examples are shown, one including a strong ultraviolet continuum, as is observed in M87 and another with a weak ultraviolet continuum. These models are referred to as the "UV-Hot" and "UV-Cold models" respectively. These diagrams make the important point that the optical–infrared colours are much bluer than expected. An interesting point is that the UV-Hot model can account for the optical colours. This accounts for a dispute which has appeared in the literature as to whether or not the optical colours of the distant radio galaxies show evidence for evolution. It plainly depends upon the form of the ultraviolet spectra which are adopted. What is incontrovertible is the fact that the infrared to optical colours require significant evolution of the stellar populations.

In our paper (Dunlop et al 1989), we describe the types of evolution models which can account for the observations. The best models are those in which most of the stars in the galaxy are created in an initial burst and there are then bursts of star formation which continue at a low level throughout the life of the galaxy. These events can account for the dispersion in colours seen in Figure 8.

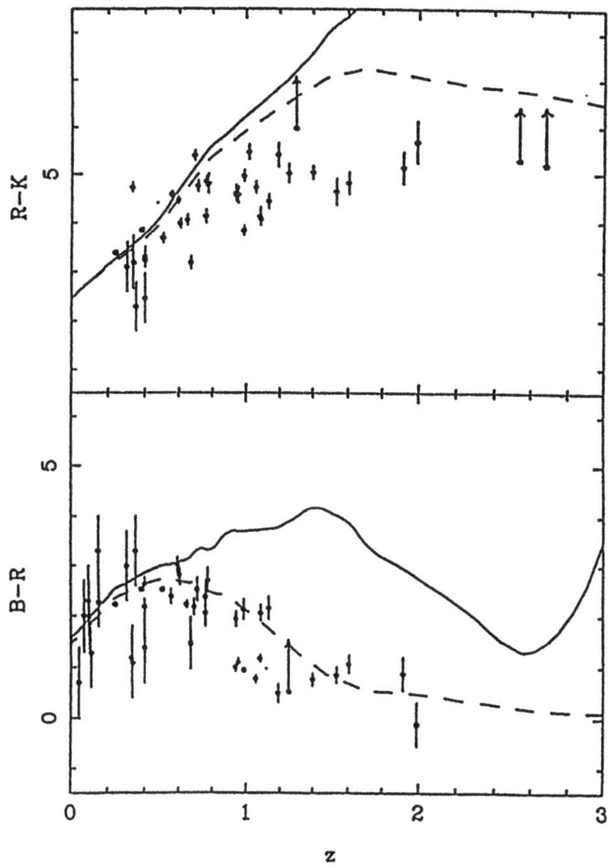

Fig. 8. The optical-infrared $(R-K)$ and optical $(B-R)$ colours of radio galaxies in the 0.1 Jy sample as a function of redshift. The lines show what would be expected if the spectra of the UV-Hot and UV-cold model galaxies were simply redshifted without any evolutionary changes in their spectra. (Dunlop et al 1989).

The next wrinkle in this story is the remarkable observation of the alignment of the giant Lyman-α clouds which are responsible for the strong narrow emission lines with the radio axes of the double radio sources and the optical images of the galaxies. I find it wholly convincing that the star formation occuring in these galaxies is stimulated by the passage of the radio jet through the ambient interstellar medium. The most recent example of this phenomenon is the radio galaxy 4C 41.17 which was studied by Chambers, Miley and van Breugal (1990)–incidently, it is also the largest redshift radio galaxy yet discovered, having redshift $z = 3.8$.

The interesting question is how much of the stellar population of the radio galaxy is created in these events. At one extreme, it might be that **all** the stars in the galaxy are formed in such events. On the other hand, these might simply be modest star formation events which have nothing to do with the formation of the bulk of the stars in the galaxy. In my view, the evidence for the alignment of the infrared image of the galaxy with the optical and radio axes is very much less convincing than the other alignments. If this view is correct, then there is not necessarily any connection between the events which create most of the stars in the galaxy and the star formation activity associated with the radio source events.

This is an important issue because it strongly affects the interpretation of the optical and infrared spectra of these galaxies. In what I will call the conventional interpretation, the infrared-optical spectrum is dominated by giant branch stars and estimates of the age of the underlying galaxy can be obtained from the degree of development of the giant branch. When Lilly (1988, 1989) undertook this analysis for the large redshift radio galaxies for which he had good optical-infrared colours, he found it difficult to explain the strength of the giant branch if the stars were younger than about $1 - 2 \times 10^9$ years. At a redshift of 3.5, this places very tight constraints upon the cosmological parameters H_0, Ω and q_0. In turn this places very important constraints upon the epoch when these galaxies could have first formed.

These observations have been interpreted in a different way by some authors (Rees 1989, Chambers and Charlot 1990). Because the time-scales for the evolution of double radio sources are believed to be much less than 2×10^9 years, attempts have been made to find models for the stellar populations which would have times scales less than this value which would account for the appearance of the strong infrared excess normally attributed to the standard giant branch. The problem with these models is that they must result in a wide dispersion in ages of the radio galaxies and hence a wide dispersion in the absolute K magnitudes of the galaxies. This is inconsistent with the observed narrow dispersion in the absolute magnitudes of the galaxies out to the largest redshifts. In addition, the need for this type of model disappears if the infrared properties of the galaxies are unrelated to their optical properties which may be the case if, as I believe, the evidence for the alignment of the infrared images of the galaxies with the radio structures is much less marked than are the optical images. In my interpretation of these remarkable observations, the radio source events indeed stimulate star formation activity in galaxies and this activity corresponds to the on-going star formation which is needed to explain the optical and infrared colours of the radio galaxies.

8 The Search for Young Galaxies

There have been many attempts to find young galaxies at very faint magnitudes. By "young galaxy" we mean galaxies in the process of formation. In one very promising approach, searches have been carried out for protogalaxies in the optical and near infrared waveband, basically searching for objects which look like gigantic HII regions (Cowie et al 1988). The elegant feature of their analysis is to look at the problem in terms of the amount of metals produced for a given number density of young objects. I refer the reader to their paper for further details. I will add two further points about searches for young objects

Deep Infrared Surveys The first of these concerns deep infrared surveys which are now possible at a wavelength of 2.2μm. With the advent of infrared arrays, it is now becoming possible to make medium deep surveys of small areas of sky. My colleagues in Edinburgh, Karl Glazebrook, John Peacock, Lance Miller and Chris Collins, have undertaken a pilot survey of 0.2 square degrees of sky which is complete to a limiting K

magnitude of about 18. The observations were made with the infrared camera IRCAM on the UK Infrared Telescope. Our prime objective was to compare the optical and infrared pictures of the sky to find out if any new classes of object might appear. If they do appear, what are they? One of the possibilities is that very red objects might be present which could be young galaxies, either because they have very large redshifts or because the young galaxies are obscured by dust.

In this pilot survey, 500 IRCAM fields have been analysed and deep optical CCD images have also been taken for all the fields. Because of the present limitation of the size of the infrared arrays, we have not attempted to go as deep as possible in the infrared waveband but rather to obtain reasonably large sky coverage. This approach contrasts with that of Lilly, Cowie and McLean in which a small region of sky was observed to a limiting K magnitude of about 20.5. The stars and galaxies in our medium-deep survey have been separated using the algorithms developed by the COSMOS team in Edinburgh. The analysis of the data is not straightforward because of the presence of ghost images. Pairing of the optical and infrared images automatically ensures that essentially all of these infrared detections are genuine.

Three preliminary results are of interest. First, the infrared galaxies found in the survey have an integral source count which maps well onto the counts of galaxies in the optical waveband. Second, there is a population of galaxies which have much steeper optical-infrared spectra than are predicted by all the standard models. The optical-infrared colours of these objects are similar to the two objects discovered by Elston, Rieke and Rieke (1988) which at one time were thought to be candidates for protogalaxies. We do not know what these objects are and have begun a programme to obtain infrared colours for them and possibly to obtain their spectra. The third preliminary result is that, of the two sources which have infrared excesses for which we now have infrared colours, one object appears relatively normal whilst the other has colours similar to what might be expected of a dust free young galaxy at a redshift of about 10. This is all very exciting but also very preliminary. The important point is that it is likely that a rather red population of infrared galaxies exists and its nature is of the greatest astrophysical interest be they young galaxies or not.

Sub-millimetre Surveys One of the oldest games in astrophysical cosmology is to work out where one can hide the primaeval galaxies. One of the hoariest old sums in this field is to work out the background radiation expected now due to the formation of a certain fraction of the heavy metals in young galaxies at redshift z. If it is necessary to make $x\%$ of heavy elements, then it is easy to show that the background radiation liberated due to the conversion of hydrogen into helium is

$$I = 8 \times 10^{-7} \frac{x\Omega_m}{(1+z)} \quad \mathrm{Wm^{-2}sr^{-1}}$$

where Ω_m is the density parameter of the baryonic content of the Universe.

It is now that the uncertainties begin. Originally, it was thought that this radiation might be released in the optical and near-infrared wavebands, as was suggested by Partridge and Peebles (1967). It might, however, all be absorbed by dust in the primaeval galaxy in which case, it might all be reradiated in the far infrared and sub-millimetre wavebands. To understand what is permissible, let us put in some concrete figures. If we adopt $\Omega_m = 0.1$, which is consistent with the constraints of primordial nucleosynthesis

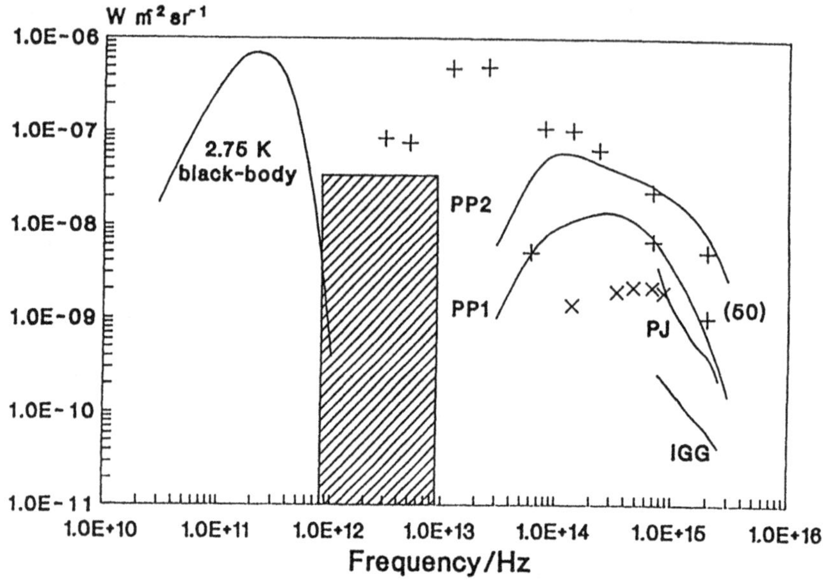

Fig. 9. The intensity of the isotropic background radiation as observed throughout the millimetre to ultraviolet waveband. In addition to the 2.75 K black-body curve, the crosses are upper limits to the background intensity in different wavebands. The crosses shows Tyson's estimate of the background due to galaxies. The figure also includes various estimates of the predicted intensity of the background due to young galaxies (PP1 and PP2 - Peebles and Partridge models; PJ - Paresce and Jakobson model; IGG - ultraviolet radiation from the hot intergalactic gas). The rectangle in the centre of the diagram is the crude estimate of the emission spectrum due to young galaxies described in the text. The reader is invited to slide this rectangle along the abscissa to find out those wavebands in which the radiation from young galaxies could be hidden. For a summary of the data, see Longair (1990).

and would make all the metals in galaxies as well as some baryonic dark matter, $x = 3$ and $z = 5$, we expect a background intensity of

$$I = 4 \times 10^{-8} \qquad \mathrm{Wm^{-2}sr^{-1}}$$

This is the background intensity which one would expect to observe spread over, say, a decade in frequency. It is intriguing to compare this intensity with the background radiation, observations of which I surveyed recently (Longair 1990). In Figure 9, I show recent observations of the background radiation in the form νI_ν. I have also shown the predicted background intensity at an arbitratry frequency and spread over a decade in frequency. The reader is invited to slide this histogram along the abscissa to find out at what wavelength the primordial galaxies could be hidden. My crude calculation is clearly consistent with the early models of Peebles and Partridge. It is evident that the galaxies could still contribute background radiation in the optical, infrared and submillimetre wavebands without exceeding the background limits. This was one of the motivations for undertaking our deep infrared survey.

The other interesting point is what happens if the young galaxies have continuum spectra similar to those of IRAS galaxies which are known to be the galaxies in which the greatest amounts of star formation are taking place now. Provided the radiation is not emitted at too large redshifts, all the background radiation could be hidden in the sub-millimetre waveband. It is an interesting calculation to work out how easy it would be to observe an intense IRAS source at a large redshift in the sub-millimetre waveband. The answer is interesting and non-intuitive. Because the submillimetre spectra of the IRAS galaxies are very steep, some of them being as steep as $I_\nu \propto \nu^4$ the effects of redshifting this spectrum into the submillimetre waveband overwhelms the effect of the cosmological inverse square law. For example, I find that an IRAS source 100 times more luminous than M82 should be quite readily detectable at a redshift of 4 at a flux density of about 0.1 Jy at 450 μm.

This is an important result because it means that with the next generation of sub-millimetre bolometer arrays such as that which is currently being built for the James Clerk Maxwell Telescope in Hawaii, a camera known as SCUBA, there is the real possibility of detecting large redshift star-forming galaxies. These calculations need to be repeated with better assumptions about the properties of young galaxies.

References

1 Barthel, 1986. In Quasars, IAU Symposium No. 119, (eds. G. Swarup and V.K. Kapahi), page 181, D. Reidel and Company.

2 Birkinshaw, M., 1990. In The Cosmic Microwave Background: 25 Years Later, (eds N. Mandolesi and N. Vittorio), 77, Kluwer Publications, Dordrecht.

3 Butcher, H. and Oemler, A., 1978. Astrophys. J., 219, 18.

4 Broadhurst, T.J., Ellis, R.S., Koo, D.C. and Szalay, A.S., 1990. Nature, 343, 726.

5 Chambers, K.C. and Charlot, S, 1990. Astrophys. J., 348, L1.

6 Chambers, K.C., Miley, G.K. and van Breugel, W.J.M., 1990. Astrophys. J. in press.

7 Cowie, L.L., Lilly, S.J., Gardner, J. and McLean, I.S., 1988. Astrophys. J., 332, L29.

8 Davies, R.D., Lasenby, A.N., Watson, R.A., Daintree, E.J., Hopkins, J., Beckman,J., Sanchez-Almeida, J. and Rebolo, R., 1987. Nature, 326, 455.

9 Dunlop, J.S., Guiderdoni, B., Rocca-Volmerange, B., Peacock, J.A., and Longair, M.S., 1989. Mon. Not. R. astr. Soc., 240, 257.

10 Dunlop, J and Peacock, J.A., 1990. Mon. Not. R. astr. Soc., 247, 19.

11 Efstathiou. G., 1990. In Physics of the Early Universe, (eds J.A. Peacock, A.F. Heavens and A.T. Davies), 361, SUSSP Publications, University of Edinburgh.

12 Ellis, R., 1987. In Observational Cosmology, IAU Symposium No. 124, (eds A. Hewitt, G. Burbidge and L.Z. Fang), 367, D. Reidel and Company.

13 Elston, R., Rieke, G.H. and Rieke, M.J., 1988. Astrophys. J. Letts., 331, L77.

14 Frenk, C.S., 1986. Phil. Trans. R. Soc. Lond. A., 330, 517.

15 Geller, M. and Huchra, J., 1988. In Large-Scale Motions in the Universe, (eds V.C. Rubin and G.V. Coyne), 3, Pontificia Academia Scientiarum.

16 Gott, J.R., 1987. Observational Cosmology, IAU Symposium No. 124, eds. A. Hewitt and G.R. Burbidge, 433.

17 Gunn, J.E., 1978. In Observational Cosmology, 8th Advanced Course, Swiss Society of Astronomy and Astrophysics, Saas-Fee, Geneva Observatory Publications.

18 Kapahi, V., 1987. In Observational Cosmology, op cit., 251.

19 Lilly, S.J., 1988. Astrophys. J., 333, 161.

20 Lilly, S.J., 1989. Astrophys. J., 340,77.

21 Lilly, S.J. and Longair, M.S., 1984. Mon. Not. R. astr. Soc., 211, 833.

22 Longair, M.S., 1989 (referred to in test as GF). In Evolution of Galaxies: Astronomical Observations, (eds. I. Appenzeller, H.J. Habing and P.Lena), 1, Springer-Verlag.

23 Longair, M.S., 1990. In The Galactic and Extragalactic Background Radiation, (eds. S. Bowyer and C. Leinert), 469, Kluwer Academic Publishers.

24 Mather, J.C., Cheng, E.S., Eplee, R.E. Jr., Isaacman, R.B., Meyer, S.S., Shafer, R.A., Weiss, R., Wright, E.L., Bennett, C.L., Boggess, N.W., Dwek, E., Gulkis, S., Hauser, M.G., Janssen, M., Kelsall, T., Lubin, P.M., Moseley, S.H. Jr., Murdock, T.L., Silverberg, R.F., Smoot, G.F.and Wilkinson, D.T., 1990. Astrophys.J., 354, L37.
25 Miley, G., 1987. In Observational Cosmology, op cit., 267.
26 Miller, L. and Mitchell, P., 1990. Mon. Not. R. astr. Soc., 244, 1.
27 Partridge, R.B. and Peebles, P.J.E., 1967. Astrophys. J., 148, 377.
28 Peacock, J.A., 1985. Mon. Not. R. astr. Soc., 217, 601.
29 Rees, M.J., 1989. Mon. Not. R. astr. Soc., 239, 1P.
30 Sargent, W.L.W., 1987. In Observational Cosmology, op.cit., 777.
31 Saunders, W., Frenk, C., Rowan-Robinson, M., Efstathiou, G., Lawrence, A., Kaiser, N., Ellis, R., Crawford, J., Xia, X-Y. and Parry, I., 1991. Nature, 349, 32.
32 Schmidt, M., 1970. Astrophys.J., 162, 371.
33 Sunyaev, R.A. (1991). Contribution to "The Infrared and Sun-millimetre Universe at Large Redshifts", COSPAR Symposium, in,press.
34 Sunyaev, R.A. and Zeldovich, Ya.B., 1980. Ann. Rev.Astr. Astrophys., 18, 537.
35 Tyson, A., 1990. In The Galactic and Extragalactic Background Radiation, op. cit., 245.
36 Warren, S.J., Hewett, P.C., Osmer, P.S. and Irwin, M.J., 1987. Nature, 330, 453.

This article was processed by the author using the TEX Macropackage from Springer-Verlag.

THE HYPOTHESIS OF THE EXPANSION OF THE UNIVERSE AND THE GLOBAL TESTS

M. Moles
Instituto de Astrofísica de Andalucía
Aptdo. 3004, 18080 Granada
SPAIN

1. Introduction.

The picture offered by the standard Cosmology is based on the interpretation of the redshift phenomenon in terms of the expansion of the universe, $1 + z = (R_0/R)$. The success of the corresponding models to explain and even predict some observational results was so impressive that that hypothesis has been most generally taken for granted and equivalent to the observational fact it has to explain, the Hubble law. The so called cosmological program was consequently defined as to get the exact values of the cosmological parameters of the model which definitively settle the universe we actually live in.

However, the efforts of more than 30 years of observational work have not yet produced well defined values for the cosmological parameters. Presently the situation is rather similar to that reviewed by Jaakkola, Moles and Vigier (1979) more than ten years ago: a Hubble constant value between 40 and 100 $kms^{-1}Mpc^{-1}$, a deceleration parameter, $-1 \leq q_0 \leq 3$, and an observed matter density well below the critical value. In fact, in many of the possible combinations allowed by the range of observed values a non-zero cosmological constant is needed to save the model.

In spite of the situation, and even if the true values of the cosmological parameters are still far from reached, the general believe on the standard Friedmann cosmology was reinforced by a number of consistency arguments, like the following:

- the Hubble law (**the fact**) is a natural consequence of the expansion of the universe (**the hypothesis**) and was predicted before it was discovered. (Note however, that the same metric could explain blueshifts as well).

- the cosmic time scale is of the right order of magnitude as compared with the ages of the oldest objects

- the abundances of the light elements and the cosmic background radiation can be *explained* within the standard frame

- the evolution seems to be one way, i.e, from matter to radiation. That would imply an universe of finite age, i.e., an evolutionary world

- no other satisfactory explanation has yet been found for the redshift phenomenon (and the other **cosmological facts**).

Those arguments are generally considered as definitive. Actually, they are very strong indeed, but not yet a proof of the rightness of the Hypothesis of the Expansion of the Universe (HEU), which constitutes the corner stone of the whole cosmological edifice. In that respect it has to be remembered that consistency proofs are necessary but not sufficient to accept a theory. In other words, an erroneous hypothesis can eventually produce right predictions. A good illustration is the theory worked out by Findlay-Freundlich (1954) on the nature of the redshift, soon after discarded on the basis of observational evidences, but nevertheless able to predict a background radiation temperature between 1.6 and 6 K.

From a methodological point of view it seems clear that the HEU has to be **directly and positively** verified or, in the meanwhile, shown to be reasonably consistent with all the available data. Thus, the question of the nature of the redshift has to be answered first on observational grounds, as stressed by many authors (see, among others, Burbidge 1973; Arp 1987; Jaakkola et al 1979, or Sandage 1987, 1988).

At that point, there are two main options to attack that problem from the observational point of view:

(i) To seek for counterexamples: non Doppler shifts, dependence of redshift on some galactic properties.

(ii) To perform global tests on the metric of the universe: to probe the effects of the metric on the observed properties.

The HEU is a well formulated hypothesis asserting that the redshift of a galaxy depends only on the space-time geometry and is independent of any other property. This substantiates option (i) to look at objects which redshift could depend on parameters other than the geometry. The effort produced a long list of candidates and originated a long controversy too many times reduced to an *oligologue* (see Arp 1987, for an account). The existence of positive relations between the redshift and some properties like compacity (Arp 1970; Bottinelli and Gougenheim 1973; Jaakkola 1973 and 1977; Collin-Souffrin et al 1974; Moles and Nottale 1981) and morphology (Jaakkola and Moles 1976; Moles and Nottale 1981; Giraud, Moles and Vigier 1982; Arp 1988) have been claimed. An excess of redshift when the light goes trough matter condensations has also been proposed (Jaakkola et al 1975) as an alternative to the standard explanation to the Rubin and Ford effect (Rubin, Ford and Rubin 1973). When some discussion was generated it always turned around the statistics or even about the evidences for objects having different redshifts being at the same distance, but the majority of these observational results have not been refuted till now.

However, it is rather difficult to consider any of such data as a proof **in itself**, since at last, it can be extremely controversial to assess what the distance to an extragalactic object actually is, or which is the dynamical status of a given system. It is nevertheless sensible to acknowledge that the accumulation of observations apparently not in agreement with the standard approach should worry the community about the rightness of the model and motivate it to try to understand the exact meaning of those observations.

On the other hand, the global approach seems to be more promising, even if, as we will discuss later, the method is not as clean as it could be thought. That approach was already pioneered by Hubble and Tolman (1935) who, apart from the original Hubble's m(z) relation, proposed the count (N(m)) and the surface brightness (SB(z)) tests. Afterwards, Hoyle (1959) discussed the potentiality of the angular size-redshift ($\theta(z)$) test (clarified by Sandage 1961). These are the so called **classical tests**. Two more tests have to be added, the look-back evolution effects and the time scale arguments, which constitute consistency tests on the evolutionary character of the universe.

The results on all those tests were reviewed by Jaakkola, Moles and Vigier (1979), LaViolette (1986) and Sandage (1987, 1988) and will be presented below.

It is very often argued to reinforce the standing of the standard model that no alternative explanation has been yet found. On that respect the remarks by Hubble and Tolman (1935) are always of use. They wrote in their quoted work: *Until further evidence is available, both the present writers wish to express an open mind with respect to the ultimately most satisfactory explanation of the nebular redshift and, in the presentation of purely observational findings, to continue to use he phrase "apparent" velocity of recession. They both incline to the opinion, however, that if the redshift is not due to recessional motion, its explanation will probably involve some quite new physical principles.*

The problem of finding alternative explanations to the redshift phenomenon is still there despite some efforts made in the last 35 years (Findlay-Freundlich 1954; Pecker, Roberts and Vigier 1973; Moles and Vigier 1974; Marič, Moles and Vigier 1976 and 1977; Moles 1978; LaViolette 1983; Pecker and Vigier 1987; Kropotkin 1988 and 1989). Those proposals are not free from controversy, some have been definitively discarded and none has received general acceptation. In that respect, we are therefore in a situation similar to that depicted by Hubble and Tolman who, nevertheless, tried to test the HEU. Now as then, the lack of alternative theories cannot be a reason to minimize the need to observationally verify the HEU: **The point here is not the confrontation between rival theories but to test the only one that does exist.**

In the present lecture we are going to review and discuss the status of the standard cosmology when it is confronted to the existing observational results on the global tests. We will consider first what the theoretical predictions are and how are they fitted by the data. Then we will consider how the consideration of non zero cosmological constant models or the introduction of evolutionary effects could change the situation.

2. The aim and formulation of the global tests.

In principle, the most direct way to positively test the HEU is the use of the global tests, which are intended to probe the space-time metric. Hubble and Tolman (1935) showed how the dependence of the surface brightness on z or the form of the function N(m) could distinguish between different types of metrics. Those, together with the (m,z) or Hubble relation and the $\theta(z)$ predictions constitute the four classical global tests.

Other than Friedmann universes we will also consider the Einstein static model. This was already done by Hubble and Tolman, who considered it as the limiting case of models where the effect of the expansion would be negligible compared to that due to an *unknown cause* of the redshift. We will show below that it is an *easy to test* model since it has essentially no free parameters. On the other hand, it is recognized that the static case has been found to fit the raw data much better than any expansion model (Jaakkola, Moles and Vigier 1979; LaViolette 1986).

2.1 The standard frame.

The standard cosmological frame is provided by the Friedmann solutions to the Einstein equations. Acceptation of the Cosmological Principle (i.e., spatial isotropy and homogeneity) implies that the metric has to be of the type

$$ds^2 = c^2dt^2 - R^2(t)[\frac{dr^2}{1-kr^2} + r^2d\theta^2 + r^2sin^2\theta d\phi^2] \quad (1)$$

The scale factor, $R(t)$, could in principle be an increasing or decreasing function of time. The fact than only redshifts are observed, together with the relation

$$1 + z = \frac{R_0}{R} \quad (2)$$

deduced from (1) for null geodesics, means that R(t) is an increasing function of time, i.e., the universe expands.

Application of the Cosmological Principle also determines the form of the matter tensor, which corresponds to that of a perfect fluid, characterized by the pressure, p, and the density, ρ. In a matter dominated universe, like the present one, the equation of state can be reduced to $p \approx 0$. The corresponding Einstein equations are

$$\frac{\dot{R}^2}{R^2} = \frac{8}{3}\pi G\rho - \frac{kc^2}{R^2} + \frac{\Lambda c^2}{3} \quad (3)$$

$$2\frac{\ddot{R}}{R} = \Lambda c^2 - \frac{kc^2}{R^2} - \frac{\dot{R}^2}{R^2} \tag{4}$$

$$\rho R^3 = cte \tag{5}$$

As mentioned in the Introduction, the general agreement on the expansion as the explanation for the redshift phenomenon implies that metric (1) and equations (3), (4) and (5) are taken for granted. Therefore, the cosmological problem reduces to the determination of the so called cosmological parameters, defined by the following relations:

$$H_0 = \frac{\dot{R}_0}{R_0} \qquad \rho_0 = \frac{3H_0^2\Omega_0}{8\pi G}$$

$$q_0 = -\frac{\ddot{R}_0}{R_0 H_0^2} \qquad \frac{\Lambda c^2}{3H_0^2} \equiv \lambda_0 = \frac{\Omega_0}{2} - q_0$$

$$\Omega_0 \equiv \frac{\rho_0}{\rho_c} = \frac{8\pi G\rho_0}{3H_0^2} \qquad \frac{kc^2}{R_0^2} = H_0^2[\frac{3\Omega_0}{2} - q_0 - 1]$$

We already indicated, that the cosmological parameters are not yet well determined (see the reviews already cited). Even if the need for an empirical verification of the HEU is in principle independent of the observational status regarding the parameters, it is clear that the lack of agreement on the type of (expansion) world we are living in has sharpened that need.

In the following we will leave aside the question of the actual values of the cosmological parameters and will concentrate on the global tests. We start with the theoretical predictions of the expansion models (with $\Lambda = 0$). Even if they have been presented many times (see, for example, Sandage 1961 and 1988), we will write them here for the sake of completeness.

The m(z) relation for the case $q_0 > 0$ reads,

$$m = 5\ log\frac{1}{q_0^2}[q_0 z + (q_0 - 1)[\sqrt{1 + 2q_0 z} - 1]] + cte \tag{6}$$

whereas for the case $q_0 = 0$ it reduces to

$$m = 5\ log[z(1 + \frac{z}{2})] + cte \tag{6'}$$

The angular size-redshift relation is given by the expression

$$log\ \theta = 2\ log(1 + z) - log[q_0 z + (q_0 - 1)(\sqrt{1 + 2q_0 z} - 1)] + cte \tag{7}$$

for $q_0 > 0$.

The N(m) relation is given by

$$N(m) = \frac{2\pi n}{QH_0^3}(1 - 2q_0)^{-3/2}(P\sqrt{1 + P^2} - sinh^{-1}P) \quad (k = -1) \tag{8}$$

$$N(m) = \frac{2\pi n}{QH_0^3}(2q_0 - 1)^{-3/2}(sin^{-1}P - P\sqrt{1 - P^2}) \quad (k = +1) \tag{8'}$$

where n is the number density, Q the covered area and (with $A = 10^{0.2m}$),

$$P = \frac{A\sqrt{k(2q_0 - 1)}}{q_0(1 + A) - (q_0 - 1)\sqrt{1 + 2A}} \quad (k = \pm 1) \tag{8''}$$

For the particular case $k = 0$, N(m) has the following expression,

$$N(m) = \frac{32\pi n A^3}{3QH_0^3}[1 + A + \sqrt{1 + 2A}]^{-3} \tag{8'''}$$

The surface brightness (in magnitudes) is related to z trough

$$SB(z) = 10\ log(1 + z) + cte \tag{9}$$

Finally, the duration of the expansion is given by

$$H_0 t_0 = \int_0^1 \sqrt{\frac{y}{(1 - \Omega_0)y + \Omega_0}}\ dy \tag{10}$$

2.2 The Einstein static case.

The metric for the static case is given by

$$ds^2 = c^2dt^2 - \frac{dr^2}{1 - r^2/R^2} - r^2d\theta^2 - r^2 sin^2\theta d\phi^2 \tag{11}$$

The corresponding proper distance has the following expression:

$$r_P = R\ sin^{-1}(\frac{r}{R}) \tag{12}$$

It is well known that the metric (11) cannot produce a redshift, which has to be introduced in an *ad hoc* way. This can be done by postulating a differential Hubble law in the form

$$\frac{d\nu}{\nu} = -\frac{c}{H} dr_P \tag{13}$$

Then, equation (12) can be rewritten,

$$r_P = \frac{H}{c}\ ln(1 + z) \tag{14}$$

and the the coordinate distance, r, is given by

$$r = R\ sin[\frac{c}{HR}\ ln(1 + z)] \tag{15}$$

These relations, together with the expression for the luminosity distance,

$$d_L = \frac{L}{4\pi l} = r\sqrt{1 + z} \tag{16}$$

suffice to express the classical tests.

We see in equation (15) that, besides H, there is a free parameter, namely R, or their combination, c/HR, which plays the role of a *curvature factor*. Note however that the influence of that factor, in realistic situations, could be very small even till large redshifts. Thus, for $\rho_0 = 10^{-30} gcm^{-3}$ and $H_0 = 75\ kms^{-1}Mpc^{-1}$ it amounts to 0.38. Therefore,

$$sin[\frac{c}{HR}\ ln(1 + z)] \approx \frac{c}{HR}\ ln(1 + z)$$

within 5% for $z \leq 5$. Then, for practical purposes we will adopt for the coordinate distance the expression

$$r = \frac{c}{H}\ ln(1 + z) \tag{15'}$$

Using equation (15′) the predictions of the static metric are parameter free and all the tests are crucial. In that sense it can be said that the static metric is easier to disprove than the expansion case.

The m(z) relation is now given by

$$m = 2.5\ log(1+z) + 5\ log[ln(1+z)] + cte \tag{17}$$

It can be verified that equation (17) is nearly indistinguishable from equation (6) for $q_0 \approx 1$.

The angular size-redshift relation is simply given by

$$log\ \theta = -log\ [ln(1+z)] + cte \tag{18}$$

The number counts as a function of z is given by

$$logN(z) = 3\ log[ln(1+z)] + cte \tag{19}$$

which, together with equation (17) gives the N(m) relation.

Finally, the surface brightness dependence on the redshift is given by the expression

$$log\ SB(z) = 2.5\ log(1+z) \tag{20}$$

It is obvious that there are no constraints on the time scale. In fact, one of the problems of the static case is to account for the agreement between the astrophysical and cosmological ages.

3. The classical tests: observational results.

From their definition, it is apparent that, to control the HEU, not all the global tests have the same probatory value. In fact, only the surface brightness one could in principle be crucial since it does not depend on the parameters of the metric but just on how the photons propagate in a given geometry. The other tests are rather adequate to set the range of allowed values of the metric parameters, and only when that range would stretch to zero could they become also crucial. Consequently, all the tests except SB(z) are consistency tests. (On the same basis have to be also considered the time scale and the look-back time evolution tests that we will discuss separately).

On the other hand, the application of the tests to real data poses a number of important problems. First, it implies the definition of well defined properties that could be considered as standards for a given family of objects. The point is then to find that class of objects, identify their standard characteristic and control that we are actually dealing with the same kind of objects at different redshift values. This is a very difficult task indeed, that cannot be verified but only shown to be consistent along all the reasoning. Thus, for example, only if a family of galaxies show a well defined m(z) relation can we accept that they are true standard candles. The same applies to metric sizes or surface brightnesses. Consequently, it is not the existence of such relations what is tested since, when the data are not fitted by the predictions it is always possible to argue that the objects are not standard candles.

The tests, at last, are only consistency arguments. That means that, in practical cases, they can disprove the theory but they cannot prove it. And to get a disproof, more than one test has to be considered at the same time; if they would demand mutually contradictory evolutionary laws or incompatible sets of cosmological parameters the hypothesis should be discarded.

There is another crucial aspect when applying the tests to empirical data. Since the HEU implies an evolutionary universe, look-back time evolution effects **have to** affect the observational parameters. This is unavoidable and, in fact, lack of evolution with z would be a proof against the HEU. In other words, as stressed by Sandage (1988), in the case of expansion models we cannot expect to find a good agreement between data and theoretical predictions in the whole z-range before taking into account the evolutionary corrections. On the other hand, these evolutionary laws have to be provided by arguments which are independent of the results of the geometrical tests. Unfortunately, there are very scarce data on the cosmic evolution, apart from those obtained from the classical tests in an obvious circular way. And they are, to say the least, totally inconclusive (see Sandage 1988). This is the reason why it has been repeatedly stated that the potentiality of the global tests is ruined by the (unknown) evolutionary aspects one has to consider.

In those respects, the Einstein metric is easier to test since there are no free parameters and all the tests become crucial. Moreover, look-back evolutionary effects are strictly forbidden in that model, so the raw data can and have to be compared with the predictions. This makes the static alternative easy to disprove. **In fact, one important virtuality of the classical tests is actually that of eliminating non-evolutionary cosmologies.**

In view of that situation, i.e., the lack of knowledge on the cosmic evolution and the possibility of disproving the static case, we will first consider the data at their face value to analyze the degree of agreement offered by the theoretical predictions. Then we will analyze the feasibility of the evolutionary laws that would be deduced when the data plus evolution had to fit the expansion model predictions.

The practical difficulties that plague the tests oblige in principle to carefully discuss and analyze the results. It is not the place here to enter the details (which are referred to the original publications and the existing reviews) and the data will be taken as they were published. The emphasis will be put not on the meaning of a particular result but on the implications of all the results concerning a given test and on the meaning of all the tests taken together.

3.1 The (m,z) relation.

The first efforts to find the value of the deceleration parameter from the m(z) relation culminated with the results by Kristian, Sandage and Westphal (1978). Their data showed the presence of a curvature in the magnitude-redshift relation for a sample of brightest cluster galaxies, supposed to be standard candles. The results, for V as well as for R magnitudes, favoured the value $q_0 \approx 1$ or slightly higher. The same kind of work, using optical band magnitudes was pursued till higher redshifts (see Spinrad and Djorgovski 1987). The results were similar and consistent with the previous analysis by Kristian, Sandage and Westphal.

It was argued that uncertainties in the extinction, K and cluster richness corrections, made the results only indicative. Thus, an important effort was devoted to establish the (m,z) relation using K magnitudes (much less affected by those problems) of the same brightest cluster galaxies. The results (Lebofsky and Eisenhardt 1986; Lillly and Longair 1984) gave however similar results, $q_0 \approx 1$ or even greater.

We already noticed that the predictions by the expansion and the static models are very similar for $q_0 = 1$. Since the data are consistent (within the errors) with precisely that value, they cannot discriminate between those models. In other words, **the (m,z) relation**, as it is obtained in the redshift range already explored, **is compatible with both, expansion and static world metrics.**

A last word on the (m,z) data. Sandage (1987) has recalled that the Hubble effect can be traced back to within 2 Mpc. It is now clear that the universe cannot be considered neither homogeneous nor isotropic down to such scales, and therefore it seems rather surprising to find there what in principle is only predicted at the scale where the Friedmann metric is applicable. Or, in other words, it is not clear from which scale down the expansion of the universe stops to manifest itself.

3.2. The (θ,z) relation.

This test is very interesting since it could be very sensitive to the model metric. Hoyle (1959) showed that, in the case of expansion, it predicts a turnover in θ (for any $q_0 > 0$) that could be detected in the already explored z-range. (Note that the minimum is obtained at $z = 1.25$ for $q_0 = 0.5$). Thus, the test could rather easily lead to strong conclusions since it is less dependent than the others on the precision of the data.

Size measurements on three different classes of sources have been considered to perform the test. We recall separately the observational results in each of the applications.

a. The largest angular sizes of double radio sources.

The first representative results of this kind were obtained by Miley (1971) with a sample of double radiosources ranging from near zero to almost 2 in redshift. To avoid projection effects, only the largest angular sizes were considered. Miley found that they follow very approximately the euclidean relation $\theta \approx z^{-1}$.

Recently Kapahi (1987) has reconsidered the same test with new data. He used the median instead of the largest angular sizes, and the result is consistent with that found by Miley. The work by Kapahi shows that the data are not completely inconsistent with the static prediction but totally out of the range covered by the expansion models.

b. The harmonic radii of cluster of galaxies.

Hickson and Adams (1979) analyzed the harmonic radii (defined within a given distance from the center) of a large sample of clusters of galaxies. The results are shown in figure 1a. It is apparent that, at their face value, the data are completely out of the range of the expansion models predictions.

Bruzual and Spinrad (1978) had used a smaller sample of clusters and a different technique to correct for the dependence of the chosen distance to scrutinize the clusters on the specific metric. Their result does not coincide with that by Hickson and Adams, but the data too cannot be accommodated by any expansion model.

On the other hand, as can be appreciated in figure 1a, the data are acceptably fitted by the static prediction.

c. The metric sise of brightest cluster galaxies.

Djorgovski and Spinrad (1981) were the first to use the parameter η defined by Petrosian (1976) for cosmological purposes. It is important to recall that it is a metric parameter, and can be directly used in the $\theta(z)$ test. Specifically they measured the angular diameter to $\eta = 2$, for a sample of 25 brightest cluster galaxies. The data, shown in figure 1b, cannot be fitted by any expansion model, whereas they are satisfactorily fitted by the static law.

Thus, from totally different samples and techniques, the data on the angular size-redshift test seem to give consistent results. Of course, they are not free of biases, but it is not at all clear how they could produce the results actually observed. In particular, for the radiosources sample, it is well known from the beginning that only the strongest sources enter the sample. Now, it is however completely unclear why the most powerful radiosources should be smaller in size as demanded by the test if the expansion model is to be saved. Regarding the harmonic radii of clusters of galaxies, again it can be argued that those radii are not necessarily standard metric sizes, or even that the clusters are not yet virialized. But until independent arguments are found, those are claims just to save the model.

Finally, the metric sizes used by Djorgovski and Spinrad don't seem affected by any bias, except that at the largest redshifts, the seeing effects become important.

Thus, the conclusion from the $\theta(z)$ test is that strong evolutionary effects or even intricated biases have to be invoked to accommodate the data within the frame of the standard expansion models. **The data**, such as they come up from the observations, **are much better fitted by the static prediction.**

3.3 The N(m) test.

Early data were shown to depart from the expansion prediction and be fitted rather well by the static case (Jaakkola, Moles and Vigier 1979). New data confirmed those results (LaViolette 1986). The empirical body (see Ellis 1987, for a review), *at its face value*, is definitively not fitted by any expansion model and, as emphasized by Sandage (1988), only the application of important evolutionary corrections even at $z \approx 0.4$, could bring the data within the range of the standard models.

3.4 The surface brightness test.

It has been repeatedly acknowledged that it is in fact a crucial test for the HEU, since no free parameters enter the relations.

To perform the test is however a difficult task. The practical problem is indeed to find a well defined metric size to which refer the surface brightness. Thus, some attempts made in the past using metric dependent sizes are not probatory. In fact, it can be easily realized that if the linear size is defined from angular sizes which do not correspond to well defined metric dimensions, the use of the model metric to translate the observed into linear sizes largely mimics (till moderately high values of z) the effect to be tested on the surface brightness.

Then, as discussed by Petrosian (1977) and Tinsley (1977), the only way to properly test the surface brightness evolution with z is to use metric sizes. Petrosian (1976) defined such a tool, showing how the evolutionary effects could also be controlled. It is then rather surprising that such test has not yet been properly intended till now. Only recently Sandage and Perelmuter (1990a, b and c) have tried to perform it.

After an exhaustive analysis of the possible bias and how to manage them, Sandage and Perelmuter were left with some objects from the sample of brightest cluster galaxies elaborated by Djorgovski and Spinrad for the $\theta(z)$ test. The results they find seem to be compatible with the HEU. However, the situation is still far from definitively settled, as stated by Sandage and Perelmuter themselves: *Yet, satisfactory as these results are for the standard model and, by consequence, for the conventional interpretation of the redshifts, the reliability of the demonstration must be questioned on several grounds.*

These questionable aspects refer to the quality of the photometry, the size of the sample, and the evolutionary corrections. One added problem is that SB depends on the absolute magnitude of the standard candle and that effect has to be reduced to normalized conditions. Thus, the data have to be corrected to standard conditions in a way which depends on the metric. This introduces some circularity in the way to handle the data.

In any case, the results are promising and, in fact, constitute the first, even if not definitive, positive evidence for the expansion models from the classical tests.

Summarizing the results on the four classical tests we can say that the situation remains almost the same as when Jaakkola, Moles and Vigier (1979) published their review: The expansion models do fit the existing (m,z) data, which are, however, fitted by the static prediction too. The (θ,z) and N(m) data are not fitted by any of the standard models, but are compatible with the static metric. Moreover, the magnitude-redshift data point to high values of q_0, whereas the (θ,z) and N(m) data require low, in fact negative, values of the deceleration parameter.

The first conclusion is then that **the static metric is not discarded by the data**. Or, stated in a positive way, the data on the global tests are fitted by the static predictions. We recall here that the static relations have no free parameters and that no look back-time evolutionary effects do exist for that world model.

Regarding the expansion models, evolutionary corrections have to be introduced to reach the agreement. It is interesting to note that both, the m(z) and N(m) tests, can be put together in agreement with the empirical data with qualitatively similar evolutionary laws. Moreover, the implied evolution is in the direction we can easily understand, i.e., the galaxies were brighter in the past. Consequently, at the qualitative level for the moment, the HEU can be also accommodated to fit the data.

In other words, and as far as the N(m) and m(z) tests are concerned, none of the two alternative metrics are discarded by the existing data.

It is the angular size-redshift test which seems to pose the worst problems to the HEU. The coincidence from three totally different elaborations of the test, favoring the static predictions, is difficult to integrate into the standard models, unless some bias or misdefinitions could produce such a *conspiracy* against the standard views and in favour of the static metric predictions.

It is then clear that the promising results by Sandage and Perelmuter on the SB(z) test are of the highest importance to settle the situation. The need to complement by new and more extensive analysis the existing meager evidences could not be overemphasized.

4. The time scale constraints.

The very existence of a (cosmic) time scale is a characteristic of the expansion models. In that respect, the relevant question is that a lower limit to that scale can be evaluated from the ages of different cosmic bodies. And the important point is that that can be done independently of any cosmological consideration.

The first step to determine that lower limit to the age of the universe is to date different kinds of objects and identify the oldest among them. We know that the best dated entities are the globular clusters in the Galaxy and, among them, the oldest have an age of at least (we take here the lowest present estimate) 15×10^9 years (including the time for the protogalaxy to collapse; Sandage 1988). Then defining $H_0 = 100h$, the age data imply $h \leq 0.65$ for $\Omega_0 = 0$ or $h \leq 0.43$ for $\Omega_0 = 1$ ($\Lambda = 0$ models). **Formally that**

implies that any value of the Hubble constant over 65 $kms^{-1}Mpc^{-1}$ is incompatible with the standard views.

The long standing controversy about the value of the Hubble constant is too well known to be repeated here. Let us then quote only some recent results. Using the relations between global properties of HII galaxies and giant extragalactic HII regions to calibrate the zero point, Melnick, Terlevich and Moles (1988) found a high value, $H_0 = 89 \pm 10$ $kms^{-1}Mpc^{-1}$. Tully (1988) and Giraud (1990) have also presented evidences, based on the Tully-Fisher distance calibration, for a high value of the Hubble constant, $H_0 = 90$ $kms^{-1}Mpc^{-1}$. On the other hand, Sandage (1988) strongly favors, from different arguments, a low value, $H_0 = 43$ $kms^{-1}Mpc^{-1}$.

Before objective reasons are advanced to reject either the high or the low H_0 values, it is clear that none of them can be discarded when discussing the consequences on the time scale.

Therefore, starting with that estimate for the age of the globular clusters, we are left with only two possibilities to save the standard model, either to demonstrate that the high H_0 values are wrong (or, conversely, that the low value is correct), or to accept that $\Lambda \neq 0$. This last possibility will be briefly discussed in the next section.

One cannot avoid the impression when analyzing the time scale question, that the situation is rather fragile. Starting with the question, it is very impressive the order of magnitude agreement between the cosmic and the astrophysical time scales. But when the details are entered, it is realized that an uncertainty of about 20% in the age of the oldest globular clusters and of a factor of 2 in the Hubble constant, which otherwise would not be completely unacceptable in the extragalactic domain, imply that a decision cannot be taken among radically different world models.

There is no time scale question with the static models, indeed. But, on the other hand, it has to be admitted that the better than order of magnitude agreement between H_0^{-1} and the age of the oldest known cosmic bodies remains a mystery for those models.

5. Effects of $\Lambda \neq 0$

The most direct indications to consider the $\Lambda \neq 0$ models would come from the time scale considerations. We already noted that the high H_0 values proposed by some authors impose a non zero cosmological constant to save the expansion hypothesis.

We have presented an analysis of the $\Lambda \neq 0$ models elsewhere (Moles 1990) and we will only summarize the results here. It was shown that, for a given density parameter, Λ has an upper limit, Λ_C. Models with $\Lambda > \Lambda_C$ are not physically permitted because either they cannot reach the present situation of the universe or they started from a non singular situation a too short time ago. A lower limit can also be considered if we want the cosmological constant to have a ponderable effect on the time scale.

Indeed, it is well known that only models with $\Lambda > 0$ and $q_0 < 0$ can result in $H_0 t_0 \geq 1$. Writing

$$\lambda = \frac{\Lambda c^2}{3H^2}$$

we call λ_1 the value corresponding to $H_0 t_0 = 1$, whereas λ_C gives an infinite time scale. For any value of the Hubble constant over 65 $kms^{-1}Mpc^{-1}$ the condition $H_0 t_0 \geq 1$ has to be satisfied. Therefore, only values of λ_0 in the range $\lambda_1 \leq \lambda_0 \leq \lambda_C$, can produce long enough time scales. What we showed is that that range is very small. For the limit case $\Omega_0 = 0$, it is $0 \leq \lambda_0 \leq 1$, whereas it is $1.85 \leq \lambda_0 \leq 2.6$ for $\Omega_0 = 1$. For the euclidean, $k = 0$, models, that range is $0.74 \leq \lambda_0 \leq 1$.

Moreover, $H_0 t_0$ departs significantly from 1 only for λ_0 approaching the critical value (see figure 2). Thus, if we take H_0 to be as high as, let's say, 80 $kms^{-1}Mpc^{-1}$, λ_0 should be equal to λ_C within a few hundreths. In other terms, should the time scale problem be posed, the value of Λ to solve it is, for a given density, almost completely determined and not far from Λ_C.

A non zero cosmological constant modifies the predictions concerning the tests. We show in figures 3a and b the fits to the data on the (m,z) and (θ, z) planes for different values of the cosmological constant. It can be seen that, within the range allowed by the precedent considerations, the predictions are not very much affected by Λ. Consequently, the fits to the data uncorrected for evolutionary effects, are not very much improved.

It is worth to note that the requisites on Λ from both sets of data are contradictory. Thus, the (m,z) data are better fitted by models with small values of Λ, whereas the (θ,z) test and the time scale arguments are better agreed by models with Λ approaching the critical value. In any case, the improvement in the (θ,z) test is only marginal since the predicted curvature has not been observed.

6. The look-back time evolution effects.

The presence of evolutionary effects is a prediction of the expansion models and its detection would be a definitive proof against the static case. However, as stated by Sandage (1988) in his review, no strong enough observational (non geometrical) evidences are yet available.

It is not our intention to rediscuss here all of those evidences, but just recall one which is considered among the strongest cases for the evolution with z, the QSOs counts.

When computing N(m) for QSOs, the corresponding slope was found to be larger that the euclidean prediction, which has the steepest possible slope. Thus, evolution in either their number density or their luminosity was invoked to explain the results. Those arguments were however critically analyzed by Jaakkola (1982). He noted first that the number count slope is steeper for the low redshift tail of the distribution, what seems contrary to the simplest expectation, i.e., to see larger effects at larger redshifts. Moreover, a morphological criterion, the point like aspect, was used together with the spectroscopical classification to select the objects. Thus, low redshift objects such as Seyfert 1 and N galaxies, which similar spectral characteristics to the QSOs, where not included in the samples. Now we know that essentially all the QSOs till the resolution limit of the ground based telescopes have in fact extended envelopes. Therefore, the morphological criterion should be dropped, and in that case, as shown by Jaakkola, the slope of N(m) for the bright QSOs samples becomes very near to the euclidean value.

This considerations serve to illustrate the difficulties to deal with cosmic evolutionary effects. To characterize them, the first step is to identify objects at different redshifts that, with respect to a given property, are identical except in that they are in different evolutionary moments. Then, since that property is only statistically defined, it is necessary to show beyond any doubt that **the differences with the redshift are larger than the scatter at a given redshift**. For example, let us suppose that we are comparing brightest cluster galaxies at different z values, and admit that the scatter in (corrected) absolute magnitude is 0.28^m and constant with z. Then at the 3σ level, the differences, once we are sure that we are dealing with similar objects, should be greater than 0.84^m to be statistically significant. This is not far from what is expected to have at $z \approx 1$.

The models of (passive) evolution are based on current ideas about stellar evolution. Then the deduced evolutionary laws are combined with the predictions of the model metric, and compared to the data. A good illustration of such an approach is that by Yosii and Takahara (1988). They found that a reasonable agreement can be reach for both the N(m) and m(z) tests together. They, rather optimistically, quoted: *..but the value of $q_0 = 0.5$ provided $z_F \approx 3 - 5$ seems to be consistent with the available observations if we take into account the luminosity evolution of galaxies.*

When trying to get information on the evolution effects one has to take into account some rather obvious constraints on the derived laws. For example, evolution in different spectral bands has to be considered and compared to the existing data. They are in fact imposed conditions. On the other hand, increasing the past luminosity has an important consequence, since that implies a corresponding increase in the amount of Helium produced locally. For the solar neighbourhood that amount is at most of 5% by mass, which places important limits on the total luminosity evolution of pur galaxy.

The results by Yosii and Takahara show that the data on two of the classical tests can be reconciled with the predictions. On the other hand, no explanation is yet available for the $\theta - z$ results. Indeed, conspiracy by observational biases to produce such coincidences are not excluded but none has been yet devised. The specific evolutionary laws for the sizes *deduced* from the observational data are, on their side, very considerable and difficult to understand (Troitskii and Gorbacheva 1988).

Finally, the value of the deceleration parameter proposed by Yosii and Takahara, $q_0 = 0.5$, implies $H_0 t_0 = 2/3$ if $\Lambda = 0$. Thus, taking as before 15 Gyr for the age of our Galaxy, only the lowest values within the observed range for the Hubble constant can produce a long enough time scale.

7. Final considerations.

Let's start with the standard expansion models. It is clear that the global tests, which are but consistency proofs, cannot provide a definitive answer until the evolutionary corrections are incorporated. At the same time, the existence of look back-time evolution effects would provide a definitive argument against static universes.

Cosmic evolution thus becomes one of the major concerns regarding physical Cosmology. The problem then is to devise how to attack the problem observationally.

We already argued about the practical difficulties to assess that look back-time evolution effects are actually being detected. Two approaches, probably complementaries, can be planned to that end. One could be to go to high enough redshifts to maximize the effects which are looked at. In that case, the problem is that it becomes harder and harder to control that the observed objects are actually similar to those at low redshift, aside from the observational difficulties to get a well sized sample.

The other approach would consist in trying to better the knowledge of the standard objects at low redshift, to reduce the observational dispersion. Then, the effects would be significant at correspondingly lower redshift and therefore, it could be easier to find similar objects and control their properties. The limit of this method is, of course, the amount of physical dispersion in the standard property.

On its side, the static model fits the global tests data rather well. As discussed before, the model is not contradicted by the observations and, in that sense, it appears as completely acceptable.

However, we cannot forget that the predictions of that model cannot even be formulated before a redshift law is postulated. And now, as when Hubble and Tolman presented their analysis, some new physics seems to be necessary to produce such a phenomenon. Besides, since there is not a hot past in that case, new explanations should be constructed for the abundances of the light elements and the microwave background radiation.

Going back to the the observational level, it has to be concluded that the problem of the nature of the redshift is still there. And whereas the N(m) and m(z) predictions of the expansion models can be made compatible with the data taking into account reasonable evolutionary laws, and the time scale can be made long enough by the effect of a non-zero cosmological constant, the angular size-redshift test results rest very intriguing and the SB(z) test awaits to be properly performed.

Thus, the most promising working directions for the next years, regarding the direct tests of the geometry of the space-time, appear to be the $\theta(z)$ and SB(z) tests, and the search for look back-time evolution effects.

References

- Arp, H.: 1970, Nature **225**, 1033.

- Arp, H.: 1987, *Quasars, Redshifts and Controversies*, Interstellar Media, California, USA.

- Arp, H.: 1988, Astron. Astrophys. **202**, 70.

- Bottinelli, L. and Gouguenheim, L.: 1973, Astron. Astrophys. **26**, 85.

- Bruzual, A. G. and Spinrad, H.: 1978, Astrophys. J. **220**, 1.

- Burbidge, G.: 1973, Mittelungen der Astronomische Gesellshaft, July-August.

- Colllin-Souffrin, S., Pecker, J.-C. and Tovmassian, H. M. : 1974, Astron. Astrophys. **30**, 351.

- Djorgovski, S. and Spinrad, H.: 1981, Astrophys. J. **251**, 417.

- Ellis, R. S.: 1987, in IAU Symp. No. 124 *Observational Cosmology*, ed. A. Hewitt, G. Burbidge and L. Z. Fang, Reidel Pub. Co., p. 367.

- Findlay-Freundlich, E.: 1954, Proc. Phys. Soc. A **67**, 192.

- Giraud, E.: 1990, Astron. Astrophys. 231, 1.

- Giraud, E., Moles, M. and Vigier, J.-P.: 1982, C. R. Acad. Sc. Paris **294**, 195.

- Hickson, P. and Adams, P. J.: 1979, Astrophys. J. Letters **234**, L91.

- Hoyle, F.: 1959, in Paris Symposium on Radio Astronomy, ed. R. N. Bracewell, Stanford Univ. Press, p. 529.

- Hubble, E. and Tolman, R.: 1935, Astrophys. J. **82**, 302.

- Jaakkola, T.: 1973, Astron. Astrophys. **27**, 449.

- Jaakkola, T.: 1977, Astrophys. and Space Sc. **49**, 99.

- Jaakkola, T.: 1982, Astrophys. and Space Sc. **88**, 283.

- Jaakkola, T. and Moles, M.: 1976, Astron. Astrophys. **53**, 389.

- Jaakkola, T., Moles, M. and Vigier, J. P.: 1979, Astron. Nachr., **300**, 229.

- Jaakkola, T., Karoji, H., Le Denmat, G., Moles, M., Nottale, L., Vigier, J.-P. and Pecker, J.-C.: 1975, Mon. Not. R. astron. Soc. **177**, 191.

- Kapahi, V. K.: 1987, in IAU Symp. No. 124 *Observational Cosmology*, ed. A. Hewitt, G. Burbidge and L. Z. Fang, Reidel Pub. Co., p. 251.

- Kristian, J., Sandage, A. R. and Westphal, J. A.: 1978, Astrophys. J. **221**, 383.

- Kropotkin, P. N.: 1988, Soviet Physics Dokl. **33**, 85.

- Kropotkin, P. N.: 1989, Soviet Physics Dokl **34**, 277.

- LaViolette, P. A.: 1983, Ph. D. thesis, Portland State University, Portland, OR.

- LaViolette, P. A.: 1986, Astrophys. J. **301**, 544.

- Lebofsky, M. and Eisenhardt, P. R. M.: 1986, Astrophys. J. **300**, 151.

- Lilly, S. J. and Longair, M. S.: 1984, Mon. Not. Royal astron. Soc. **211**, 833.

- Marič, Z., Moles, M. and Vigier, J.-P.: 1976, Astron. Astrophys. **53**, 191.

- Marič, Z., Moles, M. and Vigier, J.-P.: 1977, Lett. Nuovo Cimento **18**, 269.

- Melnick, J., Terlevich, R. and Moles, M.: 1988, Mon. Not. Royal astron. Soc. **235**, 297.

- Miley, G.: 1971, Mon. Not. R. astron. Soc. **152**, 477.

- Moles, M.: 1978, Thesis, Univ. Paris VI.

- Moles, M.: 1990, in preparation.

- Moles, M. and Nottale, L.: 1981, Astron. Astrophys. **100**, 258.

- Moles, M. and Vigier, J.-P.: 1974, C. R. Acad. Sc. Paris **278B**, 969.

- Pecker, J.-C., Roberts, A. P. and Vigier, J.-P.: 1973, Nature **241**, 338.

- Pecker, J.-C. and Vigier, J.-P.: 1987, in IAU Symp. No. 124 *Observational Cosmology*, ed. A. Hewitt, G. Burbidge and L. Z. Fang, Reidel Pub. Co., p. 507.

- Petrosian, V.: 1976, Astrophys. J. Letters **209**, L1.

- Petrosian, V.: 1977, in IAU Coll. no. 37 *Décalages vers le rouge et l'expansion de l'Univers*, ed. Ch. Balkowski, C.N.R.S., Paris, p. 213.

- Rubin, V. C., Ford, W. K. and Rubin, J. S.: 1973, Astrophys. J. Letters **183**, L111.

- Sandage, A. R.: 1961, Astrophys. J. **133**, 335.

- Sandage, A. R.: 1987, in IAU Symp. No. 124 *Observational Cosmology*, ed. A. Hewitt, G. Burbidge and L. Z. Fang, Reidel Pub. Co., p. 1.

- Sandage, A. R.: 1988, Ann. Rev. Astron. Astrophys. **26**, 561.

- Sandage, A. R. and Perelmuter, J.-M: 1990a, Astrophys. J. **350**, 481.

- Sandage, A. R. and Perelmuter, J.-M: 1990b, Astrophys. J. **361**, 1.

- Sandage, A. R. and Perelmuter, J.-M: 1990c, preprint

- Spinrad, H. and Djorgovski, S.: 1987, in IAU Symp. No. 124 *Observational Cosmology*, ed. A. Hewitt, G. Burbidge and L. Z. Fang, Reidel Pub. Co., p. 129.

- Tinsley, B. M.: 1977, in IAU Coll. no. 37 *Décalages vers le rouge et l'expansion de l'Univers*, ed. Ch. Balkowski, C.N.R.S., Paris, p. 223.

- Troitskii, V. S. and Gorbacheva, I. V.: 1988, Sov. Astron. **33**, 3.

- Tully, R. B.: 1988, Nature **334**, 209.

- Yosii, Y. and Takahara, F.: 1988, Astrophys. J. **326**, 1.

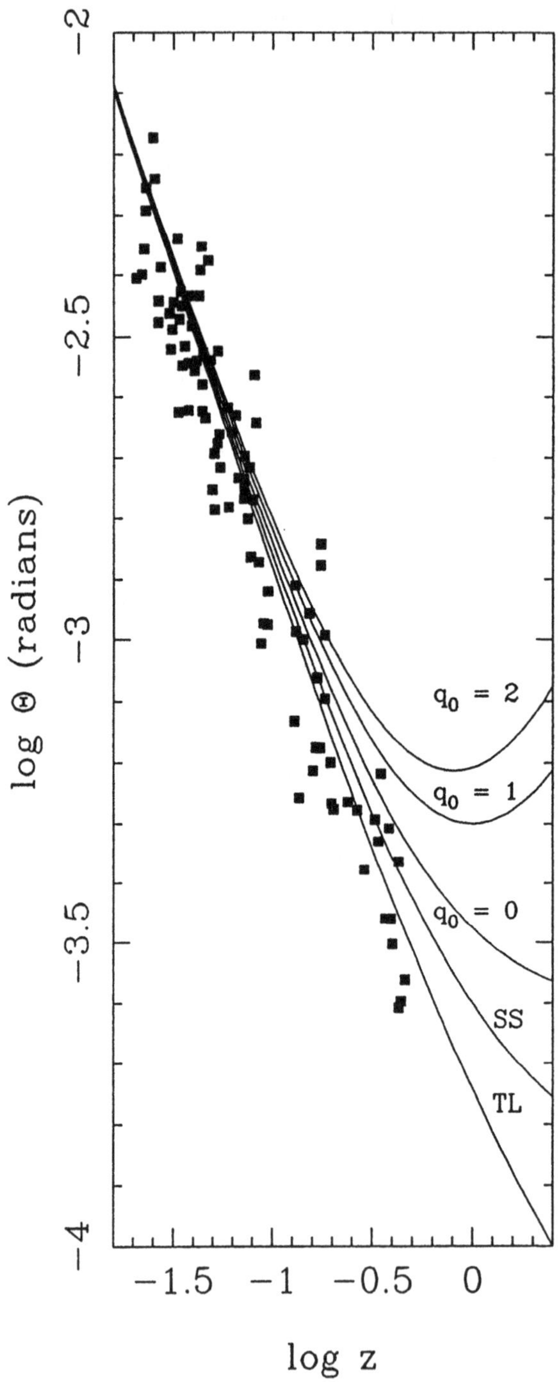

Figure 1a. Harmonic radii of clusters of galaxies as a function of the redshift. The data are from Hickson and Adams (1979). Labels are for different standard models, SS stands for Steady State and TL for tired light, static model.

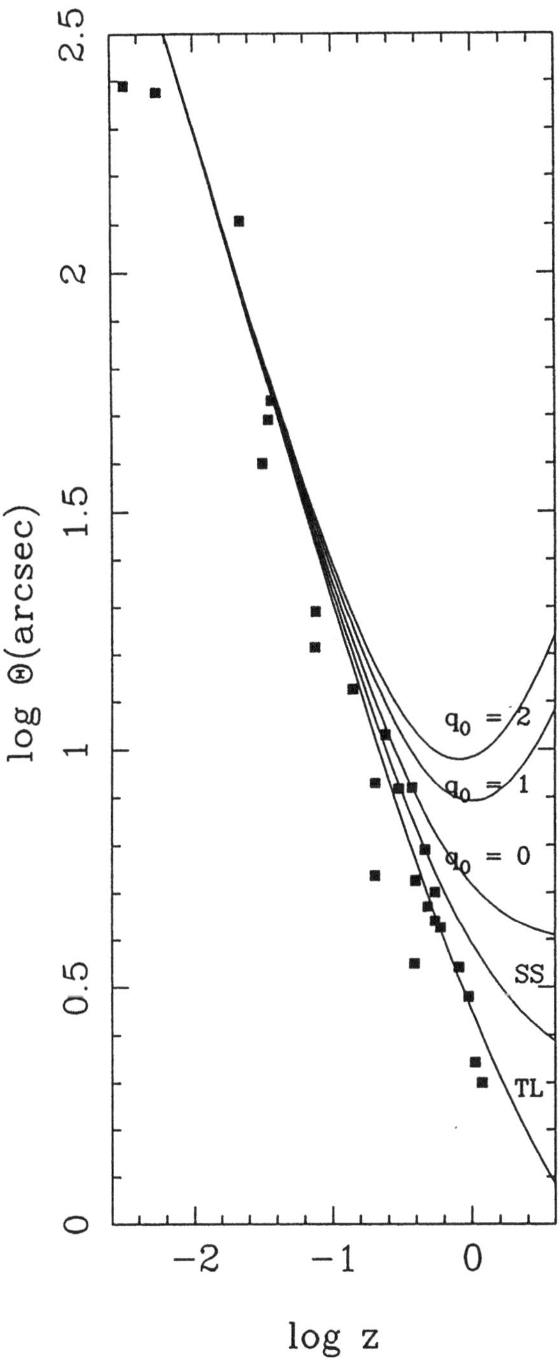

Figure 1b. The metric sizes of brightest cluster galaxies as a function of the redshift. The data are form Djorgovski and Spinrad (1981). Labels are as in figure 1a.

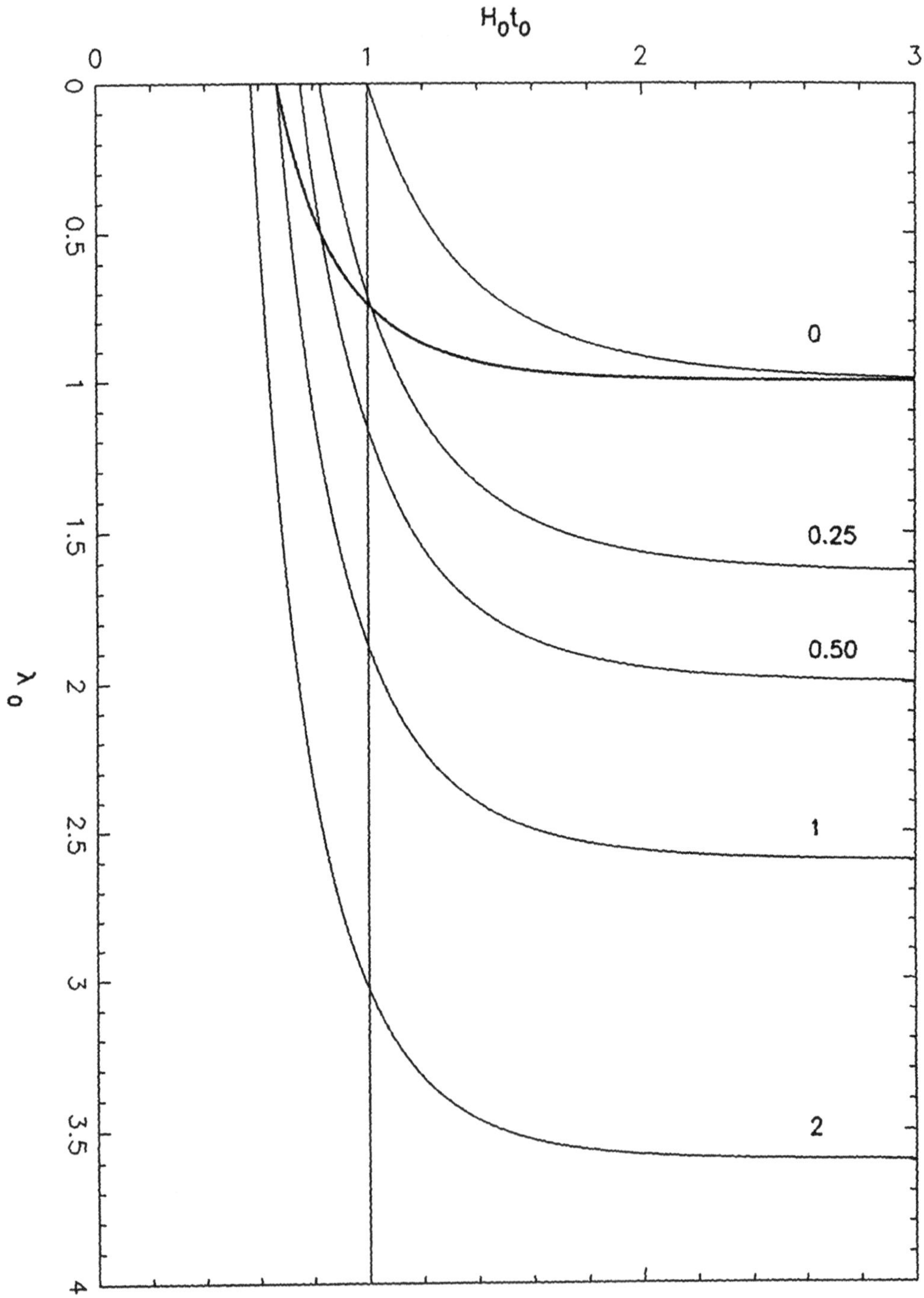

Figure 2. The value of H_0t_0 as a function of λ_0 for some values of the density parameter. Labels give the Ω_0 values. The thick line corresponds to euclidean models.

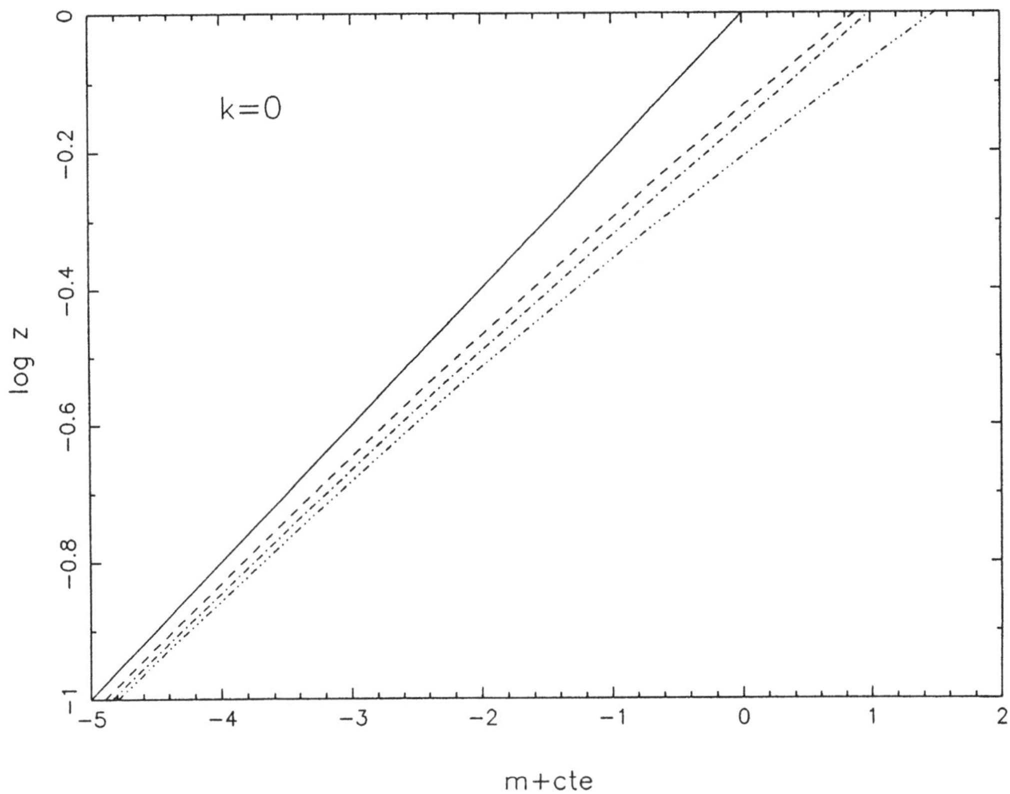

Figure 3a. The m(z) prediction of the euclidean model. The solid line corresponds to the trend shown by the observational data and the dashed line to the $\lambda_0 = 0$, $\Omega_0 = 0$ model. The other two lines are for euclidean models with $\lambda_0 = \lambda_C = 1$ (dotted-dotted-dotted-dashed line) and $\lambda_0 = \lambda_1 = 0.74$ (dotted-dashed line).

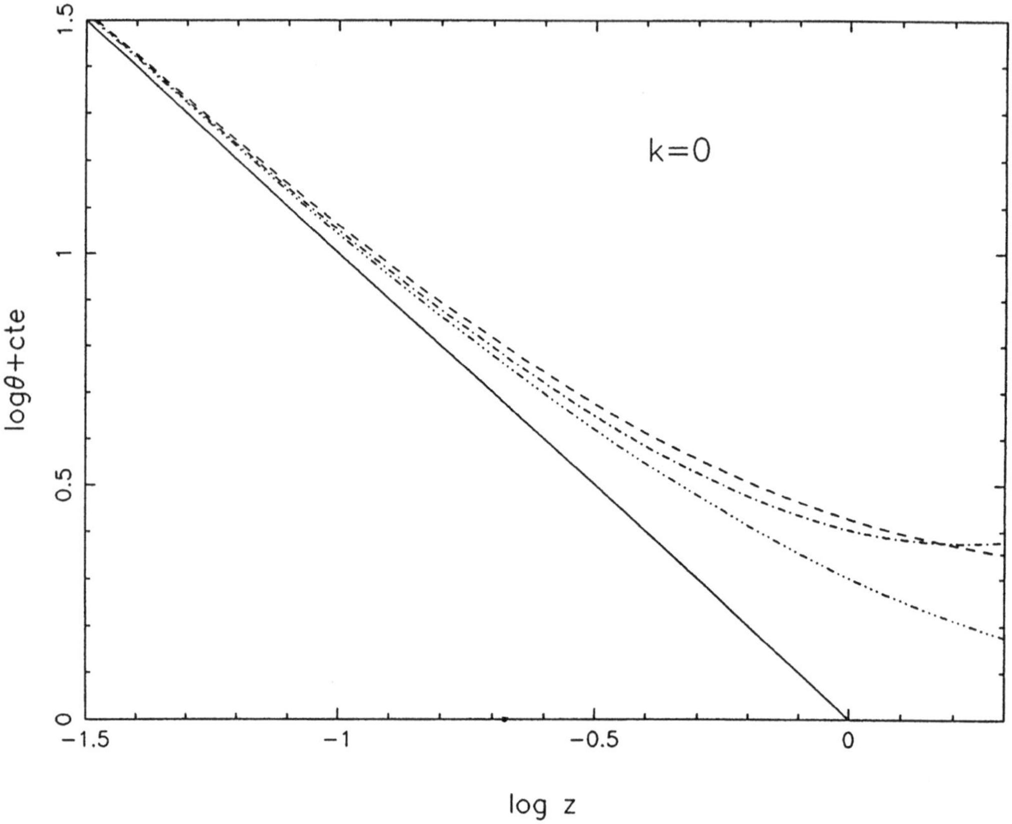

Figure 3b. The $\theta(z)$ prediction. Symbols have the same meaning as in figure 3a.

December–1990
DFFCL 12-1/1990

The Early Universe Behaviour with Non–Minimal Coupling *

P. Moniz[†]
P. Crawford
A. Barroso

Departamento de Física, Universidade de Lisboa
Ed. C1, Piso 4, Campo Grande, 1700 Lisboa
PORTUGAL

Talk presented at the XII Autumn School of Physics
The Physical Universe

1 – 5 October, 1990
Lisbon, Portugal

Abstract

We comment on the stability problem of the equilibrium states of our Universe. With a specific ansatz for the ground state we re-examine the physics of spontaneous symmetry breaking in curved space-time. Introducing a non-minimal coupling between a scalar field (a multiplet under the action of a gauge group) and gravity we show that spontaneous symmetry breaking may occur without the need of a negative value for m^2. Within this framework we calculate the Higgs mass, the value of the cosmological constant and of the effective gravitational constant.

*Work supported in part by a grant from D.F.F.C.L. – J.N.I.C.T./C.E.R.N. (projecto 900).
†Work supported in part by a J.N.I.C.T. graduate scholarship BD/138/90-RM.

1 Introduction

During the last decade theoretical research on the evolution of the Universe has undergone a considerable development. Inflationary and Quantum Cosmology represent its main achievements [1,2,3,4,5]. In this context it is legitimate to question why the Universe is as it is and how it evolved towards its actual form, with us within it. Moreover, if we suppose that we live in a stable or ground state of the Universe [6], some stability criterion (or argument) must justify the behaviour of the Universe in its very early history, before it reached the ground state where we live in. The problem of the initial conditions is the subject of Quantum Cosmology (for a review, see ref. [1,4,5]). In this field of research a large amount of work has been published and an important result is that, for a large variety of models, an Inflationary stage seems to occur quite naturally [1,5,6,7,8,9].

In the Inflationary scenario [1,2,3] the Universe undergoes several phase transitions which are driven by a primordial scalar field Φ under the influence of an effective potential, $\mathcal{V}_{eff}$, whose vacuum expectation value determines the magnitude of the cosmological constant, Λ. During these phase transitions the scalar field Φ evolves until it reaches a stable ground state. This framework is particulary relevant when considering theories with *local* or *gauge* symmetries, i.e., *space-time dependent* symmetry transformations which leave our theory invariant. The imposition of local symmetries implies the existence of *massless* vector particles. However, the phenomenology of weak interactions requires the existence of *massive* spin one gauge particles. The only mass generation mechanism that satisfies both unitarity and renormalizability is the Higgs mechanism. In such a case, we have a theory which is still invariant under the gauge symmetry but a vacuum state, which is *not* gauge invariant [10,11]. We then expect that the Universe, having started in a metastable state, will undergo sucessive phase transitions towards the stable vacuum state.

This communication is a report on previous works [12,13,14] where an analysis of the behaviour of the very early Universe is presented, assuming a *non-minimal coupling* between the scalar field Φ and gravity. We believe that this specific situation may provide important results and answers concerning the mutual influences between gravity and matter [13,14,15,16,17,18,19,20,21,22,23,24,25]. We start by reminding that there is no meaningful notion of local gravitational energy since this one is not locally defined. Thus, in curved space-time the energy density is not a safe criterion to look for equilibrium states of the system in consideration. The presence of gravity changes the situation considerably! To describe the physics of equilibrium states, either we propose assumptions concerning the ground states of the system, working them out and checking for physical results, or we perform a qualitative analysis using the full set of the differential equations of our system. This seems to be a more complete procedure, but in this work we shall restrict ourselves to the first option. As far as the other alternative is concerned, it will be the subject of future work. In section 2 we work out a particular ansatz for the ground state of the Universe at tree-level [14,15]. There we ask for a constant scalar field $\Phi = \Phi_0$ and a the space-time background that satisfy the relation

$\mathcal{R}_{\mu\nu} = \Lambda g_{\mu\nu}$, where Λ is the cosmological constant. This might be reasonable in face of the *Cosmic No-Hair Conjecture* [26,27,28,29] and of the assumption that the state of minimum excitation of the Universe must correspond to a state with a geometry of high symmetry [6,7]. For this specific choice we study the Higgs mechanism of mass generation [10,11]. This mechanism usually invoques that the mass, m, of the scalar field Φ must be *pure imaginary* ($m^2 < 0$) but we point out that with a *non-minimally coupling* between the scalar field and gravity this is not necessarily so [14]. In section 3 we present some physical consequences of the proposed ansatz. We calculate the Higgs mass and the effective gravitational constant. In section 4 we present some final remarks concerning our work. We use units $c = \hbar = 1$ and the metric signature is $(+,-,-,-)$.

2 Spontaneous Symmetry Breaking with Non –Minimal Coupling

The usual mass generation mechanism is well explained in several text books [10,11]. One introduces a coupling between the scalar field Φ and the spin one gauge field, $\mathbf{A}_\mu$, of the form $\Phi^\dagger\Phi \mathbf{A}_\mu \mathbf{A}^\mu$ which generates a mass term to $\mathbf{A}_\mu$ when the ground state of the theory is such that the vacuum expectation value of $\Phi^\dagger\Phi$ is different from zero.

To obtain this situation, the gauge invariant potential

$$\mathcal{V}[\Phi] = \mathcal{B} + m^2\Phi^\dagger\Phi + \lambda(\Phi^\dagger\Phi)^2, \tag{1}$$

which represents the self-interaction of Φ, must have a minimum for $\Phi \neq 0$. This will imply that the mass term m^2 must be negative if λ is positive, otherwise the Hamiltonian will be unbounded from below. In this picture Φ is regarded as an auxialiary field which induces the masses of $\mathbf{A}_\mu$. But can this be entirely correct? In fact, it is not, there must remain a relic of the Higgs mechanism — the *Higgs boson*. Without it the theory would violate unitarity. Hence, this tachionic character of Φ may be considered quite unfortunate.

In this section, we point that a possible solution to this problem might reside in the *non-minimal interaction* between the primordial scalar fields and gravity [14]. In doing this we are proposing to analyse the physics of a *specific* ansatz for the ground state of the Universe which is consistent with the arguments given in section 1.

The action for our system is

$$\mathcal{S} = \int d^4x\sqrt{-g}\left[\frac{M_{Pl}^2}{16\pi}\mathcal{R} - \xi\Phi^\dagger\Phi\mathcal{R} + (D_\mu\Phi)^\dagger D^\mu\Phi - \mathcal{V}[\Phi] - \frac{1}{4}\mathcal{F}^i_{\mu\nu}\mathcal{F}^{i\mu\nu}\right], \tag{2}$$

where $\mathcal{R}$ is Ricci scalar curvature, $g \equiv \det(g_{\mu\nu})$ ($g_{\mu\nu}$ is the metric tensor), M_{Pl} is the the Planck mass ($M_{Pl} = \mathcal{G}^{-1/2}$; $\mathcal{G}$ is the gravitational constant),

$$D_\mu \equiv \partial_\mu - i\ e\ \mathbf{A}_\mu \tag{3}$$

represents the covariant gauge derivative with

$$\mathrm{A}_\mu = \sum_i A^i_\mu \mathcal{T}^i \tag{4}$$

as the gauge field, $\mathcal{T}^i$ are the generators of a gauge group G, e is a coupling constant (if G is semi–simple there will be a gauge coupling constant for each factor), Φ represents a multiplet scalar field which transforms according to some representations of the gauge group G, † denotes hermitian conjugation and the last term in (2) is the usual kinetic energy of the gauge field. The action (2) includes the term

$$\xi \Phi^\dagger \Phi \mathcal{R} \tag{5}$$

which represents the *non–minimal interaction* involving directly the scalar field and gravity. This term is the only possible local scalar coupling of this sort with the correct dimensions where ξ is a dimensionless constant [30].

Let us take, for practical purposes, the case of an $SU(2) \times U(1)$ model, for which we write

$$\Phi = \begin{pmatrix} \phi^+ \\ \frac{1}{\sqrt{2}}(\vartheta + \mathcal{H} + i\phi_z) \end{pmatrix}. \tag{6}$$

Here, ϑ is the vacuum expectation value of Φ defined by

$$\langle \Phi^\dagger \Phi \rangle = \frac{\vartheta^2}{2} \equiv y^2 \tag{7}$$

and $\mathcal{H}$ denotes the Higgs field modes.

The variation of the action (2) with respect to the metric leads to the Einstein equations of motion which are

$$\mathcal{R}_{\mu\nu} - \frac{1}{2}\mathcal{R} g_{\mu\nu} + \frac{8\pi}{M^2_{pl}} T_{\mu\nu} = 0, \tag{8}$$

with

$$T_{\mu\nu} = \left(1 - \xi\frac{16\pi}{M^2_{pl}}\Phi^\dagger\Phi\right)^{-1} [(D_\mu\Phi)^\dagger D_\nu\Phi + (D_\nu\Phi)^\dagger D_\mu\Phi -$$
$$(D_\alpha\Phi^\dagger D^\alpha\Phi + \mathcal{V}[\Phi])g_{\mu\nu} + 2\xi(g_{\mu\nu}\Box - \nabla_\mu D_\nu) \mid \Phi \mid^2] + T^{(gauge)}_{\mu\nu}. \tag{9}$$

The last term is not explicitely written since it is not important for our discussion. On the other hand, the variation of $\mathcal{S}$ with respect to $\Phi^\dagger$ gives the Klein-Gordon equation

$$g^{\mu\nu}\nabla_\mu(\partial_\nu - ie\mathrm{A}_\nu)\Phi + \xi\mathcal{R}\Phi + \frac{\partial\mathcal{V}[\Phi]}{\partial\Phi} = 0, \tag{10}$$

where ∇_μ denotes the gauge *and* space-time covariant derivative which generalizes D_μ.

We now present an ansatz for the ground state [14,15]. In doing it we take into account that when gravity is present some features of the flat space-time formalism are unsuitable. In the Minkowski case the absolute stable ground state is identified with the minimum of the energy density (i.e., the Hamiltonian density). The Lorenz invariance of the vacuum demands the assumption that

$$\langle \mathbf{A}_\mu \rangle = 0 \tag{11}$$

and thus just the scalar field Φ can have a non zero vacuum expectation value. Hence, the ground state is obtained from the condition

$$\frac{\partial \mathcal{V}[\Phi]}{\partial \Phi} = 0. \tag{12}$$

However, when gravity is present the above criterion does not work, since there is no local definition of gravitational energy. As an alternative we may start by assuming that the space-time geometry of the ground state is homogeneous *both* in space *and* time [31], and as such we choose to impose the condition

$$\mathcal{R}_{\mu\nu} = \Lambda g_{\mu\nu}. \tag{13}$$

This means we have selected an Einstein space, where the Minkowski or the De Sitter spaces are particular solutions. The case of *spatial* isotropy imposes that eq.(11) remains valid and, in the ground state, the scalar field has a constant value Φ_0.

Replacing eq.(11) in eq.(9), writing $T_{\mu\nu}$ for a constant scalar field and noting that $\mathcal{R} = 4\Lambda$, it is immediate to obtain [15]

$$\Lambda = \left(1 - \xi \frac{16\pi}{M_{pl}^2} \Phi_0^\dagger \Phi_0\right)^{-1} \mathcal{V}[\Phi_0] \frac{8\pi}{M_{pl}^2}. \tag{14}$$

On the other hand, the Klein-Gordon equation implies that

$$4\xi\Lambda\Phi_0 + \left.\frac{\partial \mathcal{V}[\Phi]}{\partial \Phi}\right|_{\Phi_0} = 0. \tag{15}$$

The use of these two equations determine the stable ground state of the theory [15,16,20,21]. Notice that eq.(15) generalizes the well known condition (12) for the ground state in flat space time while eq.(14) gives the cosmological constant induced by the finite value of the potential $\mathcal{V}[\Phi]$. Replacing eqs.(1) and (14) in eq.(15) we easily obtain the equation for the ground state of Φ:

$$\left(2\lambda + \xi \frac{16\pi m^2}{M_{Pl}^2}\right) y^2 + 4\xi B \frac{8\pi}{M_{Pl}^2} + m^2 = 0. \tag{16}$$

The general solution is

$$y = \sqrt{\frac{-m^2 - 32\pi\xi B M_{Pl}^{-2}}{16\pi\xi m^2 M_{Pl}^{-2} + 2\lambda}}, \tag{17}$$

which in the limit $\xi \longrightarrow 0$, leads to the well known result

$$y^2 = -\frac{m^2}{2\lambda}, \tag{18}$$

which implies $m^2 < 0$.

However, eq.(17) shows that, for a *negative* ξ [16] and $\mathcal{B}$ positive and sufficiently large, we can have acceptable solutions for $m^2 > 0$! Let us mention that in ref.[19] it is shown that, for a *massless* scalar theory, the spontaneously symmetry breaking can occur, at *one-loop* order, only if ξ is *greater than zero*. In our case, on the contrary, we study a *massive* theory with *tree-level* breaking for *negative* ξ.

In such conditions, spontaneous symmetry breaking can occur without m^2 being negative, contrary to the standard model. The ansatz here presented just provides the suitable conditions for this particular spontaneous symmetry breaking process to happen. It is worthy to call the attention to the crucial role played in this mechanism by the *non–minimal* interaction term described by the expression (5). When $\xi < 0$, the *non–minimal coupling* term $\xi\Phi^\dagger\Phi\mathcal{R}$ can act like a negative squared mass, and thus, m^2 can stay positive.

3 Physical Consequences of the Non–Minimal Coupling

In the previous section we have proved how spontaneous symmetry breaking can occur *without* $m^2 < 0$. For such situation to happen, we needed to consider a curved space-time scenario, where the scalar field Φ was *non–minimal coupled* to gravity. This new situation suggests that some physical process, described by the standard model theory, can be drastically changed. These possible consequences are now studied and compared with the standard data.

One result of the *non-minimal coupling* of the scalar field with gravity is that the Higgs mass is changed. In fact, searching fot the $\mathcal{H}^2$ terms in the Lagrangian, it is trivial to obtain

$$m_{\mathcal{H}}^2 = m^2\left(1 + 6\lambda\frac{-m^2M_{Pl}^2 - 32\pi\xi\mathcal{B}}{16\pi\xi m^4 + 2\lambda M_{Pl}^2 m^2}\right). \tag{19}$$

Furthermore, this spontaneous symmetry breaking mechanism induces, in the Lagrangian, an extra linear term in $\mathcal{R}$ which implies that $\mathcal{G}$ is changed into an effective gravitational constant $\mathcal{G}_{eff}$ [32] given by:

$$\mathcal{G}_{eff} = \mathcal{G}\Bigg/\left(1 + \frac{m^2 + 32\pi\xi\mathcal{B}M_{Pl}^{-2}}{m^2 + \lambda(8\pi\xi)^{-1}M_{Pl}^2}\right). \tag{20}$$

To continue our analysis, let us consider the following plausible hypothesis. First, we assume that

$$(m/M_{Pl})^2 \ll \lambda\ ;\ \mid\xi\mid\approx 1.$$

Then,

$$y^2 \simeq -\frac{m^2}{2\lambda} + \frac{\mid\xi\mid}{\lambda}\frac{16\pi\mathcal{B}}{M_{Pl}^2} \tag{21}$$

which, roughly speaking, implies that for, $\mathcal{B} > (mM_{Pl})^2$, one obtains for y^2 a value of the order of m^2. For the case of a grand unified model this mechanism would require $\mathcal{B} \approx (M_X M_{Pl})^2$ (M_X^2 is the X-boson mass), which means

$$M_X^2 \approx y^2. \tag{22}$$

This, in turn, implies that the the cosmological constant, before the symmetry breaking, would be of the order of M_X^2.

According to the previous assumptions, in the Higgs mass expression (19) we can drop the first term in the denominator and obtain

$$m_{\mathcal{H}}^2 \simeq m^2 + 6\lambda y^2, \tag{23}$$

which leads to a value for the Higgs mass of the same order as the X-boson mass. With respect to the effective gravitational constant, $\mathcal{G}_{eff}$, for the values of the parameters $m, \mathcal{B}, \xi$ and λ we have assumed, the scale of the effect is of the order

$$\mathcal{G}_{eff} \simeq \mathcal{G} \Big/ \left(1 + \frac{M_X^2}{M_{Pl}^2}\right). \tag{24}$$

Another situation that is simple to study is the $\mathcal{B} = 0$ case. Then, either m^2 is negative and the situation is similar to the one in flat space-time, with a negligible correction proportional to ξ, *or*, λ is also zero and we have a free scalar field with *non-minimal coupling* to gravity. In this last case the ground state is

$$y \simeq \sqrt{-\frac{M_{Pl}^2}{16\pi\xi}}, \tag{25}$$

which, again for a negative ξ, gives a symmetry breaking of the order of the Planck scale. Now, the cosmological constant is

$$\Lambda \simeq -\frac{m^2}{4\xi} \tag{26}$$

and it can be made arbitrarily small in the limit $m \longrightarrow 0$.

4 Final Remarks

We stress the idea that the introduction of a *non-minimal coupling* between a multiplet scalar field (which transforms under some representation of a gauge group G) and the gravitational field opens the possibility of inducing spontaneous symmetry breaking without the need of choosing a negative value for m^2.

As far as the second hypothesis described in section 3 is concerned we must point out the interesting observation that this could be an example of a spontaneous symmetry breaking at the Planck scale. Within this possibility it is important to notice that it leads to a vanishing cosmological constant, as $m \longrightarrow 0$. This is done, however, at the cost of $\lambda = 0$ and renormalizability would impose the presence of the four point scalar interaction.

Aknowledgements

The authors would like to thank A.B. Henriques for his unfailling effort in providing a stimulating atmosphere throughout this Autumn school, where part of this report was iniciated. We also have benefited from useful comments from O. Bertolami, G.W. Gibbons, Y. Kubyshin and J. Mourão.

References

[1] A. Linde, *"Particle Physics and Inflationary Cosmology"*, Contemporary Concepts in Physics, Volume 5, Harwood Academic Publishers (Chur, 1990)

[2] E. Kolb and M. Turner, *"The Early Universe"*, Lecture Note Series – Frontier in Physics – 69, Addison–Wesley (Boston, 1990)

[3] Rosa Dominguez-Tenreiro and Mariano Quirós, *"An introduction to Cosmology and Particle Physics"*, World Scientific Publishing Co. (Singapore, 1988)

[4] J. V. Narlikar and T. Padmanabhan, *"Gravity, Gauge Theories and Quantum Cosmology"*, D. Reidel Publishing Co. (Dordrecht, 1986)

[5] L.Z. Fang anr R. Ruffini (eds.), *"Quantum Cosmology"*, in Advanced Series in Astrophysics and Cosmology - Vol. 3, World Scientific Publishing Co. (Singapore, 1987)

[6] J. Hartle and S.W. Hawking, Phys. Rev **D28** (1983) 2960

[7] S.W. Hawking, Nucl. Phys. **B244**, (1984) 135

[8] S.W. Hawking and P.C. Luttrell, Nucl. Phys. **B247** (1984) 250

I.G. Moss and W.A. Wright, Phys.Rev. **D29**, (1984) 1069

S.W. Hawking and Z.C. Wu, Phys. Lett. **151B** (1985) 15

P.F. Gonzalez–Dias, Phys.Lett. **159B** (1985) 19

D.N. Page in Quantum Conceptes of Space and Time, R. Penrose and C.J. Isham (eds), Claredon Press, (Oxford, 1986)

[9] G. Esposito and G. Platania, Class. Quantum Grav. 5 (1988) 937

[10] D. Bailin and A. Love, *"Introduction to Gauge Field Theory"*, Adam and Hilger (Bristol, 1986)

[11] T.P. Cheng and L.F. Li, *"Gauge Theory of Elementary Particle Physics"*, Oxford University Press, 1988.

[12] P. Crawford, *"Soluções Cosmológicas com Campos não Abelianos"*, Ph.D. Thesis, Universidade de Lisboa, 1987

[13] P. Moniz, *"Homogeneous Cosmologies with Scalar Fields"*, M.Sc. Thesis, Universidade de Lisboa, 1990

[14] P. Moniz, P. Crawford and A. Barroso, Class. Quantum Grav. **7** (1990) L143

[15] M.S. Madsen, Class. Quantum Grav. **5** (1988) 627.

[16] Y. Hosotani, Phys.Rev. **D32** (1985) 1949.

[17] L.H. Ford, Phys. Rev. **D35** (1987) 2339.

[18] J. D. Bekenstein, Found. Phys. **16** (1986) 409.

[19] G. Denardo and E. Spallucci, Nuov. Cim. **83** (1984) 35.

[20] O. Bertolami ,Phys. Lett. **86B** (1987) 161

[21] A Grib, V.M. Mostepanenko and V.M. Frolov, Theor. Math. Phys. **33** (1977) 42

[22] B.L. Hu and D.J. O'Connor, Phys. Rev. **D36** (1987) 1701

[23] A.D. Dolgov, *"An attempt to get rid of the Cosmological constant"*, in The Very Early Universe, G.W. Gibbons, S.W. Hawking and S.T.C. Siklos (eds.), Cambridge University Press (Cambridge, 1983)

[24] B. Allen, Nucl. Phys. **B226** (1983) 288

[25] R. Fakir and W.G. Unruh, Phys. Rev. **D41** (1990) 1783
R. Fakir and W.G. Unruh, Phys. Rev. **D41** (1990) 1792
R. Fakir, Phys. Rev. **D41** (1990) 3012

[26] R. M. Wald, Phys. Rev. **D28** (1983) 2118

[27] G. W. Gibbons and S. W. Hawking, Phys. Rev. **D15** (1977) 2738

[28] W. Boucher and G. W. Gibbons, *"Cosmic Baldness"* in The Very Early Universe, G. W. Gibbons, S. W. Hawking and S. T. C. Siklos (eds), Cambridge University Press (Cambridge, 1983)

[29] S. W. Hawking and I. G. Moss, Phys. Lett. **110B** (1982) 35

[30] N.D. Birrell and P.C.W. Davies, *"Quantum Fields in Curved Spaces"*, Cambridge University Press (Cambridge, 1982)

[31] M.P. Ryan and L.C. Sheppley, *"Homogeneous Relativistic Cosmologies"*, Princepton University Press (Princepton, 1975)

[32] A. Linde, Phys. Lett. **93B** (1980) 394

STABILITY OF COMPACTIFICATION IN EINSTEIN-YANG-MILLS THEORIES

O. Bertolami [1)], Yu. A. Kubyshin [2)] and J.M. Mourão [3) 4)]

1) Centro de Física da Matéria Condensada
Av. Prof Gama Pinto 2, 1699 Lisboa-Codex, PORTUGAL

2) Nuclear Physics Institute, Moscow State University,
Moscow 119899, USSR

3) Centro de Física Nuclear
Av. Prof. Gama Pinto 2, 1699 Lisboa-Codex, PORTUGAL

4) Departamento de Física, Instituto Superior Técnico
Av. Rovisco Pais, 1000 Lisboa PORTUGAL

Abstract

We study the dynamics of multidimensional Einstein-Yang-Mills cosmologies. The stability of compactification after the period of inflationary expansion of the external dimensions is analysed.

1. Introduction

In multidimensional field theories - the so-called Generalized Kaluza-Klein (GKK) theories - the extra dimensions serve the purpose of unifying different four-dimensional fields. For instance the metric and gauge fields that we observe in four-dimensions could be different manifestations of one single multidimensional object - the metric [1-3]. Alternatively a gauge field in a multidimensional space-time could lead in four-dimensions to a gauge field and a Higgs field with a self interacting symmetry breaking potential [4-9].

For a GKK theory to be in agreement with our everyday experience we must admit that the extra (or internal) dimensions have a very small characteristic size. As proposed by Cremmer, Sherk and Luciani [10] this could be the result of a spontaneous symmetry breaking phenomenon in the multidimensional theory. They found that in a multidimensional Einstein-Yang-Mills (EYM) theory the symmetry between all the dimensions can be spontaneously broken by the existence of solutions corresponding to a factorization of the space-time in

$$E^{4+d} = M^4 \times I^d \quad , \tag{1.1}$$

M^4 being the four-dimensional Minkowski space-time and I^d a d-dimensional compact space with a size of the order of the Planck length, $L_{pl} = \sqrt{16\pi k} \approx 10^{-33}$ cm, where k is the gravitational constant [2,7-10].

However, for a given compactifying solution to correspond to the ground state of the theory it must be stable both with respect to classical and quantum fluctuations. It has been shown that some of the compactifying solutions in EYM systems are stable against symmetric [7-9] and general [11-13] small classical fluctuations.

In a cosmological setting the multidimensional space-time is assumed to have, in large scales, the form

$$E^{4+d} = \mathbb{R} \times G^{ext}/H^{ext} \times G^{int}/H^{int} \tag{1.2}$$

admitting local coordinates $x^{\hat{\mu}} = (t, x^i, \xi^m)$, where $\hat{\mu} = 0, 1, \dots, 3 + d$; $i = 1, 2, 3$; $m = 4, \dots, d + 3$, $\mathbb{R}$ denotes a time-like direction and G^{ext}/H^{ext} (G^{int}/H^{int}) the space of external (internal) spatial dimensions realized as a coset space of the external (internal) isometry group G^{ext} (G^{int}). In this approach spontaneous compactification occurs if, as a result of the cosmological evolution, the scale factor - a (t) - of the external space increases up to its observed today macroscopic value while the scale factor - b(t) - of the internal space is kept static or slowly varying and very small (see e.g. [9,14-17]). The classical stability at zero temperature of some solutions of this type in EYM theories has been proved in refs. [9,16,17]. In ref. [16] these solutions were found to survive semiclassically a period of inflationary expansion of the external dimensions for the different models of inflation: old, new, extended and chaotic.

We shall assume that temperature can be introduced in a multidimensional EYM cosmological model by considering gauge fields with non-vanishing external space components of the strength tensor. These components are assumed to generate, after inflation, the radiation energy density which dominates the Universe and fixes its temperature.

In Section 2 the dynamical equations of multidimensional EYM cosmologies are derived and some of their properties are studied. In Section 3 we show that the effective potential for the dilaton field depends crucially on the temperature. At zero temperature the solution corresponding to compactified internal dimensions is (for suitable values of the cosmological constant of the multidimensional theory [9]) classically stable but semiclassically unstable, however this solution for a non-vanishing temperature becomes the true ground-state of the dilaton field.

2. Classical dynamical equations

In this section we shall obtain the equations describing the dynamics of a multidimensional EYM cosmological model. An important point to make is that since we are interested in the post inflationary period, for which the three-dimensional external Universe is radiation dominated, we shall not (as in refs. [9,16,17]) set the external-space components $\hat{A}_i$ of the gauge field to zero.

We shall consider that the large scale dynamics of the post inflationary Universe is dominated by the bosonic sector of the multidimensional theory. Though this may not seem a very good assumption especially for the matter dominated period ($t_{Univ} > 10^{12}$ sec.) we believe that the inclusion of fermions will not lead to any qualitative change in the analysis of the stability of the vacuum.

Following the original Kaluza-Klein idea (see [1] and e.g. [2,3]) we would expect the bosonic sector of the multidimensional theory to be that of pure gravity with dynamics described by the action

$$S_{gr}[\hat{g}_{\hat{\mu}\hat{\nu}}] = \frac{1}{16\pi\hat{k}} \int_{E^{4+d}} d\hat{x} \sqrt{-\hat{g}} \left(\hat{R} - 2\hat{\Lambda} \right), \tag{2.1}$$

where $\hat{g} = \det(\hat{g}_{\hat{\mu}\hat{\nu}})$, $\hat{R}$ is the scalar curvature, $\hat{k}$ and $\hat{\Lambda}$ are respectively the gravitational and cosmological constants in 4+d dimensions. Assuming that the space-time has the factorized form (1.2) and that the space of internal dimensions G^{int}/H^{int} is a compact space with a very small size, then it follows that the theory (2.1) leads to an effective four-dimensional Einstein-Yang-Mills theory with a gauge group K ($K \subset G^{int}$) and a multiplet of scalar fields [1-3]. However, this theory has important difficulties such as the lack of stable compactifying solutions and the impossibility of obtaining proper chiralities and masses for the fermions in the reduced theory. One must therefore consider either alternative theories of gravity or theories with a more complicated bosonic sector including besides the metric, gauge and other bosonic fields in the multidimensional space-time. These are the GKK theories referred above. This procedure raises, of course, doubts on the motivation of the GKK theories. If, for consistency with observations, one needs in D = 4 + d dimensions a theory almost as complicated as the four dimensional one, why then to consider the internal dimensions at all? It can be argued however, that the multidimensional EYM theory is the bosonic sector (or part of it) of a superstring theory. Furthermore, since the existence of extra dimensions is not in contradiction with observations one should study this possibility and try to understand the reason of why the dimensionality of the macroscopic space-time is four and not any other number.

As mentioned previously we shall consider multidimensional EYM theories. In these theories the basic difficulties of pure Einstein theories can be naturally circumvented [7-17]. Let the gauge group $\hat{K}$ of the (4 + d)-dimensional theory be a simple compact Lie group. The action is given by:

$$S[\hat{g}_{\hat{\mu}\hat{\nu}}, \hat{A}_{\hat{\mu}}, \hat{\chi}] = S_{gr}[\hat{g}_{\hat{\mu}\hat{\nu}}] + S_{gf}[\hat{A}_{\hat{\mu}}, \hat{g}_{\hat{\mu}\hat{\nu}}] + S_{inf}[\hat{\chi}, \hat{A}_{\hat{\mu}\hat{\nu}}, \hat{g}_{\hat{\mu}\hat{\nu}}], \tag{2.2}$$

where $S_{gr}[\hat{g}_{\hat{\mu}\hat{\nu}}]$ is given by (2.1),

$$S_{gf}[\hat{A}_{\hat{\mu}}, \hat{g}_{\hat{\mu}\hat{\nu}}] = \frac{1}{8\hat{e}^2} \int_{E^{4+d}} d\hat{x} \sqrt{-\hat{g}} \quad \mathrm{Tr}\, \hat{F}_{\hat{\mu}\hat{\nu}} \hat{F}^{\hat{\mu}\hat{\nu}} \tag{2.3a}$$

$$S_{inf}[\hat{\chi}, \hat{A}_{\hat{\mu}\hat{\nu}}, \hat{g}_{\hat{\mu}\hat{\nu}}] = -\int_{E^{4+d}} d\hat{x} \sqrt{-\hat{g}} \left(\frac{1}{2}(\partial_{\hat{\mu}} \hat{\chi})^2 + \hat{U}(\hat{\chi}) \right), \qquad (2.3b)$$

$\hat{F}_{\hat{\mu}\hat{\nu}} = \partial_{\hat{\mu}}\hat{A}_{\hat{\nu}} - \partial_{\hat{\nu}}\hat{A}_{\hat{\mu}} + [\hat{A}_{\hat{\mu}}, \hat{A}_{\hat{\nu}}]$, $\hat{e}$ is the gauge coupling constant, $\hat{\chi}$ is the inflaton field responsible for the inflationary expansion of the external space and $\hat{U}(\hat{\chi})$ is the potential for $\hat{\chi}$. It is assumed that the potential $\hat{U}(\hat{\chi})$ is bounded from below, has a global minimum and that without loss of generality $\hat{U}_{min} = 0$.

To study cosmological models associated with action (2.2) we must restrict ourselves to spatially homogeneous and (partially) isotropic field configurations, which are symmetric under the action of the group $G^{ext} \times G^{int}$. To find these configurations we shall use the theory of symmetric fields [3-9,12,13,18]. Let us for definiteness consider the case with gauge group $\hat{K} = SO(N)$, $N \geq 3 + d$ and

$$E^{4+d} = \mathbb{R} \times S^3 \times S^d , \qquad (2.4)$$

where S^3 (S^d) is the three (d) dimensional sphere. The group of spatial homogeneity and isotropy is, in this case

$$G^{HI} = SO(4) \times SO(d+1) , \qquad (2.5a)$$

while the group of spatial isotropy is

$$H^I = SO(3) \times SO(d) , \qquad (2.5b)$$

which is in agreement with the alternative realization of E^{4+d} as

$$E^{4+d} = \mathbb{R} \times SO(4)/SO(3) \times SO(d+1)/SO(d) =$$

$$= \mathbb{R} \times (SO(4) \times SO(d+1))/(SO(3) \times SO(d)) . \qquad (2.6)$$

In the theory of symmetric fields a central rôle is played by the so-called Cartan one-form which, in the present case is defined as follows. Let $\sigma(y) \in SO(4)\times SO(d+1)$, $y \in (SO(4)\times SO(d+1))/(SO(3)\times SO(d)) = S^3 \times S^d$, be some choice of representatives in the cosets y, such that $[\sigma(y)] = y$, i.e. a (local) section in the principal bundle

$SO(4)xSO(d+1) \rightarrow S^3xS^d$. The Cartan one-form on S^3xS^d is defined as the pull back of the canonical left-invariant form on the group SO(4)xSO(d+1) [19]:

$$\omega_{(y)} = \sigma^{-1}(y)\, d\sigma(y)\,. \tag{2.7}$$

The form ω takes values on $\mathfrak{so}(4) \oplus \mathfrak{so}(d+1)$ = Lie(SO(4)xSO(d+1)) - the Lie algebra of the group SO(4)xSO(d+1) - and therefore can be decomposed as

$$\omega = \sum_{\alpha=1}^{d+3} \omega^{\alpha} T_{\alpha} + \sum_{1 \le i<j \le 3} \omega^{ij} \frac{T_{ij}^{(4)}}{2} + \sum_{1 \le m<n \le d} \tilde{\omega}^{mn} \frac{\tilde{T}_{mn}^{(d+1)}}{2}\,, \tag{2.8}$$

where $\{T_{ij}^{(4)}, 1 \le i<j \le 4, \tilde{T}_{mn}^{(d+1)}, 1 \le m<n \le d+1\}$ is a basis in the Lie algebra $\mathfrak{so}(4) \oplus \mathfrak{so}(d+1)$ of G^{HI} and

$$T_{\alpha} = \frac{T_{\alpha 4}^{(4)}}{2}\,, \qquad \text{for } \alpha = 1, 2, 3$$

$$T_{\alpha} = \frac{\tilde{T}_{\alpha-3d+1}^{(d+1)}}{2}\,, \qquad \text{for } \alpha = 4, \ldots, d+3\,. \tag{2.9}$$

with the following commutation relations

$$[T_{ij}^{(4)}, T_{i'j'}^{(4)}] = \delta_{ij'} T_{ji'}^{(4)} + \delta_{ji'} T_{ij'}^{(4)} - \delta_{ii'} T_{jj'}^{(4)} - \delta_{jj'} T_{ii'}^{(4)} \tag{2.10a}$$

$$[T_{ij}^{(4)}, \tilde{T}_{mn}^{(d+1)}] = 0 \tag{2.10b}$$

$$[\tilde{T}_{mn}^{(d+1)}, \tilde{T}_{m'n'}^{(d+1)}] = \delta_{mn'} \tilde{T}_{nm'}^{(d+1)} + \delta_{nm'} \tilde{T}_{mn'}^{(d+1)} - \delta_{mm'} \tilde{T}_{nn'}^{(d+1)} - \delta_{nn'} \tilde{T}_{mm'}^{(d+1)}\,. \tag{2.10c}$$

The one-forms ω^{α}, $\alpha = 1, \ldots, d+3$ in (2.8) form a (local) moving coframe in S^3xS^d. In this coframe the components of a SO(4)xSO(d+1)-invariant metric on S^3xS^d are independent of the local coordinates (x^i, ξ^m). Moreover the most general form of a SO(4)xSO(d+1)-invariant metric in E^{4+d} reads

$$\hat{g} = -\tilde{N}^2(t)\, dt^2 + \tilde{a}^2(t) \sum_{i=1}^{3} \omega^i \omega^i + b^2(t) \sum_{m=4}^{d+3} \omega^m \omega^m\,, \tag{2.11}$$

where $\tilde{a}(t)$, $b(t)$ and the lapse function $\tilde{N}(t)$ are arbitrary nonvanishing functions of time. Notice that in (2.11), $\sum_{i=1}^{3} \omega^i \omega^i$ and $\sum_{m=4}^{d+3} \omega^m \omega^m$ coincide with the standard metrics $d\Omega_3^2$ and $d\Omega_d^2$ in the three and d-dimensional spheres respectively. For later purpose, we make the following conformal change of the variables that characterize the four-dimensional part of the metric:

$$\tilde{N}^2(t) = \left(\frac{b_0}{b(t)}\right)^d N^2(t) \tag{2.12a}$$

$$\tilde{a}^2(t) = \left(\frac{b_0}{b(t)}\right)^d a^2(t) \quad . \tag{2.12b}$$

The SO(4) x SO(d+1)-*invariant* ansatz for the inflaton field $\hat{\chi}$ reads:

$$\hat{\chi}(t, x^i, \xi^m) = \hat{\chi}(t) \quad . \tag{2.13}$$

To fix a sector of SO(4) x SO(d+1)-*symmetric* gauge fields one must choose an homomorphism λ of the isotropy group SO(3) x SO(d) to the gauge group SO(N) [4-9]:

$$\lambda : \ SO(3) \times SO(d) \rightarrow SO(N) \quad . \tag{2.14}$$

We choose λ to be the simplest nontrivial embedding defined by the branching rule

$$\underline{N}\downarrow_{\lambda(SO(3) \times SO(d))} = (\underline{3}\,,\underline{1}) + (\underline{1}\,,\underline{d}) + (N\text{-}3\text{-}d)\,(\underline{1}\,,\underline{1})\,. \tag{2.15}$$

Then the SO(4) x SO(d+1)-symmetric ansatz for the gauge field is [4-9,18] :

$$\begin{aligned}\hat{A}(t) = {} & \frac{1}{2}\sum_{p,q=1}^{N-3-d} B^{pq}(t)\, T^{(N)}_{3+d+p\,3+d+q}\, dt + \frac{1}{2}\sum_{1\le i<j\le 3} T^{(N)}_{ij}\,\omega^{ij} + \frac{1}{2}\sum_{4\le m<n\le 3+d} T^{(N)}_{mn}\,\tilde{\omega}^{m-3\,n-3} + \\ & \sum_{i=1}^{3}\left(\frac{1}{4} f_0(t) \sum_{j,k=1}^{3} T^{(N)}_{jk}\,\varepsilon_{jik} + \frac{1}{2}\sum_{p=1}^{N-3-d} f_p(t)\, T^{(N)}_{i\,d+3+p}\right)\omega^i + \\ & \sum_{m=4}^{d+3}\left(\frac{1}{2}\sum_{q=1}^{N-3-d} g_q(t)\, T^{(N)}_{m\,d+3+q}\right)\omega^m ,\end{aligned} \tag{2.16}$$

where the functions $f_p(t)$, $p = 0, \dots, N\text{-}3\text{-}d$, $g_q(t)$, $q = 1, \dots, N\text{-}3\text{-}d$, $B^{pq}(t)$, $1 \le p < q \le N\text{-}3\text{-}d$ are arbitrary and $T^{(N)}_{pq}$, $1 \le p < q \le N$ are the generators of the gauge group SO(N).

By substituting (2.11) - (2.13) and (2.16) into action (2.2) we obtain an one-dimensional effective action for the functions of time t that parametrize the symmetric field configurations:

$$S^{eff} = S^{eff}[a, \psi, f_o, \vec{f}, \vec{g}, \chi, N, \hat{B}] =$$
$$= 16\pi^2 \int_{t_1}^{t_2} dt\, Na^3 \left\{ -\frac{3}{8\pi k}\frac{1}{a^2}\left(\frac{\dot{a}}{N}\right)^2 + \frac{3}{32\pi k}\frac{1}{a^2} + \frac{1}{2}\left(\frac{\dot{\psi}}{N}\right)^2 + \frac{1}{2}\left(\frac{\dot{\chi}}{N}\right)^2 + \right.$$
$$e^{d\beta\psi}\frac{3}{4e^2}\frac{1}{a^2}\left(\frac{1}{2}\left(\frac{\dot{f}_o}{N}\right)^2 + \frac{1}{2}\left(\frac{\vec{\mathcal{D}}_t f}{N}\right)^2\right) + e^{-2\beta\psi}\frac{d}{4e^2}\frac{1}{b_o^2}\frac{1}{2}\left(\frac{\vec{\mathcal{D}}_t g}{N}\right)^2$$
$$\left. - W(\psi, a, f_o, \vec{f}, \vec{g}, \chi) \right\} \tag{2.17}$$

where $k = \dfrac{\hat{k}}{v_d b_o^d}$, $e^2 = \dfrac{\hat{e}^2}{v_d b_o^d}$, $\beta = \sqrt{\dfrac{16\pi k}{d(d+2)}}$, v_d is the volume of S^d for $b = 1$, with $\psi = \beta^{-1} \ln (b/b_0)$ and $\chi = \sqrt{v_d b_o^d}\,\hat{\chi}$ denoting the dilaton and the inflaton fields respectively. In (2.17) the dots denote time derivative and $\mathcal{D}_t$ is the covariant derivative with respect to the SO(N-3-d) gauge field $\hat{B}(t)$ in $\mathbb{R}$:

$$\vec{\mathcal{D}}_t f(t) = \frac{d}{dt}\vec{f}(t) + \hat{B}(t)\vec{f}(t) \tag{2.18a}$$
$$\vec{\mathcal{D}}_t g(t) = \frac{d}{dt}\vec{g}(t) + \hat{B}(t)\vec{g}(t), \tag{2.18b}$$

where $\vec{f} = \{f_p\}$, $\vec{g} = \{g_p\}$, $p = 1, \dots, N\text{-}3\text{-}d$ and $\hat{B}$ is the (N-3-d)x(N-3-d) antisymmetric matrix $\hat{B} = (B_{pq})$. The potential W in (2.17) is given by:

$$W = e^{-d\beta\psi}\left\{-e^{-2\beta\psi}\frac{1}{16\pi k}\frac{d(d-1)}{4}\frac{1}{b_o^2} + e^{-4\beta\psi}\frac{1}{b_o^4}\frac{d(d-1)}{8e^2}V_2(\vec{g}) + \frac{\Lambda}{8\pi k} + U(\chi)\right\} +$$
$$e^{-2\beta\psi}\frac{1}{a^2 b_o^2}\frac{3d}{4}\frac{1}{8e^2}(\vec{f}\,\vec{g})^2 + e^{d\beta\psi}\frac{1}{a^4}\frac{3}{4e^2}V_1(f_o, \vec{f}), \tag{2.19}$$

where $\Lambda = v_d b_o^d \hat{\Lambda}$, $U(\chi) = v_d b_o^d\, \hat{U}(\hat{\chi}/\sqrt{v_d b_o^d})$ and

$$V_1(f_o, \vec{f}) = \frac{1}{8}\left[\left(f_o^2 + \vec{f}^{\,2} - 1\right)^2 + 4f_o^2\,\vec{f}^{\,2}\right] , \qquad (2.20a)$$

$$V_2(\vec{g}) = \frac{1}{8}\left(\vec{g}^{\,2} - 1\right)^2 . \qquad (2.20b)$$

Notice that the Lagrangian in (2.17) does not depend on the time derivatives of N and $\hat{B}$. This means that these variables play the rôle of Lagrange multipliers associated with local symmetries of (2.17). The lapse function N is associated with the invariance of S^{eff} with respect to arbitrary time reparametrizations while $\hat{B}$ is related with local SO(N-d-3)-invariance [18]. The equations of motion can be easily obtained by applying the variational principle to (2.17). These equations are identical to the equations that one obtains by substituting the ansätze (2.11) - (2.13) and (2.16) directly in the multidimensional equations of motion (i.e. the ansätze are consistent). In the "gauge" N=1, $\hat{B}$ = 0 one finds the following equations:

(i) Friedmann equation:

$$\left(\frac{\dot{a}}{a}\right)^2 = -\frac{1}{4a^2} + \frac{8\pi k}{3}\left\{\frac{\dot{\psi}^2}{2} + \frac{\dot{\chi}^2}{2} + \frac{1}{8e^2}e^{d\beta\psi}\frac{6}{a^2}\left(\frac{\dot{f}_o^{\,2}}{2} + \frac{\dot{\vec{f}}^{\,2}}{2}\right) + \frac{d}{4e^2}\frac{1}{b_o^2}e^{-2\beta\psi}\frac{\dot{\vec{g}}^{\,2}}{2} + W\right\}, \qquad (2.21)$$

(ii) Klein-Gordon equation for the dilaton field:

$$\ddot{\psi} + 3\frac{\dot{a}}{a}\dot{\psi} = -\frac{\partial W}{\partial\psi} + \beta\,\frac{3d}{4e^2}\frac{1}{a^2}e^{d\beta\psi}\left(\frac{\dot{f}_o^{\,2}}{2} + \frac{\dot{\vec{f}}^{\,2}}{2}\right) - \beta\,\frac{d}{2e^2}\frac{1}{b_o^2}e^{-2\beta\psi}\frac{\dot{\vec{g}}^{\,2}}{2} \qquad (2.22)$$

(iii) Klein-Gordon equation for the inflaton field:

$$\ddot{\chi} + 3\frac{\dot{a}}{a}\dot{\chi} = -e^{-d\beta\psi}\frac{\partial U}{\partial\chi} , \qquad (2.23)$$

(iii) Yang-Mills equations:

$$\ddot{f}_o + \frac{\dot{a}}{a}\dot{f}_o + d\,\dot{\psi}\,\dot{f}_o = -\frac{1}{a^2}\frac{\partial V_1}{\partial f_o} \qquad (2.24a)$$

$$\ddot{\vec{f}} + \frac{\dot{a}}{a}\dot{\vec{f}} + d\,\dot{\psi}\,\dot{\vec{f}} = -\frac{1}{a^2}\frac{\partial V_1}{\partial\vec{f}} - e^{-(d+2)\beta\psi}\frac{1}{4b_o^2}(\vec{f}\,\vec{g})\,\vec{g} \qquad (2.24b)$$

$$\ddot{\vec g} + 3\frac{\dot a}{a}\dot{\vec g} - 2\dot\psi\dot{\vec g} = -\frac{1}{b_o^2}e^{-(d+2)\beta\psi}\frac{d-1}{2}\frac{\partial V_2}{\partial \vec g} - \frac{1}{a^2}\frac{3}{4}(\vec g\,\vec f\,)\,\vec f \qquad (2.24c)$$

$$3b_o^2\, e^{(d+2)\beta\psi}(\dot f_p f_q - \dot f_q f_p) + d\, a^2 (\dot g_p g_q - \dot g_q g_p) = 0 \qquad (2.24d)$$

Let us now study the solutions of equations (2.21) - (2.24) corresponding to static configurations of the gauge and inflaton fields. The following static configuration

$$\chi = \chi^v \quad , \quad f_o = f_o^v \quad ,$$
$$\vec f = \vec f^{\,v} \quad \text{and} \quad \vec g = \vec g^{\,v} \qquad (2.25)$$

is solution of the equations (2.23) and (2.24) if

$$\frac{\partial U}{\partial \chi}\Big|_{\chi = \chi^v} = 0 \quad , \quad \frac{\partial V_1}{\partial f_p}\Big|_{f_o = f_o^v\, ,\ \vec f = \vec f^{\,v}} = 0 \qquad p = 0, \ldots, N-3-d$$
$$\frac{\partial V_2}{\partial g_p}\Big|_{\vec g = \vec g^{\,v}} = 0 \qquad p = 1, \ldots, N-3-d \quad , \qquad (2.26a)$$

and

$$(\vec f^{\,v}\,\vec g^{\,v}) = 0 \ . \qquad (2.26b)$$

Substituting the static configuration (2.25) in (2.21) and (2.22) the equations for a(t) and $\psi(t)$ become

$$\left(\frac{\dot a}{a}\right)^2 = -\frac{1}{4a^2} + \frac{8\pi k}{3}\left\{\frac{\dot\psi^2}{2} + \Omega(\psi, a)\right\} \qquad (2.27)$$

$$\ddot\psi + 3\frac{\dot a}{a}\dot\psi = -\frac{\partial\Omega}{\partial\psi} \ , \qquad (2.28)$$

where

$$\Omega(\psi, a) = e^{-d\beta\psi}\left\{-e^{-2\beta\psi}\frac{1}{16\pi k}\frac{d(d-1)}{4}\frac{1}{b_o^2} + e^{-4\beta\psi}\frac{1}{b_o^4}\frac{d(d-1)}{8e^2}\,\upsilon_2 + \frac{\Lambda}{8\pi k}\right\} +$$
$$e^{d\beta\psi}\frac{1}{a^4}\frac{3}{4e^2}\,\upsilon_1 \ , \qquad (2.29)$$

with $\upsilon_1 = V_1(f_o^v, \vec f^{\,v})$, $\upsilon_2 = V_2(\vec g^{\,v})$ and we have assumed that $U(\chi^v) = 0$.

During inflation the scale factor a(t) of the external dimensions grows exponentially and therefore one can neglect the last term in the effective potential (2.29) for the dilaton. Let us briefly recall this situation which was studied in detail in refs [9] and [16]. If $\upsilon_2 = 0$, as in the case of pure gravity, there are no stable compactifying solutions, i.e. solutions for which $b = b^v \approx \sqrt{16\pi k}$. If $\upsilon_2 > 0$ then the shape of the potential $\Omega_\infty(\psi) = \Omega(\psi, \infty)$ depends on the value of the cosmological constant $\hat{\Lambda}$ of the multidimensional theory [9]. Notice that although in our case the only extremum of $V_2(\vec{g})$ - see (2.20b) - with $\upsilon_2 > 0$ is the unstable local maximum $\vec{g}^v = 0$, models with an absolute minimum of V_2 for which $\upsilon_2 > 0$ can be easily found by considering either non-regular embeddings λ or internal spaces G^{int}/H^{int} with a non-simple isotropy group H^{int}. For $\Lambda > c_2/16\pi k$ ($c_2 = \frac{(d+2)^2(d-1)}{d+4}\frac{e^2}{16\upsilon_2}$) there are no compactifying solutions and for

$$\frac{c_1}{16\pi k} < \Lambda < \frac{c_2}{16\pi k} \tag{2.30}$$

($c_1 = d(d-1)\frac{e^2}{16\upsilon_2}$) a compactifying solution exists which is classically stable but semiclassically unstable. Nevertheless it has been shown [16] that this solution survives the inflationary period without tunneling to the true decompactified vacuum. A value of $\Lambda < \frac{c_1}{16\pi k}$ leads to a negative value of the effective four-dimensional cosmological constant $\Lambda^{(4)}$ (the value of $\Omega_\infty/8\pi k$ at the local minimum). Since the four-dimensional cosmological constant must satisfy the bound

$$|\Lambda^{(4)}| \lesssim 10^{-120}\frac{1}{16\pi k}, \tag{2.31}$$

it then follows that the multidimensional cosmological constant $\hat{\Lambda} = \frac{\Lambda}{v_d b_o^d}$ has to be fine tuned in such a way that

$$|\Lambda^{(4)}| = |v_d b_o^d \hat{\Lambda} - \frac{c_1}{16\pi k}| \lesssim 10^{-120}\frac{1}{16\pi k}. \tag{2.32}$$

In the next section we shall use the analog of eq. (2.28) to study the dynamics of the dilaton field in the radiation dominated period that follows inflation.

3. Stability of compactification after inflation

After the period of accelerated expansion of the external dimensions and the subsequent reheating the Universe became radiation dominated. We shall model this situation by considering that the main contribution to the temperature comes from non-vanishing external-space components of the gauge field $\hat{F}_{\mu\nu}$ ($\mu = 0, 1, 2, 3$) and that therefore (see (2.29) and recall that $a \sim 1/T$) the effective potential for ψ at a non-vanishing temperature becomes:

$$\tilde{\Omega}(\psi, T) = e^{-d\beta\psi} \left\{ \frac{2e^2}{d(d-1)} \frac{1}{\upsilon_2 (16\pi k)^2} (e^{-2\beta\psi} - 1)^2 + \frac{\Lambda^{(4)}}{8\pi k} \right\} + T^4 e^{d\beta\psi} , \quad (3.1)$$

where we have set $b_0^2 = 4\pi k \frac{d(d-1)}{e^2} \upsilon_2$ and have assumed that $\upsilon_2 > 0$. For this choice of b_0 the point $\psi = 0$, i.e. $b = b_0$, is very close to the minimum of (3.1) corresponding to a solution of spontaneous compactification (see (3.2a)). Notice that unlike the zero temperature case ($\tilde{\Omega}(\psi, 0)$) the potential (3.1) has now an infinite potential barrier for large ψ. After inflation one has that $T < \frac{10^{-5}}{\sqrt{16\pi k}}$ and $|\Lambda^{(4)}| << 16\pi k T^4$. In this case the potential (3.1) is of the double well type and it has two minima ψ_- and ψ_+. The first one is very close to zero

$$\psi_- = \frac{\sqrt{d(d+2)}\, d^2(d-1)\, \upsilon_2}{16e^2} (16\pi k)^{3/2}\, T^4 \quad (3.2a)$$

and accordingly

$$\tilde{\Omega}_- = \tilde{\Omega}(\psi_-, T) = T^4 . \quad (3.2b)$$

For ψ_+ we have

$$\psi_+ = \frac{1}{d} \sqrt{\frac{d(d+2)}{16\pi k}} \ln \left(\frac{1}{16\pi k T^2} \sqrt{\frac{2e^2}{d(d-1)}} \right) \quad (3.2c)$$

and

$$\tilde{\Omega}_+ = \tilde{\Omega}(\psi_+, T) = \frac{T^2}{16\pi k} \sqrt{\frac{2e^2}{d(d-1)}} . \quad (3.2d)$$

Therefore since $\tilde{\Omega}_- << \tilde{\Omega}_+$, $\psi = \psi_-$ is for $16\pi k |\Lambda^{(4)}| << (16\pi k)^2 T^4 << 1$ the true ground-state. On the other hand, this corresponds to the situation on which the internal dimensions are compactified, that is:

$$b_- = \sqrt{\frac{4\pi k\, d(d-1)\, \upsilon_2}{e^2}}\ \exp\left(\sqrt{\frac{16\pi k}{d(d-1)}}\ \psi_-\right) \cong \sqrt{\frac{4\pi k\, d(d-1)\, \upsilon_2}{e^2}} \qquad (3.3)$$

The false vacuum $\psi = \psi_+$ corresponds to the situation on which the internal dimensions are decompactified in the limit $T \to 0$:

$$b_+ = \sqrt{\frac{4\pi k\, d(d-1)\, \upsilon_2}{e^2}}\ \exp\left(\sqrt{\frac{16\pi k}{d(d-1)}}\ \psi_+\right) \to \infty \quad . \qquad (3.4)$$

We see then that for $0 < 16\pi k\, \Lambda^{(4)} \ll (16\pi k)^2\, T^4 \ll 1$ (with the corresponding $\hat{\Lambda}$) the decompactified internal dimensions are semiclassically unstable! In this situation even if after inflation the internal dimensions were decompactified, the multidimensional Universe would tend to tunnel to the solution for which the internal dimensions were compactified. This result concerning the metastability of ψ_+ is much stronger than the one related with the semiclassical stability of the compactifying solutions.

Acknowledgments

One of us (Yu. K.) would like to thank GTAE, CFNUL and CFMC (Lisbon) for financial support and Prof. A. Barbosa Henriques and the Gravity Group for the warm hospitality extended to him during his stay in Lisbon where the present paper was prepared.

References

[1] T. Kaluza, Sitzungsber. Preuss. Acad. Wiss. Phys. Math. **K1** (1921) 996;
O. Klein, Z. Phys. **37** (1926) 895.

[2] M.J. Duff, B.E.W. Nilsson and C.N. Pope, Phys. Rep. **130** (1986) 1;
J. Rayski, Acta Phys. Polon., **27**(1965) 947; **28** (1965) 87;
R. Kerner, Ann. Inst. H. Poinc., **A9** (1968) 143;
A. Trautman, Rep. Math. Phys, **1** (1970) 29;
Y.M. Cho, J. Math. Phys. **16** (1975) 2029.

[3] R. Coquereaux and A. Jadczyk, "Riemannian geometry, fiber bundles, Kaluza-Klein theories and all that..." (World Scientific Lecture Notes in Physics, Vol. 16, Singapore, 1988).

[4] N.S.Manton, Nucl. Phys. **B158** (1979) 141;
P. Forgacs and N.S. Manton, Commun. Math Phys. **72** (1980) 15.

[5] I.P Volobujev and G. Rudolph, Theor. Math. Phys. **62**(1985) 261;
I.P. Volobujev and Yu.A. Kubyshin, Theor. Math. Phys **68** (1986) 788; **68** (1986) 885;
Yu.A. Kubyshin, J.M. Mourão and I.P. Volobujev, Int. J. Mod. Phys. **A4** (1989) 151; Theor. Math. Phys. **78** (1989) 41; **78** (1989) 191;
G. Rudolph and I.P. Volobujev, Nucl. Phys. **B313** (1989) 95.

[6] F.A. Bais, K.J. Barnes, P. Forgacs and G. Zoupanos, Nucl. Phys. **B263** (1986) 557;
N.G. Kozimirov and I.I. Tkachev, Z. Phys **C36** (1987) 83;
K. Farakos, G. Koutsoumbas, M. Surridge and G. Zoupanos, Nucl. Phys. **B291** (1987) 128;
K. Farakos, D. Kapetanakis, G. Koutsoumbas and G. Zoupanos, Phys Lett. **211B** (1988) 322.

[7] I.P. Volobujev, Yu.A. Kubyshin, JETP Letters **45** (1987) 581; Theor. Math. Phys. **75** (1988) 509;
Yu.A. Kubyshin, J.M. Mourão, G. Rudolph and I.P. Volobujev, "Dimensional Reduction of Gauge, Theories, Spontaneous Compactification and Model Building (Lecture Notes in Physics Vol. 349, Springer-Verlag, 1989).

[8] M. Surridge, Z. Phys. **C37** (1987) 77;
Yu.A. Kubyshin, J.M. Mourão and I.P. Volobujev, Phys. Lett **203B** (1988) 349; Nucl. Phys. **B322** (1989) 531;

A. Nakamura and K. Shiraishi "Spontaneous compactification and structure of gauge vacua in six-dimensional gauge theory", preprint INS-Rep. 795 (1989).

[9] Yu.A. Kubyshin, V.A. Rubakov and I.I. Tkachev, Int. J. Mod. Phys. **A4** (1989) 1409.

[10] E. Cremmer and J. Scherk, Nucl. Phys. **B118** (1977) 61;
J.F. Luciani, Nucl. Phys. **B135** (1978) 111.

[11] S. Randjbar-Daemi, A. Salam and J. Strathdee, Nucl Phys. **B214** (1983) 491; Phys. Lett. **124B** (1983) 345.

[12] A.N. Schellekens, Nucl. Phys. **B248** (1984) 704; Phys. Lett. **143B** (1984) 121.

[13] P. Forgacs, Z. Horvath and L. Palla, Phys. Lett. **147B** (1984) 311.

[14] P.G.O. Freund, Nucl.Phys. **B209** (1982) 146;
V.A. Rubakov and M.E. Shaposhnikov, Phys. Lett. **125B** (1983) 139;
Q. Shafi and C. Wetterich. Phys. Lett. **129B** (1983) 387; Phys. Lett. **152B** (1985) 51;
E.W. Kolb and R. Slansky, Phys. Lett. **135B** (1984) 378;
A.B. Henriques, Nucl. Phys. **B277** (1986) 621;
A.B. Henriques, A.R. Liddle and R.G. Moorhouse, Nucl. Phys, **B311** (1989) 719.

[15] E.W. Kolb. M.J. Perry and T.P. Walker, Phys. Rev. **D33** (1986) 869;
K. Maeda, Phys. Lett. **186B** (1987) 33.

[16] L. Amendola, E.W. Kolb, M. Litterio and F. Occhionero, Phys. Rev. **D42** (1990) 1944.

[17] G. Clement, Class. Quantum Grav. **5** (1988) 325.

[18] O. Bertolami, J.M. Mourão, R.F. Picken and I.P. Volobujev, "Dynamics of Euclideanized Einstein-Yang-Mills Systems with Arbitrary Gauge Groups" Lisbon Preprint, IFM-9/90, to appear in the Int. J. Mod. Phys. A;
P.V. Moniz and J.M. Mourão, "Homogeneous and isotropic closed cosmologies with a gauge sector" Lisbon Preprint IFM-11/90.

[19] A. Salam and J. Strathdee, Ann. Phys. (N. Y.) **141** (1982) 316.

Multivariate Analysis and Pattern Recognition Methods: A Short Review and Some Current Directions

F. Murtagh[1]
Space Telescope – European Coordinating Facility
European Southern Observatory
Karl-Schwarzschild-Str. 2
D-8046 Garching (Germany)

Email: murtagh@dgaeso51.bitnet

1 Introduction

The version of this article presented at the XII Autumn School in Lisbon in October 1990 differs from the hard-copy version in the following respects. Firstly, in the oral presentation, we begun by commenting on the way in which statistical analysis of data is carried out, and on the increasing importance of dynamic, interactive environments. Secondly, making use of IRAS (Infrared Astronomical Satellite, operational in 1983) Point Source Catalog data, the principal components analysis and k-means partitioning methods were discussed. Thirdly, and finally, some well-known studies such as Bhavsar and Ling (1988) were discussed. Further material on the first point is to be found in Murtagh (1990). Comprehensive background material on the second point is to be found in Murtagh and Heck (1987). Finally, the reference indicated above for the third point should be consulted.

In this article, we aim to overview background aspects of a few important topics in the area of multivariate data analysis. In the case of graph-theoretic methods, these data analysis approaches have been widely used on astronomical data in recent years. Clustering by mode detection, partitioning and hierarchical clustering methods are briefly reviewed.

Approaches based on graph theoretic methods (section 2) and those based on variance minimization (section 3) have different objectives in many cases. The former *may* be more appropriate for detecting filamentary-type structures, whereas the latter *may* be found more effective for the general problem of condensing or summarizing data.

1.1 Bibliography

1. S.P. Bhavsar and E.N. Ling, "Are the filaments real?", *The Astrophysical Journal*, **331**, L63–L68, 1988.

2. F. Murtagh and A. Heck, *Multivariate Data Analysis*, Kluwer, Dordrecht, 1987.

3. F. Murtagh and Ph. Nobelis, "Statistical software", in C. Jaschek and F. Murtagh (Eds.), *Errors, Bias and Uncertainties in Astronomy*, Cambridge University Press, Cambridge (U.K.), 1990, pp. 245–260.

---

[1] Affiliated to Astrophysics Div., Space Science Dept., European Space Agency

2 Graph-Theoretic Methods

2.1 The Problem

Since a graph representation captures the notion of relative separation, it provides a powerful tool for clustering-type problems. A number of such methods will be looked at, together with their applicability for processing point patterns. The latter do not need to be in the plane, and methods described below generalize easily to higher dimensions.

2.2 Terminology

A graph G is a set (V, E) of vertices and edges. The set of edges is often weighted, and one may therefore have

$$E : V \times V \longrightarrow \mathbb{R}^+$$

i.e. E is the set of all pairs (i, j) such that $i, j \in V$, and the associated weight $w(i, j)$ is a positive real. If the graph is *directed*, then nodes and arcs replace vertices and edges.

A *planar graph* is one which can be represented in the plane, without cross-overs of edges. A planar graph can be shown to have less than or equal to $3n - 6$ edges, when n vertices are in question. This provides a bound on such structures as the Delauney triangulation, to be looked at below.

A *path* in a graph is a sequence of distinct edges of the form $(x_1, x_2), (x_2, x_3), \ldots$. A *circuit* is a path whose final vertex is the same as its first. A *Hamiltonian* path, or circuit, is a path or circuit which traverses every vertex once and once only (save for the last vertex in the circuit case). The Hamiltonian path or circuit problem is not easily solvable. The decision as to whether or not a given graph has a Hamiltonian circuit belongs to the difficult class of problems which are known as NP-complete. A polynomial-time algorithm for this problem is therefore not thought to exist. The *travelling salesperson problem* (TSP) is an alternative term for the Hamiltonian circuit problem. A salesperson must visit each and every customer, and not double back on any occasion. The graph of this problem may not be *complete*: there may be less than $n(n-1)/2$ vertices, where $n = \mid V \mid$, the cardinality of the vertex set. There is a relationship between the TSP and the minimal spanning tree (MST), which is looked at in the next section. In fact, as discussed by Garey and Johnson (1979, p. 131), the MST can be used in a clever way to provide an approximate TSP where the total length of the latter is no worse than twice the total weight of the MST.

2.3 Minimal Spanning Tree

A *tree* is a graph which does not contain a circuit: it is therefore a "skeleton" graph. A *spanning tree* is one which includes every vertex. A *minimal spanning tree* (MST) has minimal totalled edge weights, when compared to any other possible spanning tree. The MST has proven to be a useful data structure for picking out plausible clusters of points (vertices) or for detecting outlier points.

Fig. 1 shows aspects of the construction of a MST on five vertices. Read the set of boxes on the left hand-hand side in a top-to-bottom direction, noticing that edges are drawn if the associated weights are less than a threshold, which is being raised from box to box. The vertices are considered as five points in the usual Euclidean plane. Thus the edge weights are identical in this case to Euclidean distances between appropriate points. The closely related data structures known as single linkage hierarchical clustering, which we will return to below, is also shown. The thresholded graphs define *components*, i.e. connected subsets of the set of vertices. The MST can always be determined in $O(n^2)$ time. In fact, for various types of data, and using various speed-up tactics, the MST may be determined in an even more favourable order-of-magnitude time. A number of references for efficient algorithms for the MST are given in Murtagh (1985).

The MST is used for detecting separated clusters of points, or even clusters of different densities by studying the $n-1$ edge weights in the MST. Note that although Fig. 1 showed a 2-dimensional example, the MST can easily be constructed in 3-dimensional space or any arbitrary m-dimensional one. If we find some very large edge weight in the MST, then probably by cutting this edge of the MST, we are left with two clearly separated clusters. Zahn (1981) defines an *inconsistent edge* as one whose weight is significantly greater than the average of nearby edge weights at either extremity. Nearby edges may be defined, e.g., as all edges within path-length 2 of the end-vertices of the examined edge.

An operation on the MST to avoid the effects of anomalous and otherwise noisy points is *pruning*. "Hairs" are removed from the MST by removing edges with the following property: vertices of degree 1 connected to vertices of degree 3 or greater.

If we find that some of the MST's edges are of small weight, and some of large weight, using e.g. a histogram of edge weights, then this signals clusters of differing densities.

Some examples of point patterns where the MST can perform well are shown in Fig. 2.

2.4 Other Graph Methods for Analyzing Point Patterns

The MST is a robust and flexible method which does not require excessive computation time or storage space. Other data structures have also been proposed for the analysis of point patterns.

The *Relative Neighborhood Graph* (RNG) is defined as follows. Points i and j are relative neighbors if

$$d(i,j) \leq \max\{d(i,k), d(j,k)\}$$

for all $k = 1,\ldots,n, k \neq i,j$. Points i and j are therefore at least as close to one another as they are to any other point. This implies that relative neighbors are such that their *lune* (see Fig. 3) is empty. The RNG is the graph for which edges are defined between relative neighbors. The RNG results in a greater number of edges than does the MST.

Another graph structure is the *Gabriel Graph* (GG) or least-squares adjacency graph. An edge in graph GG exists whenever

$$d^2(i,j) \leq d^2(i,k) + d^2(j,k)$$

for $k \neq i,j$. The region of exclusion, here, is defined by a disk (Fig. 3).

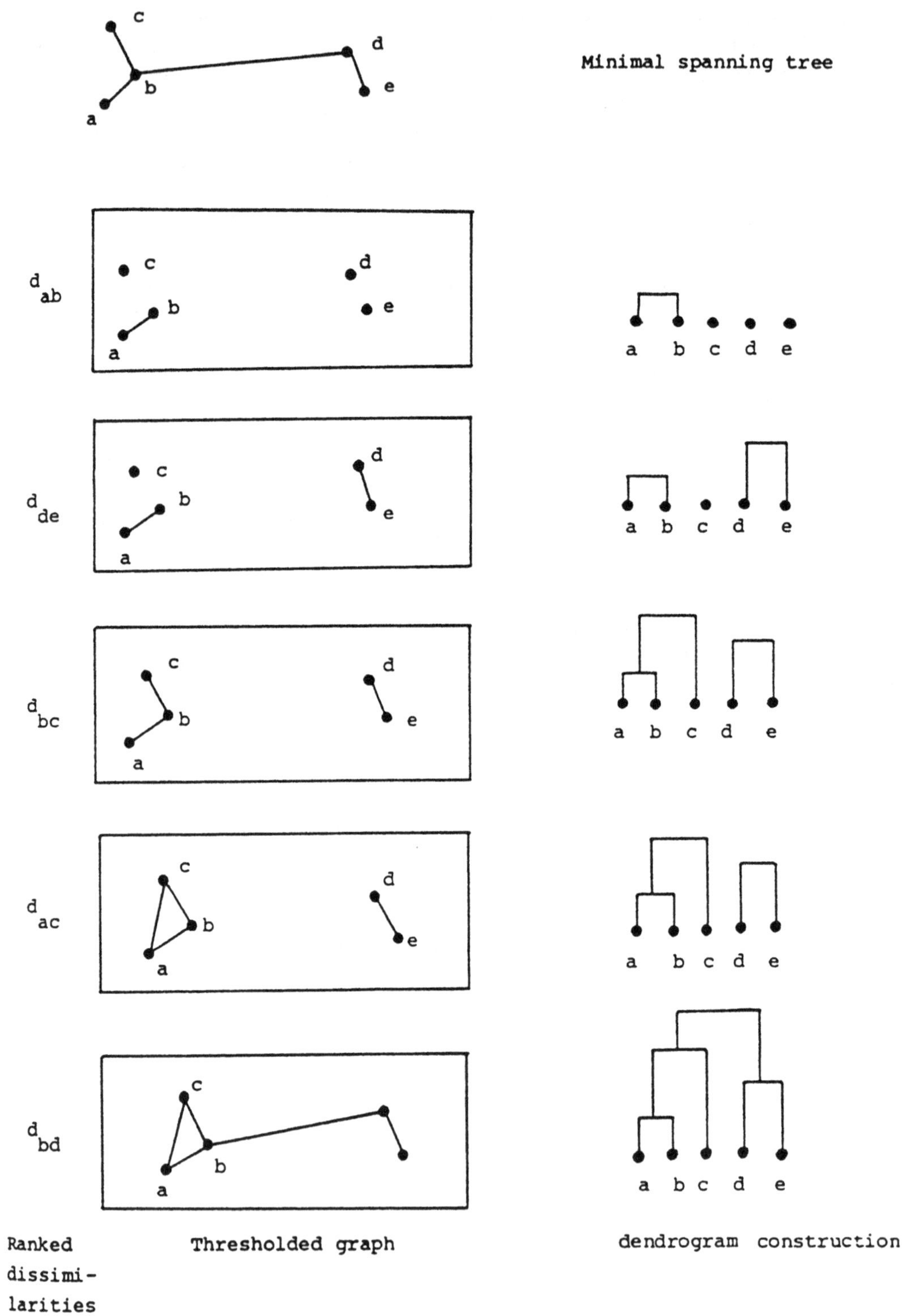

Figure 1: The minimal spanning tree and single link hierarchy.

Well-separated, compact groups.

Elongated clusters.

Linked globular groups.

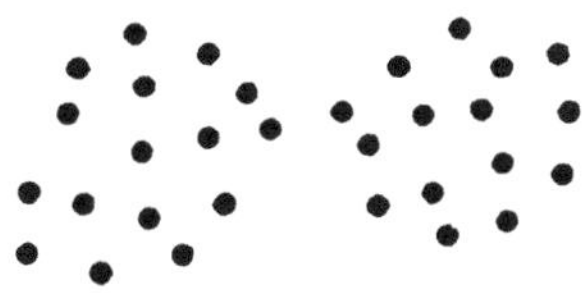

Touching globular groups.

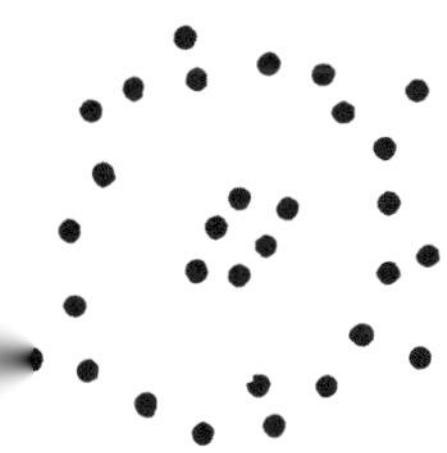

Concentric groups.

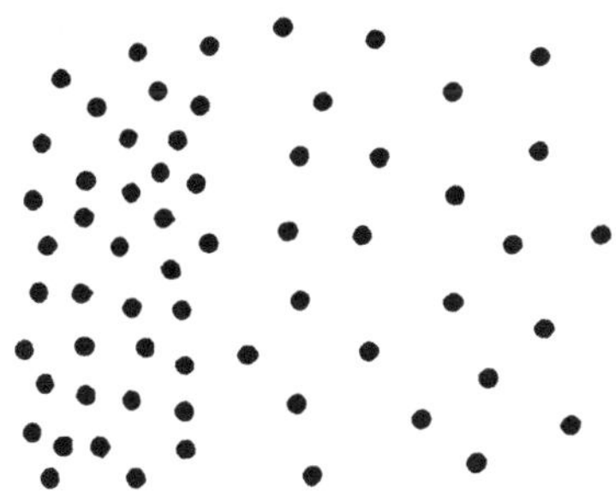

Groups characterised by differing densities.

Figure 2: Point patterns which the MST can aid in analyzing.

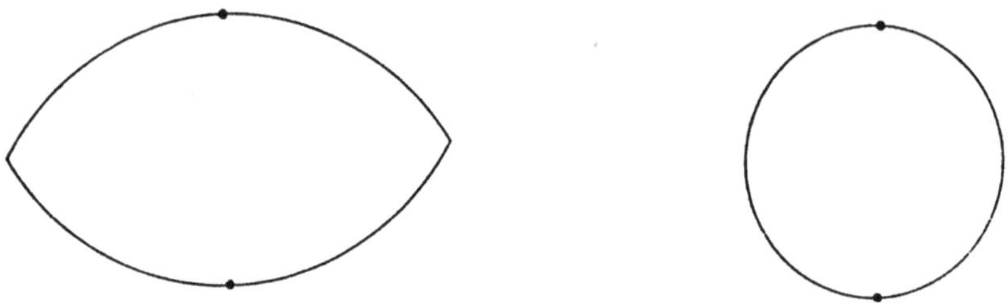

Figure 3: A pair of relative neighbors, indicating their *lune*; and a vertex-pair in a Gabriel Graph indicating their *disk*.

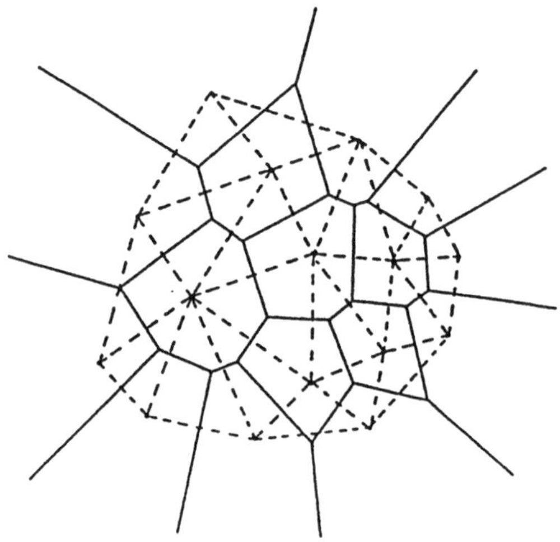

Figure 4: The complete lines show the Voronoi diagram of 16 points; the hatched lines show the Delauney triangulation.

If DT is the Delauney triangulation (to be looked at below), we have the following result:

$$MST \subseteq RNG \subseteq GG \subseteq DT$$

i.e. an edge in the MST is necessarily an edge in the RNG of the same graph, but the reverse does not need to be valid; and such a relationship holds for the other graph structures also.

The *Voronoi diagram* or *Dirichlet tesselation* is defined as follows. Consider the convex polygonal region

$$\{x \mid d(x,i) < d(x,j)\} \quad \text{for all points } i \neq j$$

where i and j belong to a finite set of n points in the plane or in a space of higher dimension. Such polygons may be bounded or may extend to infinity (Fig. 4). They are constructed from the intersection of the perpendicular bisectors of the line joining point i with point j. The polygon could be described as the "territory" of point i: it is the part of the plane nearer to i than to any other point j.

A Voronoi diagram on n points has at most $2n-5$ vertices and $3n-6$ edges. The *dual* of the Voronoi diagram is known as the *Delauney triangulation* (Fig. 4).

Interest in the Voronoi diagram has arisen from the possible meaning which can be attributed to the polygons surrounding each point. In ecology, animal territorial rights might prevail. The Voronoi diagram has also been used as a model encompassing information relating to pancakes, filaments, voids and clusters of galaxies. It has been used, in an interesting fashion, as a basic framework within which the basic laws of galaxy evolution take place.

The Voronoi diagram has been generalized in various ways: to metrics other than the Euclidean one; to the Voronoi diagram of circles and lines rather than points; k-order Voronoi diagrams are based on k-nearest neighbors; weighted Voronoi diagrams define tiles in terms of weighted distances, i.e. the regions of influence defined by the polygons are not due to the same "expansion factors" in all tiles (Aurenhammer and Edelsbrunner, 1984); and finally the problem of Voronoi diagrams in 3-dimensional space ("Voronoi foam") and higher dimensions has been tackled. These areas are briefly surveyed in Lee and Preparata (1984). For computational issues in 2-dimensional space, see Preparata and Shamos (1985).

2.5 Bibliography

1. Graph theory: e.g. A. Tucker, *Applied Combinatorics*, Wiley, New York, 1980.

2. Minimal spanning tree: the seminal article is by C.T. Zahn, "Graph-theoretical methods for detecting and describing Gestalt clusters", *IEEE Transactions on Computers*, **C-20**, 68–86, 1971.

   V. Di Gesù and C. Zahn, "Experiments with a general dotted curve finder", SLAC Computation Research Group technical report CGTM No. 161, Oct. 1974.

   J.D. Barrow, S.P. Bhavsar and D.H. Sonoda, "Minimal spanning trees, filaments and galaxy clustering", *Monthly Notices of the Royal Astronomical Society*, **216**, 17–35, 1985.

   S.P. Bhavsar and E.N. Ling, "Are the filaments real?", *The Astrophysical Journal*, **331**, L63–L68, 1988.

   M.R. Garey and D.S. Johnson, *Computers and Intractability. A Guide to the Theory of NP-Completeness*, W.H. Freeman, San Francisco, 1979.

   F. Murtagh, *Multidimensional Clustering Algorithms*, COMPSTAT Lectures Volume 4, Physica-Verlag, Vienna and Würzburg, 1985.

3. Voronoi diagrams: V. Icke and R. van de Weygaert, "Fragmenting the universe. I. Statistics of two-dimensional Voronoi foams", *Astronomy and Astrophysics*, **184**, 16–32, 1987.

   V. Icke and R. van de Weygaert, "Voronoi cosmology", *Quarterly Journal of the Royal Astronomical Society*, 1990 (in press).

   F.P. Preparata and M.I. Shamos, *Computational Geometry*, Springer-Verlag, New York, 1985.

D.T. Lee and R.P. Preparata, "Computational geometry – a survey", *IEEE Transactions on Computers*, **C-33**, 1072–1101, 1984.

N. Ahuja, "Dot pattern processing using Voronoi neighborhoods", *IEEE Transactions on Pattern Analysis and Machine Intelligence*, **PAMI-4**, 336–343, 1982.

S.K. Bhaskar, A. Rosenfeld and A. Wu, "Models for neighbor dependency in planar point patterns", *Pattern Recognition*, **22**, 533–559, 1989.

F. Aurenhammer and H. Edelsbrunner, "An optimal algorithm for constructing the weighted Voronoi diagram in the plane", *Pattern Recognition*, **17**, 251–257, 1984.

4. Other graph-theoretic structures:

R. Urquhart, "Graph theoretical clustering based on limited neighborhood sets", *Pattern Recognition* **15**, 173–187, 1982; "Erratum", *Pattern Recognition* **15**, 427, 1982.

D.W. Matula and R.R. Sokal, "Properties of Gabriel Graphs relevant to geographic variation research and the clustering of points in the plane", *Geographical Analysis*, **12**, 205–222, 1980.

3 Clustering and Mode-Seeking Methods

3.1 The Problem

The number of ways of assigning n objects to g clusters is known in combinatorics as a Stirling number of the second kind. Since it grows rapidly, exhaustive enumeration is ruled out for any sizable value of n. Thus, other strategies must be used in order to get a partition which satisfies some criterion. Heuristics which produce suboptimal solutions, such as the minimum distance or exchange methods described below, work well in practice.

3.2 Partitioning

Given that variance expresses compactness, a possible criterion to be optimized by a partitioning method is

$$\min_P \sum_{\mathbf{p} \in P} \sum_{\mathbf{i} \in \mathbf{p}} d^2(\mathbf{i}, \mathbf{p}).$$

Here, $\mathbf{p}$ is a cluster of partition P, represented by its centre of gravity or mean vector; i is one of the set of n objects considered, and, as shorthand notation, if $\mathbf{i} \in \mathbf{p}$ then object $\mathbf{i}$ belongs to the cluster represented by $\mathbf{p}$. To avoid the trivial result of each cluster being a singleton, we impose the constraint of a set number of clusters, k. This type of algorithm is referred to as *k-means*. Numerous implementation details give rise to variants of k-means, but the broad lines of such algorithms will be sketched out here.

One approach to this problem is to carry out $n - k$ agglomerations in a hierarchical clustering. An alternative approach uses iterative refinement:

Arbitrarily define a set of k cluster centers. Assign each object to the cluster centre to which it is closest. Redefine cluster centers on the basis of the current cluster memberships.

*If the totalled within-class variances (*see expression above*) is smaller than previously, then reiterate.*

We have omitted mention of a test for convergence. "Closeness" may be defined using the Euclidean, or other, distance. The initial cluster centers may be chosen by averaging some object-vectors; or on the basis of prior knowledge of the data; or as random vectors. Each iteration of the above algorithm may be studied with regard to time and space requirements, but the number of iterations is indeterminate. Nonetheless, even simple iterative algorithms of this sort usually have good convergence properties. The criterion which we set out to minimize may not in fact be minimized: the minimum distance algorithm briefly described above is a sufficient but not a necessary prescription for an optimal partition. Thus, given this suboptimality, the above algorithm would normally be run a number of times over in order to select the best outcome. The starting configuration would be varied, and the best result selected as a probable near-to-optimal outcome.

An alternative algorithm is based on the *exchange method*:

Arbitrarily choose an initial partition. For each i, *see if the criterion to be minimized is bettered by relocating* i *to another cluster. If this is the case, relocate* i *to the cluster which allows the largest decrease in criterion to be brought about. When all objects have been considered, reiterate if further decreases in the criterion are possible.*

With partitioning methods of the sort described, one must know the number of clusters beforehand. For large datasets, there may be little alternative to using such methods. Using solar seismology data (courtesy of M. Fofi, Rome), the clustering of 9 dopplergram-related values for 16384 pixels took about 8 hours CPU time on a VAX 8600 using Ward's minimum variance method. For a run using a partitioning method on the same machine, just over 8 minutes were required.

3.3 Mode Detection

In mode seeking, it is attempted to find dense regions of parameter space, or peaks in intensity. The latter case arises in image processing. In the former case, we define vertex weights in addition to given or derivable edge weights. The edge weights would often be Euclidean distances between points. The vertex weights may be defined, for example, in the following ways:

1. $w(i) = \mid \{j \mid d(i,j) \leq r\} \mid$ i.e. the weight of point i is the number of other points j within a set radius of i.

2. $w(i) = \frac{1}{k} \sum_j d(i,j)$ where j is a member of the set of k nearest neighbours of i.

3. $w(i) = \sum_j e^{(-d^2(i,j)/W^2)}$ for an appropriate normalizing constant, W.

4. In image processing, we may consider intensities or multispectral vectors of intensity, f_i, at i.

A possible mode seeking algorithm would implement a 'tree climbing' strategy by forming a directed link – an arc – between a vertex and a neighboring vertex with the steepest gradient. 'Neighboring' has a straightforward definition if a radial threshold is used; or if the 4 or 8 nearest pixels are used in a digitized image. When the tree is complete, directed paths can be traced out to find the modal vertices. A change of direction along a path indicates a valley region between two or more modes. It is clear, therefore, that the directed trees which are the data structures produced by this general class of algorithms are amenable to the same type of processing as was already seen for the MST.

Note that contiguity can, if desired, play a central role in this type of clustering. In addition to computational efficiency, mode seeking is therefore also particularly appropriate for many types of spatial data.

3.4 Hierarchical Clustering Methods

Hierarchical clustering constitutes an important and widely-used class of methods. A *dendrogram*, or tree diagram, of a set of objects is constructed. The n objects are specified, each, by an m-dimensional vector. By means of a distance (e.g. the Euclidean distance), the following algorithm requires precisely $n-1$ *agglomerations* in total:

Find the two least-distant objects; agglomerate them by taking these two objects out of the list of objects, and by substituting them with a cluster vector. Ensure that this cluster vector is now in the list of objects. Determine the next two least-distant objects. Continue, as before, until only one object remains.

We have ignored ties, but these can usually be arbitrarily handled. More importantly, we have not specified how distance was defined between a newly-created cluster (i.e. agglomerated objects) and the previously-existing objects. In fact, this might not be a distance at all, but rather a *dissimilarity.* A distance satisfies the following axioms, for distinct objects i, j and k (which usually are associated with vectors in $\mathbb{R}^m$):

$d(i,j) = d(j,i)$ (symmetry)

$d(i,j) \geq 0;\ d(i,i) = 0$ (positive definiteness)

$d(i,j) \leq d(i,k) + d(k,j)$ (triangular inequality)

If the last property, the triangular inequality, is not satisfied, then the measure of (dis)association is a *dissimilarity.* The triangular inequality has a transparent meaning in Euclidean space: one cannot go from i to j more quickly through any intermediary k (different from i, j).

The second oversight above – that of defining singleton-cluster and cluster-cluster dissimilarity – is at the heart of the half-dozen widely-used hierarchical clustering methods. Consider for the moment that i, j, and k are singletons, and that the "union" symbol defines an agglomeration. The most popular hierarchical methods include those based on the following definitions.

- $d(i \cup j, k) = \min\ \{d(i,k), d(j,k)\}$ (the single linkage method);
- $d(i \cup j, k) = \max\ \{d(i,k), d(j,k)\}$ (the complete linkage method);
- $d(i \cup j, k) = \alpha_i d(i,k) + \alpha_j d(j,k) + \beta d(i,j) + \gamma \mid d(i,k) - d(j,k) \mid$ (a general formula using parameters α, β and γ which give rise to a particular method; this is the Lance-Williams formula).

In the Lance-Williams formula, it may be verified (by pencil and paper, if necessary) that the single link method is none other than: $\alpha_i = \alpha_j = \frac{1}{2}$, $\beta = 0$, and $\gamma = -\frac{1}{2}$. The complete link method uses the same values, save for the value of γ which is positive.

The single link method gives rise to clusters which are connected components, or simply *components*, of the graph of vertices and edges. The complete link method gives rise to clusters which are *complete subgraphs*, also known as *cliques*.

Ward's minimum variance method defines

$$\alpha_i = \frac{n_i + n_k}{n_i + n_j + n_k} \qquad \beta = \frac{n_k}{n_i + n_j + n_k} \qquad \gamma = 0$$

where n_i is the cardinality of cluster i. The single and complete linkage methods have a clear interpretation, but the minimum variance method offers a clearer definition of the center of any cluster. It is: $(n_i\mathbf{g}_i + n_j\mathbf{g}_j)/(n_i + n_j)$ where $\mathbf{g}_i$ is the mean of cluster i. The dissimilarity between cluster centers is: $(n_in_j)/(n_i+n_j)d^2(\mathbf{g}_i, \mathbf{g}_j)$, where d is the Euclidean distance.

It can be shown that the minimum variance method simultaneously aims at compact clusters which are as separated as possible. On the other hand, the single link method frequently gives rise to the "friends of friends" syndrome: i is close to j, j is close to k, therefore all three find themselves in one cluster, even though i might not be close to k. The complete link method often gives rise to properties which are fairly similar to the minimum variance method.

Consider a dendrogram constructed by any means, and consider points i, j and k. Let us define "dendrogram distances" as follows. $\delta(i,j)$ is the value of the criterion which is stepwise optimized in constructing the hierarchy. That is to say, it is the value resulting from the application of the Lance-Williams formula, with appropriate parameters, at a particular stage. Such distances derived from a dendrogram satisfy an even stronger condition than the triangular inequality. This condition is the *ultrametric inequality*:

$$\delta(i,j) \leq \min\{\delta(i,k), \delta(j,k)\}.$$

It can be shown that the single link method produces the *maximal inferior ultrametric*: this means that the dendrogram distances, δ, yielded by this method are as close as possible, but always less than or equal, to the given set of initial distances. The complete link method, on the other hand, produces one possible set of *minimal superior ultrametrics*.

The assessment of clustering results may be carried out using discriminant analysis methods. These sometimes are referred to as, respectively, unsupervised and supervised classification. Fisher's linear discriminant attempts to find the best linear separation between two predefined classes in a multidimensional space. Multiple discriminant analysis generalizes this by seeking new coordinate axes which optimally discriminate between two or more predefined classes. Using Bayesian principles, and assuming Gaussian properties for the classes, can lead to linear, quadratic or more general forms of discriminating surfaces. Finally, non-parametric forms of discrimination such as the k-nearest neighbors approach offer very flexible decision surfaces. Further aspects of discrimination methods may be found in Murtagh and Heck (1987) or Hand (1981).

3.5 Bibliography

1. Partitioning: H. Späth, *Cluster Dissection and Analysis: Theory, Fortran Programs, Examples*, Ellis Horwood, Chichester (U.K.), 1985.

2. Mode detection: F. Murtagh, "Algorithms for contiguity-constrained clustering", *The Computer Journal*, **28**, 82–88, 1985.

3. Hierarchical clustering: F. Murtagh, *Multidimensional Clustering Algorithms*, Physica-Verlag, Würzburg and Vienna, 1985.

   F. Murtagh and A. Heck, *Multivariate Data Analysis*, Kluwer Academic Publishers, Dordrecht, 1987.

4. Discriminant analysis: F. Murtagh and A. Heck (1987), reference above.

   D.J. Hand, *Discrimination and Classification*, Wiley, New York, 1981.

4 Open Problems: Messy Data

"Messy" data, as it has been characterized, may have quite varied causes. Mixed quantitative/qualitative data, missing data, data with errors, and so on, are usually considered under this heading. It is common practice in the physical sciences to consider error estimates with measured quantities. Methods for exploring data do not always take such errors into account. There is now general realization, for instance, that ordinary linear regression produces false results if both (or all) variables have error estimates (see references in Murtagh, 1990). The frequent availability of data with upper limits has also been extensively discussed in Feigelson (1990) and other articles. Such censoring may arise for example in the case of known sources where certain flux measurements cannot be obtained because they are below the background noise level, or they are otherwise obscured. What can be measured is some value below which is the real value.

IRAS PSC data indicates well that mixtures of error estimates need to be considered in practice. Gaussian errors of different qualities (good and moderate) are present with censored observations. The enhancement of currently used methods to accomodate such data is an open problem at the present time.

4.1 Bibliography

1. F. Murtagh, "Linear regression with errors in both variables: a short review", in C. Jaschek and F. Murtagh (Eds.), *Errors, Bias and Uncertainties in Astronomy*, Cambridge University Press, Cambridge (U.K.), 1990, pp. 385–391.

2. E.D. Feigelson, "Censored data in astronomy", in C. Jaschek and F. Murtagh (Eds.), *Errors, Bias and Uncertainties in Astronomy*, Cambridge University Press, Cambridge (U.K.), 1990, pp. 213–226.

3. S.P. Bhavsar, "Bootstrap, data permuting and extreme value distributions – getting the most out of small samples", in C. Jaschek and F. Murtagh (Eds.), *Errors, Bias and Uncertainties in Astronomy*, Cambridge University Press, Cambridge (U.K.), 1990, pp. 107–122.

COOLING OF NEUTRON STARS

C. J. Pethick

Nordita, Blegdamsvej 17, DK-2100 Copenhagen Ø, Denmark
and
Department of Physics, University of Illinois at Urbana-Champaign,
1110 West Green Street, Urbana, IL 61801, U. S. A.

ABSTRACT. A review is given of the important neutrino emission processes in neutron star interiors, and their consequences for the cooling of neutron stars.

1. Introduction

The subject of neutron stars is now a vast one, which it is difficult to survey briefly. In this article a discussion is given of various aspects of the cooling of neutron stars, a subject where there have been a number of recent developments. Emphasis will be placed on the basic physical processes, rather than on detailed numerical estimates. For supplementary reading, the reader is referred to an extended treatment of various aspects of neutron stars in the book by Shapiro and Teukolsky (1983), and to a recent review by Baym (1991) of the equation of state of matter at high densities and the properties of neutron stars. A review of matter at subnuclear densities is given by Pethick and Ravenhall (1991a).

One of the ways in which it is hoped to obtain more information about the internal constitution of neutron stars is by study of thermal radiation from their surfaces. In the early life of a neutron star, the main mechanism for energy loss is emission of neutrinos from the interior. Neutrinos at energies less than 10 MeV can escape from a neutron star in a time of order seconds or less, and they are therefore very effective in removing energy despite the fact that neutrinos are generated by weak interaction processes. Later in the life of a neutron star, the neutrino generation rate falls as the temperature falls, and the primary mechanism for heat loss is thermal emission from the stellar surface.

Neutrino emission rates depend on the properties of dense matter, and therefore measurements of the surface emission hold out the promise of enabling one to distinguish between a number of possible states of dense matter. The simplest neutrino-emitting processes one can envisage are beta decay of the neutron

$$n \rightarrow p + e^- + \bar{\nu}_e \tag{1}$$

and electron capture on protons,

$$p + e^- \rightarrow n + \nu_e. \tag{2}$$

In beta equilibrium, the neutron, proton, and electron chemical potentials, μ_n, μ_p, and μ_e satisfy the condition

$$\mu_n = \mu_p + \mu_e, \tag{3}$$

which corresponds physically to the requirement that it cost no energy to convert a neutron into a proton plus an electron, or vice versa. When this condition is satisfied, there is, at zero temperature, no phase space available for neutrinos and antineutrinos in the final states of reactions (1) and (2). For most of the life of a neutron star, temperatures in the interior are much less than the Fermi temperatures of the constituents, typically of order 100 MeV in energy units, which corresponds to a temperature of order $10^{12}K$. Matter is therefore degenerate, which implies that particles participating in reactions (1) and (2) must have energies which lie within $\sim k_BT$ of their respective Fermi energies.

In the mid 1960's, when cooling of neutron stars was first studied in detail, it was argued that the reactions (1) and (2) could not occur. The reasoning for this was that the neutrons, protons and electrons participating in these processes must have momenta close to their respective Fermi momenta, which we denote by p_{Fn}, p_{Fp} and p_{Fe} respectively, and therefore, since the neutrino momentum is of order k_BT/c, which is small compared with the Fermi momenta of the other participating particles, for momentum to be conserved in the reactions, the condition

$$p_{Fe} + p_{Fp} \geq p_{Fn} \tag{4}$$

must be satisfied. Since the density of particles of species i is given by

$$n_i = p_{Fi}^3/3\pi^2\hbar^3, \tag{5}$$

and in dense matter the proton fraction is typically of order a few percent, it was argued that momentum could not be conserved. We now reexamine this conclusion in the light of our current knowledge of nuclear physics.

2. The Direct Urca Process

First, we estimate the minimum proton fraction for the reactions (1) and (2), which we refer to collectively as the direct Urca process, to proceed. (The discussion in this section follows closely that of Lattimer et al., 1991.) If matter consists only of neutrons, protons and electrons, the condition for charge neutrality is $n_p = n_e$, or $p_{Fp} = p_{Fe}$. Thus the threshold condition (4) becomes

$$p_{Fp} \geq p_{Fn}/2, \tag{6}$$

or

$$n_p \geq n_n/8. \tag{7}$$

Thus the proton fraction $x = n_p/n$, where $n = (n_n + n_p)$ is the total baryon density, is given by

$$x \geq \frac{1}{9} \approx 11.1\%.$$

If the electron chemical potential exceeds the muon rest mass, $m_\mu c^2 \approx 105.7\ MeV$, muons will also be present in dense matter, and this will increase the threshold proton concentration. If $\mu_e >> m_\mu c^2$, the threshold proton concentration is ≈ 0.148, and for smaller values of μ_e, the threshold concentration lies between $\frac{1}{9}$ and this value. At densities typical of the central regions of neutron stars, the calculated proton concentration of matter is very sensitive to the choice of physical model, and for many physically plausible models it exceeds the threshold value, as Lattimer *et al.* discuss. However, it is not possible to say with certainty whether or not the direct Urca process can occur because our present knowledge of interactions between nucleons at high density is insufficiently precise to allow one to estimate proton concentrations accurately. As calculations by Wiringa, Fiks and Fabrocini (1988) demonstrate, a major source of uncertainty is lack of knowledge of the three-nucleon interaction, and, in particular, of its isospin dependence. Various aspects of the nuclear physics of matter at densities in excess of that of nuclear matter are discussed by Pethick and Ravenhall (1991b).

If the proton concentration exceeds the threshold value for the reaction (1), the rate of emission of antineutrino energy per unit volume may be estimated using Fermi's Golden Rule, and it is given by

$$L_\beta = \frac{2\pi}{\hbar} 2 \sum G_F^2 cos^2\theta_C (1 + 3g_A^2) n_1 (1 - n_2)(1 - n_3) \epsilon_4 \delta^{(4)}(p_1 - p_2 - p_3 - p_4), \tag{8}$$

where n_i is the Fermi function, and the subscripts $i = 1 - 4$ refer to the neutron, proton, electron and antineutrino, respectively. The p_i are four momenta, and ϵ_4 is the antineutrino energy. The sum over states is to be performed only over three-momenta $\vec{p}_i$, and the prefactor 2 takes into account the initial spin states of the neutron. The beta decay matrix element, squared and summed over spins of final particles and averaged over angles is $G_F^2 cos^2\theta_C(1 + 3g_A^2)$, where $G_F \approx 1.436 \times 10^{-49} erg\ cm^3$ is the weak coupling constant, θ_C is the Cabibbo angle and $g_A \approx -1.261$ is the axial vector coupling constant. Final electron and proton states must be vacant if the reaction is to occur, and this accounts for the blocking factors $1 - n_2$ and $1 - n_3$. The electron-capture process (2) gives the same luminosity as process (1), but in neutrinos, and therefore the total luminosity of the direct Urca process is twice Eq. (8). The integrals may be calculated straightforwardly, since the neutrons, protons and electrons are very degenerate, and one finds

$$\begin{aligned} L_{Urca} &= \frac{457\pi}{10080} \frac{G_F^2 cos^2\theta_C (1 + 3g_A^2)}{\hbar^{10} c^5} m_n\, m_p\, \mu_e (k_B T)^6\ \Theta_t \\ &\approx 4 \times 10^{27} (Y_e n/n_s)^{1/3}\ T_9^6\ \Theta_t\ erg\ cm^{-3}\ s^{-1}, \end{aligned} \tag{9}$$

where $n_s = 0.16 fm^{-3}$ is the density of nuclear matter. Here Θ_t is the threshold factor $\Theta(p_e + p_p - p_n)$, which is +1 if the argument exceeds zero, and is zero otherwise. T_9 is the temperature measured in units of $10^9 K$.

Particle interaction effects change this result in a number of ways. First, the neutron and proton densities of states are determined by effective masses rather than bare masses. Second, the effective weak interaction matrix elements can be modified by the medium. However, these effects are unlikely to reduce the luminosity by more than a factor of 10.

The temperature dependence of the direct Urca emissivity may easily be understood from phase space considerations. The neutrino or antineutrino momentum is $\sim k_B T/c$, and thus the phase space available in final states is proportional to $(k_B T/c)^3$. The participating neutrinos, protons and electrons are degenerate, and therefore for the reaction to occur they must have energies that lie within $\sim k_B T$ of the energies at the Fermi surfaces, and thus each degenerate particle contributes a factor $\sim k_B T$. (The fact that only 3 of the 4 particle energies are independent might be expected to reduce the powers of T by one, but this effect is compensated by the fact that we are interested in the rate of emission of neutrino energy).

Whether or not the direct Urca process occurs in reality, the above calculation is an instructive one, since neutrino emissivity from most of the other neutrino emission processes considered in dense matter may be understood in terms of it.

3. The Modified Urca Process

Until 1990, it was tacitly assumed that proton concentrations in dense matter lie below the threshold for the direct Urca process. In this case, the simplest weak interaction processes that can occur are (1) and (2) modified by the addition in initial and final states of a nucleon, whose sole purpose is to enable the condition of momentum conservation to be satisfied. The processes are thus

$$n + n \rightarrow n + p + e + \bar{\nu}_e \tag{10}$$

and

$$n + p + e \rightarrow n + n + \nu_e. \tag{11}$$

These were first discussed by Chiu and Salpeter (1964), and detailed estimates of their rates have subsequently been made by Finzi (1964), Bahcall and Wolf (1965), and Friman and Maxwell (1977).

The process is basically the process (1) and (2), with the modification that a nucleon in an initial or final state interacts, via the nucleon-nucleon strong interaction, with a bystander nucleon. There is a total of 5 degenerate particles participating in these processes, and therefore, on the basis of the phase space arguments given above, one would expect the emissivity to vary as T^8, a conclusion borne out by the detailed calculations.

The magnitude of the modified Urca luminosity may be estimated in terms of that for the direct Urca process. The lowest order matrix element for the modified Urca process has an extra strong interaction matrix element, V and an extra energy denominator (typically of order of a Fermi energy) compared with the usual weak interaction matrix element, and their ratio is thus $\sim V/E_F$. The additional neutrons in the initial and final states each contribute a factor proportional to the number of states accessible, of order the density of states at the Fermi energy, $\sim n_n/E_F$, times the thermal energy. Thus the modified Urca luminosity is of order $(V/E_F)^2(n_n k_B T/E_F)^2$ times that of the direct Urca process. Since neutron matter is a strongly interacting system, a typical neutron potential energy, $n_n V$, is comparable to the Fermi energy, and therefore the modified Urca rate is of order $(k_B T/E_F)^2$ times the characteristic direct Urca rate. Since Fermi energies are of order 100 MeV, which corresponds to a temperature of order $10^{12}K$, this factor is of order $10^{-6}T_9^2$, an estimate in good agreement with what is found from detailed calculations.

The time for the core of a neutron star to cool to a given temperature may be estimated by equating the rate of loss of thermal energy to the neutrino luminosity. The thermal energy is of order $n(k_B T)^2/E_F$, and one finds the characteristic times τ, are given by

$$\tau_{Urca} \sim \frac{1\ minute}{T_9^4}, \tag{12}$$

and

$$\tau_{modUrca} \sim \frac{1\ year}{T_9^6}. \tag{13}$$

Thus if the direct Urca process cannot occur, the core temperature will exceed $\sim 10^8 K$ for $\sim 10^6$ yrs, while if the direct Urca process can occur, the temperature will be that high only for about a week.

4. Pion Condensation

In 1965 Bahcall and Wolf suggested that the presence of a Bose condensation of pions in dense matter could lead to enhanced neutrino emission compared with the modified Urca process, which has been regarded as the "standard" process for neutrino emission in the intervening quarter of a century. In the presence of a pion condensate, the nuclear excitations become a mixture of neutrons and protons, which we denote by f. It is therefore possible for the processes

$$f \to f + e^- + \bar{\nu}_e \tag{14}$$

and

$$f + e^- \to f + \nu_e \tag{15}$$

to occur. Excitations which have energies within $\sim k_BT$ of the neutron and proton Fermi energies can participate in the reactions, and these can have momenta of magnitude comparable to the neutron Fermi momenta. Since the f quasiparticles in the initial and final states may both have momenta of order the neutron Fermi momenta, momentum conservation considerations do not prohibit the reactions, as they do for the case of the direct Urca process for nucleons, Eqs.(1-2). More detailed studies of pion condensation carried out in the 1970's indicate that it will occur, if it does at all, at a finite wavelength. However the finite momentum imparted to nucleons by scattering from the pion condensate may be shown not to alter the conclusion that the reactions (14-15) can occur.

The rate of the process (14-15) may be calculated by arguments similar to these for the nucleon direct Urca process, except that the weak interaction matrix element that enters is that between two f quasiparticles, rather than between a neutron and a proton, and the phase space is different because the momentum of the condensate, k, enters the momentum conservation condition (Maxwell et al., 1977). The strength of the pion condensate is measured by an angle θ_π, and for small θ_π one finds that the square of the beta decay matrix element summed over spins is $(\theta_\pi^2/4)(1+(g_A k/p_{Fe})^2)$. Since $k > p_{Fe}$, one finds that the phase space is reduced by a factor p_{Fe}/k compared with the rate for the direct Urca process for nucleons, and therefore

$$L_\pi = L_{Urca}(\theta_\pi^2/4)(1+(g_A k/p_{Fe})^2)p_{Fe}/k. \tag{16}$$

For typical estimates of parameters, one finds that the rate of emission of neutrino energy from a pion condensate is more than an order of magnitude less than the direct Urca rate for nucleons. However, unlike the direct Urca process, the Urca process in the presence of a pion condensate will occur for arbitrary proton concentrations. However, more recent studies indicate, that, because of the strong spin-isospin repulsion, pion condensation is unlikely in neutron stars.

5. Kaon Condensation

Another possible ground state of dense matter is a kaon condensate (Kaplan and Nelson, 1986). The $SU(3)\times SU(3)$ symmetry of the strong interactions leads to an attractive interaction between kaons and nucleons proportional to the baryon density, and therefore at sufficiently high density one has the possibility that the kaon energy vanishes, and a Bose condensate of kaons appears. Such a state may be regarded as a strange version of a pion condensate; a pion condensate may be thought of as a pairing of neutrons and proton holes in a state with the quantum numbers of the pion, whereas a kaon condensate may be considered as a pairing of neutrons and Σ^- holes (for a K^+ condensate). In quark language, a pion condensate corresponds to a finite value of $\langle \bar{d}u \rangle$, while a K^+ condensate corresponds to a finite value of $\langle \bar{s}u \rangle$. Here $\langle \ldots \rangle$ denotes an expectation value, and u, d, and s denote the up, down and strange quark fields, respectively. In contrast to pion condensation, which would be expected to occur at a finite wavelength because of the repulsive pion-nucleon s-wave interaction, kaon condensation is expected to occur in a spatially uniform state.

The rate of neutrino emission from a kaon condensate may be calculated in a fashion similar to that for pion condensation. The chief differences are that the condensate is spatially uniform, and the weak interaction matrix elements contain a factor $sin\theta_C$, rather than $cos\theta_C$, because there are strange particles involved in the process. The final result is (Brown, Kubodera, Page and Pizzochero, 1988)

$$L_K = L_{Urca} \frac{{\theta_K}^2}{8} sin^2\theta_C, \tag{17}$$

where θ_K is the kaon condensation angle, analogous to θ_π for pion condensation. This result shows that for $\theta_K^2 \approx 0.1$, the neutrino luminosity of a kaon condensate is about one thousandth of the typical nucleon Urca rate.

6. Quark Matter

Yet another possibility for the state of matter at high densities is quark matter, in which quarks can move around essentially as free particles, rather than being bound together as colour singlet entities, such as nucleons and pions. Neutrino emission from such a system was considered by Iwamoto (1980, 1982), and the basic processes are the quark analogues of the nucleon processes (1-2),

$$d \rightarrow u + e^- + \bar{\nu}_e, \tag{18}$$

and

$$u + e^- \rightarrow d + \nu_e. \tag{19}$$

The condition for beta equilibrium is

$$\mu_d = \mu_u + \mu_e, \tag{20}$$

which is the analogue of Eq.(3) for nucleons. If quarks and electrons are treated as massless non-interacting particles, this condition is

$$p_{Fd}c = p_{Fu}c + p_{Fe}c, \tag{21}$$

which is precisely the threshold condition for the direct Urca process to occur. At threshold the momenta of the u quark, the d quark and the electron must be parallel. However, as Iwamoto pointed out, the weak-interaction matrix element for this case vanishes. If quark-quark interactions are taken into account, the expression for the quark chemical potential contains interaction corrections, which to lowest order in the QCD coupling constant, α, are

$$\mu_i = (1 + \frac{8}{3\pi}\alpha)p_{Fi}c, \quad i = u, d, \tag{22}$$

and the direct Urca process is kinematically allowed for quarks and electrons which are not collinear. The calculation of the neutrino and antineutrino emission rates proceeds in essentially the same way as for the nucleon process, and the overall result for the luminosity is

$$L_q = \frac{914}{315} \frac{G_F^2 cos^2\theta_C}{\hbar^{10} c^7} \alpha\, p_{Fd}\, p_{Fu}\, \mu_e (k_B T)^6. \tag{23}$$

This has a form similar to the nucleon Urca rate (9), but there are some significant differences. First, there is a factor α, which reflects the fact mentioned above that the weak interaction matrix element vanishes for collinear relativistic particles, whereas for non-relativistic nucleons the corresponding matrix element is essentially independent of angle. The second difference is that the quantities p_{Fu}/c and p_{Fd}/c take the place of the nucleon masses. Third, the numerical coefficient is somewhat smaller. However, since p_{Fu}/c and p_{Fd}/c are expected to be less than m_n and α is less than or of the order of unity, the neutrino luminosity from quark matter is expected to be rather less than the characteristic rate for the nucleon process. However, it is important to note that for quark matter the electron fraction is uncertain. For instance, if u, d, and s quarks may be treated as massless, the electron fraction vanishes identically. Detailed estimates of the composition of quark matter for various models are given by Duncan, Shapiro, and Wasserman (1983) and Alcock, Farhi, and Olinto (1986).

7. Summary and Observational Aspects

From the above discussion, it is clear that if any of the many variants of the direct Urca process can occur, the early cooling of a neutron star will be much more rapid than if only the modified Urca process occurs. If central temperatures of neutron stars can be determined from observation, this offers the possibility of distinguishing among some possibilities for the state of dense matter in neutron stars.

Attempts to deduce internal temperatures of neutron stars have been made by searching for the thermal X-ray emission from the surfaces of neutron stars. If the outer layers of a neutron star are in a steady state thermally, the central temperature may be deduced from the surface temperature, assuming the thermal transport properties and equation of state of matter in the outer layers are known. Depending on the surface temperature, the central temperature is estimated to be 10 - 100 times greater than the surface temperature. Detailed calculations of neutron star envelopes are given by Urpin and Yakovlev (1979), by Gudmundsson, Pethick and Epstein (1983), and by Hernquist and Applegate (1984).

The assumption of a thermal steady state in the outer layers of the star is a reasonable one provided the thermal relaxation time of the outer layers of the star is short compared with the timescale for variations of the core temperature. Thermal relaxation times for central temperatures of order $10^9 K$ are typically of order 100 years, and depend on the thickness of the outer layers, which is sensitive to the mass and radius of the neutron star. (Nomoto and Tsuruta, 1981; Brown et al., 1987). Thus one can see that if the core cools by any of the processes we have considered above, except the modified Urca one, the core will have cooled to a temperature of order $10^8 K$ or less before the surface layers of the star can respond thermally. Thus X-ray emission seen from such neutron stars which are less than 100 years old would be provided by the thermal energy residing in the outer layers of the star, and not from heat transported from the core. With current and planned X-ray telescopes it is unlikely that one could detect X-ray emission from such a star unless it were less than about 100 years old.

There are a number of effects which make the picture more complicated than the one presented above. Some of these are observational. First, in observations of known neutron stars it is difficult to be sure that any X-rays detected are the result of thermal emission from the surface, rather than some other process, such as emission by a surrounding nebula or by the electromagnetic mechanisms responsible for pulsar emission. In a second type of observation one searches for a compact source of X-ray emission in a supernova remnant. If no radiation is seen, one does not know whether this indicates the presence a cool neutron star, or no neutron star at all. On the theoretical side, superfluidity of nucleons in the core would reduce the neutrino emissivity, and consequently the core would cool more slowly. Under these conditions, emission of neutrino-antineutrino pairs by the bremsstrahlung process could be the dominant source of energy loss. (The heat capacity is also reduced, but by a smaller factor than the neutrino emissivity because electrons remain normal). Calculations of even rather gross features such as the equation of state of dense matter at densities typical of neutron star interiors are subject to considerable uncertainties, and it is even more difficult to determine whether or not matter is superfluid or superconducting. There are also uncertainties in the physics of the outer parts of neutron stars, among which are the effects of strong magnetic fields (See, e.g., Romani, 1987; Miller, 1990). For a review of neutron star cooling calculations, see Tsuruta (1986).

Within the next few years one expects a number of X-ray telescopes to be launched, and new searches for thermal emission from neutron stars will be made. This gives added incentive for making further theoretical studies. Our discussion above shows that if fast cooling of neutron stars is observed, this could be due either to matter composed of neutrons, protons and electrons, with a sufficiently high proton concentration, or to an exotic state such as a pion condensate, kaon condensate or quark matter.

In the course of preparing this review, I have profitted from discussions with many colleagues, especially G. Baym, F. K. Lamb, J. M. Lattimer, M. Prakash, and D. G. Ravenhall. The work was supported in part by U. S. National Science Foundation grant PHY86-00377, and NASA grant NAGW-1583.

References

Alcock, C., Farhi, E., and Olinto, A. 1986, *Ap. J.* **310**, 261.

Bahcall, J.N., and Wolf, R.A. 1985, *Phys. Rev. Lett.* **14**, 343; *Phys. Rev.* **140**, B1452.

Baym, G. 1991, in *Neutron Stars: An Interdisciplinary Field*, (D. Pines and J. Ventura, eds.), Kluwer, Boston, in press.

Brown, G.E., Kubodera, K., Page, D., and Pizzochero, P. 1988, *Phys. Rev.* **D 37**, 2042.

Chiu, H.Y., and Salpeter, E.E. 1964, *Phys. Rev. Lett.* **12**, 413.

Duncan, R. C., Shapiro, S. L., and Wasserman, I. 1983, *Ap. J.* **267**, 358.

Finzi, A. 1964, *Ap. J.* **139**, 1398.

Friman, B.L., and Maxwell, O.V. 1979, *Ap. J.* **232**, 541.

Gudmundsson, E. H., Pethick, C. J., and Epstein, R. I. 1983, *Ap. J.* **272**, 286.

Hernquist, L., and Applegate, J. H. 1984, *Ap. J.* **287**, 244.

Iwamoto, N. 1980, *Phys. Rev. Lett.* **44**, 1637.

Iwamoto, N. 1982, *Ann. Phys.* **141**, 1.

Kaplan, D.B., and Nelson, A.E. 1986, *Phys. Lett. B* **175**, 57; **179**, 409.

Lattimer, J.M., Pethick, C.J., Prakash, M., and Haensel, P. 1991, to be published.

Miller, M. C. 1990, *Ph. D. thesis*, Caltech.

Nomoto, K., and Tsuruta, S. 1981, *Ap. J. Lett.* **250**, L19.

Maxwell, O., Brown, G.E., Campbell, D.K., Dashen, R.F., and Manassah, J.T. 1977, *Ap. J.* **216**, 77.

Pethick, C.J., and Ravenhall, D.G. 1991a, in *Neutron Stars: An Interdisciplinary Field*, (D. Pines and J. Ventura, eds.), Kluwer, Boston, in press.

Pethick, C.J., and Ravenhall, D.G. 1991b, *Proceedings of the Texas/ESO-CERN Meeting on Relativistic Astrophysics, Brighton, 1990*, Annals of the New York Academy of Sciences, in press.

Romani, R. W. 1987, *Ap. J.* **313**, 718.

Shapiro, S.L., and Teukolsky, S.A. 1983, *Neutron Stars, White Dwarfs and Black Holes*, Wiley, New York.

Tsuruta, S. 1986, *Comm. Astrophys. and Spa. Sci.* **11**, 151.

Urpin, V. A., and Yakovlev, D. G. 1979, *Astrofizika* **15**, 647.

Wiringa, R. B., Fiks, V., and Fabrocini, A. 1988, *Phys. Rev.* **C38**, 1010.

EXPERIMENTS IN HIGH ENERGY PHYSICS: A BRIEF INTRODUCTION

Mário Pimenta
LIP/IST

Laboratório de Instrumentação e Física Experimental de Partículas
Av. Elias Garcia 14 1^0 1000 Lisboa, Portugal

ABSTRACT

A brief description of a "standard" High Energy Physics experiment focussing on what are the main measurements and how they are obtained is herein reported. The historical example of the discovery of neutral currents by the Gargamelle experiment in 1973 is presented to show the long path between the data collection and the final results.

A "standard" experiment in High Energy Physics

The main idea in High Energy Physics is to collide two objects (particles), observe what comes out and try to guess what was in. To collide two particles we need accelerating and focusing devices, to observe we need detectors and to interpret the results we need models and theories [1].

According both to the beam and to the target, experiments can be classified as: fixed target experiments (an accelerated particle beam collides with a target at rest in the laboratory, fig.1a), collider experiments (two accelerated particle beams collide with each other, fig.1b) and cosmic rays experiments [2] (the target and the detector are waiting in the Laboratory for a cosmic ray to come, fig.1c).

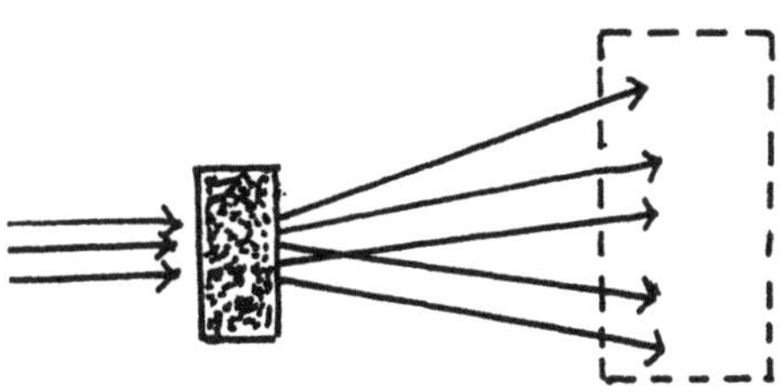

Fig. 1a) Fixed Target Experiment

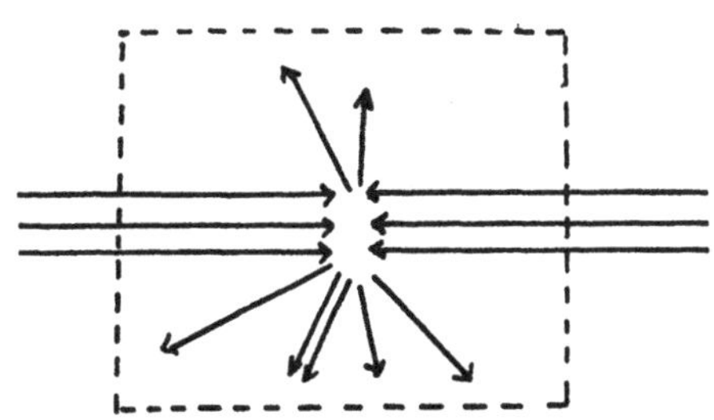

Fig. 1b) Collider Experiment

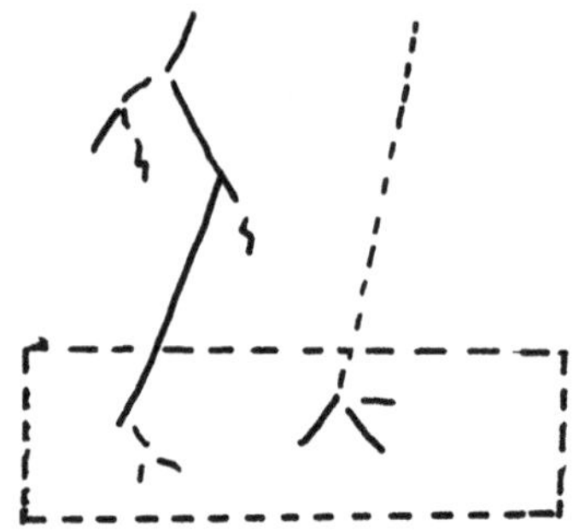

Fig. 1c) Cosmic Ray Experiment

The main building blocks of an accelerator are :the source, the electric field and the magnetic field. The source can either be primary if the particles are the constituints of the ordinary matter (protons, electrons, ions) or secondary if the particles are produced in a previous collision (anti-protons, positrons, pions, kaons,...). The electric field accelerates the beam and the magnetic field guides and focus it. The centre of mass energy provided by accelerators has been increasing in an exponential way since the thirties (1 Mev) till nowadays (1 TeV).

Detectors are mainly used to measure the momentum and the velocity of charged tracks and the energy of both charged and neutral tracks.

Charged tracks going through a thin layer of matter ionizes it. The ionization energy can be collected as visible energy (emission of γs by the excited nucleous in scintillators, fig. 2a) or directly as kinetic energy (in the wire or in the drift chambers the ions are accelerated by an electric field and collected in the cathod causing an electric current fig.2b). Either way the ionization is a mechanism that enables us to know where the particle has gone through and even if we have precise enough detectors, the energy that has been deposited. Using several layers of ionization detectors we are able to reconstruct the charged particles trajectories.

To measure the momentum of charged particles we can profite from the well known fact that, within constant magnetic fields, charged particles have circular trajectories and there is a trivial relationship between the radius of the trajectory and the momentum of the particle.

Fig. 2a) Scintillator

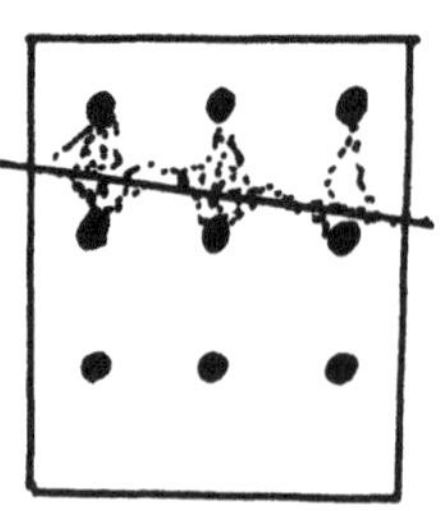

Fig. 2b) Wire Chamber

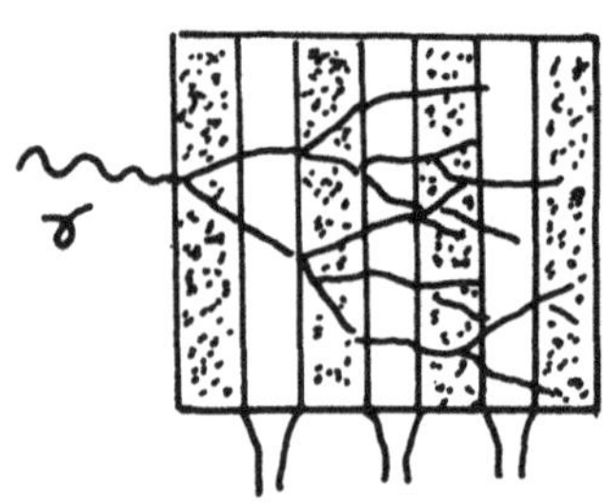

Fig. 3) Calorimeter

The energy of the particles, both charged and neutral, can be measured by stopping the particles inside a block of matter and collecting a fraction of the deposited energy. For instances in the detector shown in fig.3 (sampling calorimeter : sandwich of heavy material and scintillator) the particle goes through the first layer where it interacts with the atoms of the heavy material and it originates several secondary particles. These particles deposit some ionization energy in the first active layer and afterwards they go through the second layer where new interactions can take place. All the process is repeated in the next layers until all the energy is spent.

The measurement of the velocity of charged particles is usually made using the Cerenkov effect. A description of this measurement can be found elsewhere in these proceedings [2].

Some detectors give a "visual" image of the interaction (charged tracks only). A good example of these detectors is the bubble chamber (fig.4);the bubble chamber is a large vessel containing a liquid under pression and close to the boiling point. When the beam enters the chamber there is a sudden decompression and the boiling temperature is exceeded. A charged tracks leaves ions through its path that make little bubles. The charged tracks are, therefore, seen as a path of small bubbles.

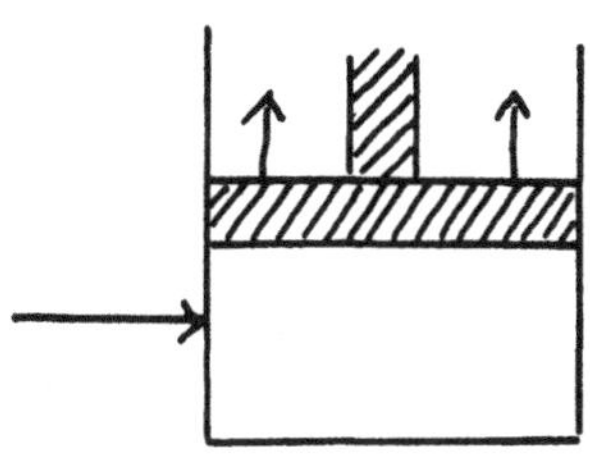

Fig. 4a) Beam entering a Bubble Chamber

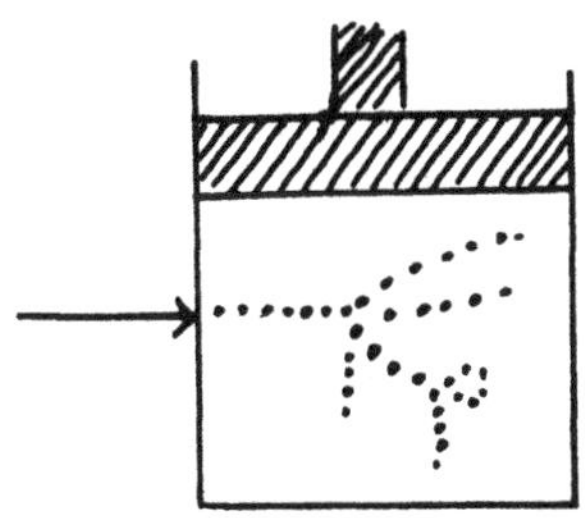

Fig. 4b) Interaction in a Bubble Chamber

Of course there are a lot of different detectors and they work in a much more subtle way than the oversimplified version that has been presented so far. In fig.5 a picture of DELPHI detector (one of the four LEP experiments) can be seen just to show how huge and complicated these detectors can be.

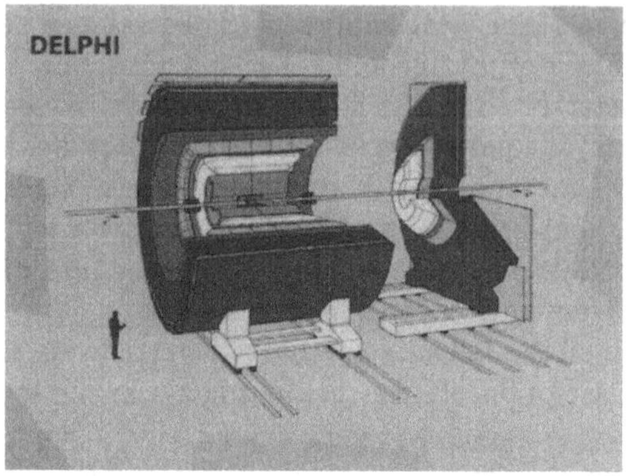

Fig. 5) Delphi Detector

An example : Experiments with neutrino beams in the Laboratory and the discovery of neutral currents

The story of neutrinos is a good example of the close interaction between experiment and theory. Several good reviews can be found elsewere [3],and here we will just focus briefly on some of the steps.

In the 1920's, while observing the continuos spectrum of the emited electron it was discovered, that the Beta decay apparently did not conserve either energy or angular and linear momentum. In 1930, Pauli suggested the existence of a low mass, low interaction particle which was called later on "neutrino" (by this time Niels Bohr was suggesting the possible violation of energy conservation), and soon afterwards Fermi proposed the so called Fermi Theory of the Weak Interaction which gave a good framework to compute the observed quantities in the following twenty years. But the first direct proof of neutrino existence was only obtained by Crown and Reines in 1956, using neutrinos from a nuclear reactor, and the first neutrino beam was produced in the beginning of the 60's from the decay of pions and kaons. A large number of experiments with neutrino beams have since then been performed and very important results have been obtained such as the existence of several neutrino families, the discovery of neutral currents, the test of quark and standard model and so forth.

Neutrino's interaction can be mediated by the charged W bosons (charged currents) or by the neutral Z^0 boson (neutral currents)- fig.6 . However a neutral current event is very difficult to observe for the incoming and the outcoming neutrinos can not be seen (they are neutral). The only sign in the detector is, in leptonic interactions, one electron, or, in semi-leptonic interactions, some hadrons that must be distinguished from the semi-leptonic neutrinos charged currents final states (some hadrons and a lepton). Detectors for neutrinos experiments must then be both massive (the interaction rate is very low) and very accurate (good final state identification).

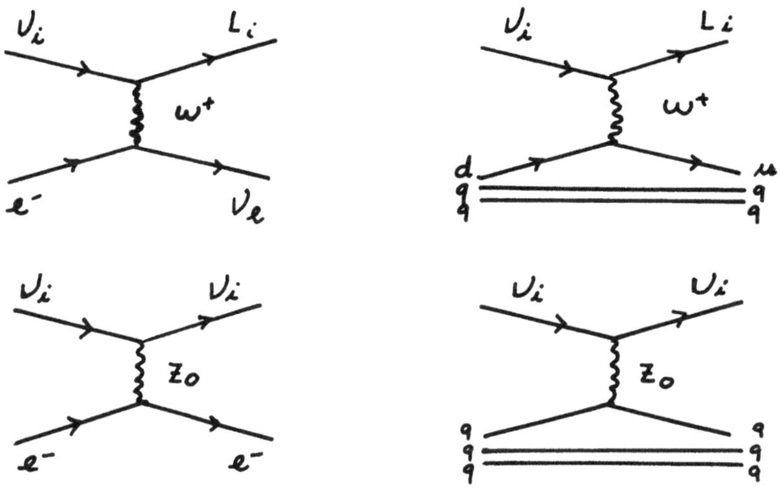

Fig. 6) Leptonic and semi-leptonic neutrino interactions

The neutral currents were first discovered [4] in Gargamelle experiment at CERN (1973). Gargamelle was an enormous bubble chamber ,5 meters long with 10 tons of liquid freon at 10 to 25 Atomspheres. More than 3 millions pictures were taken during its lifetime. In that famous experiment only one event over every 700.000 events had the characteristic signature of leptonic neutral current- an electron in the final state (fig.7).

Fig. 7) The first neutral current event

Before computing a result we need to know more than just the number of the events that satisfy our experimental cuts. We have to study carefully the inefficiency of the method (the probability to loose a good event), the expected background (the number of events from all the other channels that can also satisfy the cuts), and of course to determine the number of particles in the beam and in the target (luminosity). Our result can then be expressed as a cross section (the number of observed events minus the expected background divided by the efficiency and by the luminosity) which is the usual form of presentation of results in a "standard" high energy physics experiment.

REFERENCES

1) There are many good books in elementary particle physics . The following references are therefore a small set of books I considered useful to a first introduction on the subject.

- Introduction to Elementary Particles, David Griffiths, John Wiley and Sons.
- The Experimental Foundations of Particle Physics, Robert N. Cahen and Gerson Goldhaben, Cambridge University Press
- Detectors for particle radiation, Konrad Kleinknecht, Cambridge University Press
- Statistics for nuclear and particle physicists, Louis Lyons, Cambridge University Press

2) - Experiments with neutrinos , Paula Bordalo , in these proceedings

3) - Weak Interactions of Leptons and Quarks , E.D. Commins and P.H. Bucksbaum, Cambridge University Press
- Weak interactions, D.Bailin, Adam Hilger Ltd

4) - Search for elastic Muon-Neutrino electron scatering, Hasert et al., PL 46B 121(73).
- Observation of neutrino like interactions without Muon or electron, PL 46B 148(73).

NUCLEOSYNTHESIS IN BIG-BANG MODELS

F.D. Santos

Centro de Física Nuclear da Universidade de Lisboa
Av. Gama Pinto 2, 1699 Lisboa, Portugal

I. Introduction

The synthesis of the light elements in the first few minutes of the Universe is a unique event to probe our understanding of the very early Universe. In particular the comparison between the predicted abundances of D, ^{3}He, ^{4}He and ^{7}Li, from models of primordial nucleosynthesis, to those inferred from observation provides information on the baryon density of the Universe. Detailed comparisons between theory and observation show that the present density of nucleons is such that the Universe cannot be "closed" by nucleons alone.

In order to obtain more constraining bounds to cosmology and particle physics from primordial nucleosynthesis it is essential to take into account the most recent theoretical and observational data. Here I will give an overview of the current status of big-bang nucleosynthesis [1], including a discussion of those aspects of the nuclear physics which are relevant in the quest for more accurate cosmological knowledge.

II. Thermodynamics of the Early Universe

The presence of a uniform radiation field with a temperature [2] of $T_\gamma = 2.735 \pm 0.06$ implies a hot singularity which means an ever increasing temperature as the cosmic time $t \to 0$. A crucial assumption in all cosmological models of the early Universe is the existence of a phase of complete thermodynamic equilibrium at some very early epoch where $T = 10^{11}$K or earlier. The hot big-bang picture results naturally from a Friedmann-Lemaitre (FL) model in which the expansion factor R(t) satisfies the equation [3,4]

$$\left(\frac{\dot{R}}{R}\right)^2 = \frac{8\pi}{3} G\rho - \frac{K}{R^2} + \frac{\Lambda}{3}$$

where R(t) approaches zero towards the limit $t \to 0$. During the radiation dominated phase the energy density ρ in FL models evolves proportionally to R^{-4} or equivalently, proportionally to T^4. At very early times the curvature term KR^{-2} and the cosmological constant Λ become negligible compared with the potential energy term proportional to ρ. This implies

$$\dot{R}^2 = \frac{8\pi}{3} G \rho \, R^2.$$

Since ρR^4 = const. the expansion factor R(t) satisfies $(\dot{R}R)^2$ = const., $R \propto t^{1/2}$ and

$$\rho = \frac{3}{32\pi G} t^{-2}. \tag{2.1}$$

The existence of thermodynamic equilibrium implies an enormous simplification of the physics in the early Universe: all quantities of interest depend only on the temperature and the chemical potentials μ_i of the particles present. For bosons $|\mu_i| \leq m_i c^2$ and for fermions there are no restrictions. Photons have $m_i = 0$, hence $\mu_i = 0$, and a number density $n_\gamma = (\pi/13)\ [kT/(hc)]^3$.

It is reasonable to assume that the densities corresponding to the electron-lepton and μ-lepton numbers n_{L_e}, n_{L_μ} are small and of the same order as the baryon density n_B. This kind of lepton assymmetry arises naturally in GUT theories.

With these assumptions the chemical potentials of all particles are equal to zero. Consequently only those with $m_i c^2 << kT$ can be present in appreciable numbers. These particles can be treated to a good approximation as ultrarelativistic $c|\vec{p}| >> mc^2$. For bosons and fermions

$$n_B = (g_B/2) n_\gamma \qquad \rho_B = (g_B/2)\rho_\gamma$$

$$n_F = \frac{3}{8} g_F n_\gamma \qquad \rho_F = \frac{7}{16} g_F \rho_\gamma .$$

The total energy density in a radiation dominated phase is then

$$\rho = \rho_B + \rho_F = \frac{1}{2} g(T) \rho_\gamma$$

where

$$g(T) = g_B + \frac{7}{8} g_F$$

is the number of statistical degrees of freedom.

At temperatures below 10^{12} K (kT ≅ 100 MeV) only the leptons $\mu^{\pm}$, $e^{\pm}$, ν_{μ}, $\bar{\nu}_{\mu}$, ν_e, $\bar{\nu}_e$ and the photons exist in appreciable numbers in equilibrium. The number of nucleons is considerably smaller $n_B \cong 10^{-9}\, n_{\gamma}$. For temperatures above 10^{10}K the neutrino interactions with the other leptons can mantain a thermal equilibrium. At lower temperatures the neutrinos drop out of thermal equilibrium and even after the decoupling follow a thermal Fermi distribution. After the annihilation of the $\mu^{\pm}$ and before $e^{\pm}$ annihilation i.e. for T > 1 MeV, g(T) = 43/4 and therefore

$$\rho = \frac{43}{8}\,\frac{aT^4}{c^2} \tag{2.2}$$

where $a = 7.56 \times 10^{-15}$ erg cm^{-3} K^{-4} is the radiation density constant. Inserting eq. (2.2) into (2.1) we obtain

$$t = \left(\frac{3c^2}{172\pi G a T^4}\right)^{1/2} = 0.99\, T_{10}^{-2}\ \mathrm{s} \tag{2.3}$$

where $T_{10} = T/10^{10}$. Below T ≅ 5×10^9 K (kT < 0.5 MeV) electrons and positrons start to annihilate into photons and this annihilation produced an increase in the temperature T_{γ} of the photons relative to the temperature T_{ν} of the neutrinos. The requirement that the entropy is constant leads to

$$T_{\gamma} = \left(\frac{11}{4}\right)^{1/3} T_{\nu} = 1.401\, T_{\nu}\,.$$

For all lower temperatures T_{γ} and T_{ν} are still proportional to R^{-1}. The decoupling of photons from matter at T ≅ 10^4K does not appreciably change this law since the number of hidrogen atoms is very small. For this temperature range

$$t = 3.16\ T_{\gamma,10}^{-2}\ \mathrm{s}.$$

where $T_{\gamma,10}$ is the photon temperature in units of 10^{10}K.

III. Primordial Nucleosynthesis

The ratio of neutrons to protons immediatly before nucleosynthesis determines the amount of heavier elements that can be formed. Protons are transformed into neutrons and vice-versa through the reactions

$$p + e^- \leftrightarrow n + \nu_e$$

$$n + e^+ \leftrightarrow p + \bar{\nu}_e \, .$$

The corresponding cross sections can be computed from the (V-A) charged-current interaction of the weak forces. The ratio of the numbers of neutrons N_n to the total number of nucleons

$$X_n = \frac{N_n}{N_n + N_p}$$

is determined from the balance between the $p \rightarrow n$ and $n \rightarrow p$ reactions

$$-\frac{dX_n}{dt} = \lambda(n \rightarrow p)\, X_n - \lambda(p \rightarrow n)\,(1 - X_n)$$

where λ are the reaction rates. For temperatures $kT >> Q = (m_n - m_p)c^2 = 1.293$ MeV, $T=T_\nu$,

$$\frac{\lambda(p \rightarrow n)}{\lambda(n \rightarrow p)} = e^{-(Q/kT)} \, ,$$

$$X_n = [1 + \exp(Q/kT)]^{-1} \, . \qquad (3.1)$$

Notice that $T \rightarrow \infty$ implies $X_n \rightarrow 1/2$. Furthermore using the weak interaction cross section we obtain

$$\lambda(n \rightarrow p) = \lambda(p \rightarrow n) = \frac{7}{15}\, \pi^3 \left(G_V^2 + 3G_A^2\right)\, (kT)^5 = 0.36\, T_{10}^5\, s^{-1}$$

As long as the lifetime τ for the $p \rightarrow n$ and $n \rightarrow p$ reactions is much smaller than the expansion time of the Universe the system is in thermodynamic equilibrium. For $\tau = \frac{1}{\lambda} \leqslant 0.1\, t$, eq. (2.3) gives $T > 3 \times 10^{10}$K. At this temperature eq. (3.1) implies that $X_n = 0.38$. As the temperature drops below $T = 3 \times 10^9$ K ($kT \cong 1$ MeV) the weak interactions can no longer establish thermal equilibrium against the expansion rate H(t). In fact for $T = 3 \times 10^9$ K the mean lifetime is $\tau = 4.2\, t$ and there is no longer enough time for thermal equilibrium, between neutrinos and nucleons, to be established. The ratio X_n freezes out at a temperature T_f and at lower temperatures is changing only by the β - decay of the neutrons. The time evolution of X_n, resulting from β decay, is now described by

$$X_n(t) = X_n(0) \exp(-t/\tau_n)$$

with $\tau_n = 889.8 \pm 4.4$ s [5].

The freeze-out temperature T_f and the time from freeze-out to nucleosynthesis will be very important. When the temperature dropped to approximately 10^9K (KT $\cong$ 0.1 MeV) the nucleosynthesis ocurred very rapidly. Since the density is relatively low only two-body reactions are responsible for the production of the relevant nuclei. The abundances depend only on the nucleon flux determined by the nucleon density and on the cross sections for the various processes. To characterize the nucleon density n_B it is convenient to use the parameter $\eta_{10} = 10^{10}\eta$ where $\eta = n_B/n_\gamma$. Furthermore it is useful to compare the nucleon density to the critical density $\rho_c = 3\, H^2/(8\pi G)$. Thus we can write

$$\Omega_B H_{50}^2 = 1.47 \times 10^{-2} \left(\frac{T_\gamma}{2.735}\right)^3 \eta_{10} \qquad (3.2)$$

where $\Omega_B = \rho_B/\rho_c$ and H_{50} is H in units of 50 Kms^{-1} Mpc^{-1}. For $2.675 < T_\gamma < 2.795$ K, $0.8 < H_{50} < 2$ and $3 < \eta_{10} < 10$ we obtain $0.01 < \Omega_B < 0.25$. Part of the uncertainty in Ω_B results from the lack of a precise value for the Hubble parameter. However the comparison betweeen the predicted and observed abundances can reduce the uncertainty in η_{10} and also is Ω_B.

IV. Reaction Network Calculations

The crucial reaction to start the synthesis of complex nuclei is $p + n \to D + \gamma$. At very early times, because the Universe is very hot, there are many high energy gamma rays present to dissociate the deuterons by the inverse reaction $D + \gamma \to p + n$. The nucleosynthesis starts only when the radiactive capture competes favorably with photo-dissociation and this occurs for $kT \lesssim kT_D \cong 0.1$ MeV and $t \gtrsim t_D \cong 100$ s. When the abundance of deuterons is at a sufficiently high level ^{3}H, ^{3}He and ^{4}He are synthesized very quickly. Fig.1 shows the evolution of n, p, D, ^{3}H, ^{3}He and ^{4}He in the early Universe predicted by the standard Big - Bang model. Heavier elements are not produced in significant amounts because at mass numbers A = 5 and A = 8 there is no stable nucleus. A small amount of ^{7}Li and ^{7}Be is produced through the reactions $^3H(\alpha,\gamma)^7Li$, $^3He(\alpha,\gamma)^7Be$ and $^7Be(e^-,\nu_e)^7Li$. Since ^{4}He is the more tightly bound of the light nuclei almost all neutrons present before nucleosynthesis starts are incorporated into ^{4}He nuclei. Therefore the ^{4}He mass fraction $Y_p \cong 2X_n$. Effects that increase the value of X_n such as, for instance, increasing the number of neutrino

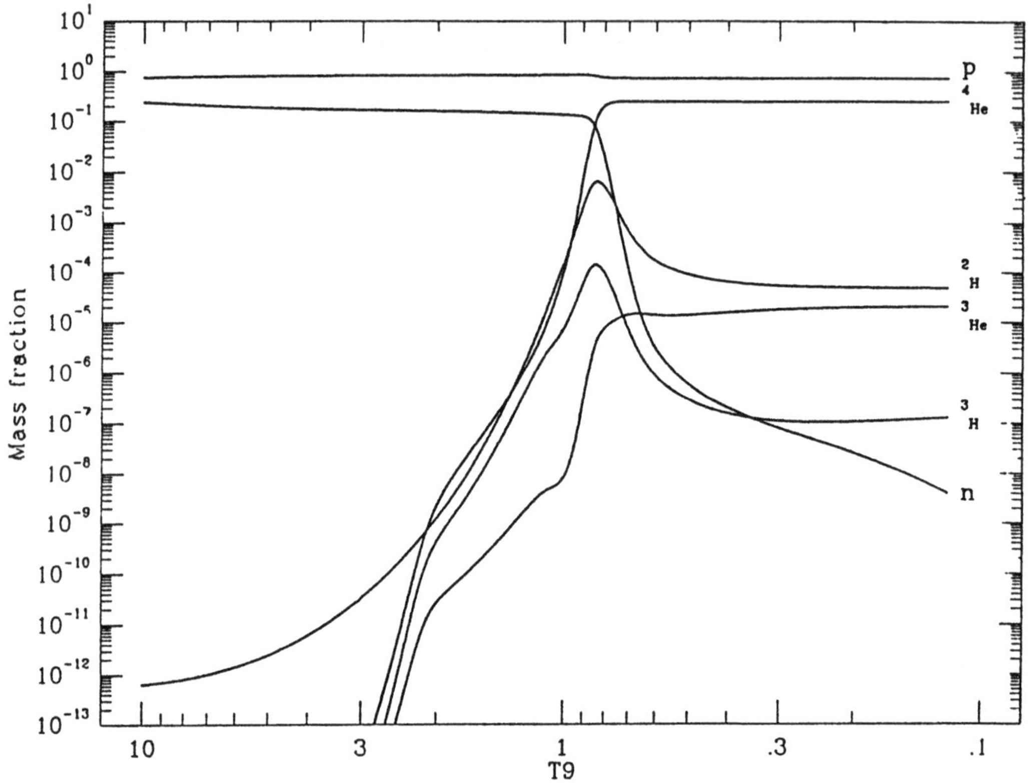

Figure 1- Evolution of the abundances in the early Universe predicted by the standard Big-Bang model. Mass fractions are shown as a function of T_9 the temperature in units of 10^9 degrees [10].

families lead to an increase in the predicted Y_p. Thus, whereas the predicted abundance of ^4He is mostly determined by X_n, the abundances of D, ^3He and ^7Li are sensitive to the nucleon density parameter η_{10}.

By increasing η_{10} we increase the cross sections of the reactions involving deuterons and therefore the predicted abundances of D and ^3He decrease. For $1 \leq \eta_{10} \leq 10$ a good fit to the detailed nucleosynthesis calculations [6] gives

$$\left(\frac{D}{H}\right)_P = 4.6 \times 10^{-4}\ \eta_{10}^{-5/3} \tag{4.1}$$

Over the same range of values of η_{10}, $(^3He/H)_P$ decreases from 3×10^{-5} to 0.9×10^{-5} [6]. Although the predicted ^4He mass fraction has a weak dependence on η_{10}, Y_p is very sensitive to X_n and therefore to the number of neutrino families N_ν and the neutron half-life $\tau_{1/2}$. From a fit to the calculated Y_p for $2.5 \leq \eta_{10} \leq 10$, $2 \leq N_\nu \leq 4$ and $10.1 \leq \tau_{1/2} \leq 10.5$ min. Olive et al. [5] obtain

$$Y_p = 0.228 + 0.010 \ln \eta_{10} + 0.012 (N_\nu - 3) + 0.018 (\tau_{1/2} - 10.28) \tag{4.2}$$

The abundance of ^7Li has a rather complex behaviour as a function of η_{10}. Direct formation dominates at small η and formation via ^7Be dominates at large η. As a result the predicted abundance has a minimum for $\eta_{10} \cong 3$.

There are still some uncertainties in the cross sections of some of the nuclear reactions in the network of the primordial nucleosynthesis. The most important are those associate with the production of ^7Li. Recently Deliyannis et al. [7] performed a systematic study of the uncertainty in the production of ^7Li introduced by uncertainties in the nuclear reaction data. Table 1 lists the uncertainties associated with each reaction in the ^7Li producing chain

Table 1

Reaction	Uncertainty (%)
$^1H(n,\gamma)^2H$	10
$^2H(p,\gamma)^3He$	16
$^2H(d,n)^3He$	10
$^2H(d,p)^3H$	10
$^3He(\alpha,\gamma)^7Be$	6
$^3H(d,n)\,^4He$	10
$^3H(\alpha,\gamma)\,^7Li$	17.5
$^7Be(n,p)^7Li$	20
$^7Li(p,\alpha)^4He$	8

A substantial improvement in the quality of the basic nuclear reaction data is clearly needed.

Recently it was realized [8,9] that the reaction $^2H(d,\gamma)^4He$, in the energy range where primordial nucleosynthesis occurs, has a cross section which is a factor of approximately 35 higher than the previous value used in nucleosynthesis calculations. However the new value has little effect on the predicted abundances [10] because the other reactions that destroy deuterons, form ^4He and proceed through the strong nuclear force, have larger cross sections.

V. Comparison with Observations

It is certainly not easy to obtain reliable observational data on the abundances to compare with the predictions from primordial nucleosynthesis calculations. ^4He can be observed throughout the Universe via its emission lines in HII regions. In this method the amount of neutral ^4He must be estimated by modelling. Furthermore we have to consider that some of the observed ^4He results from nucleosynthesis produced in stars by hydrogen burning. Thus to extract the primordial abundance we need to correct for stellar and galactic evolution. The fact that the inferred abundance of ^4He is almost everywhere in the Universe close to 0.25 by mass is one of the most impressive confirmations of the hot Big-Bang. Y_p is slightly smaller in ionized gas clouds with low metallicity where the material has probably suffered less astration. Recent ^4He determinations [5] in high excitation, low metallicity HII regions gives

$$Y_p = 0.23 \pm 0.01 \tag{5.1}$$

Deuterium is very quickly burned away when incorporated into stars and it cannot be significantly synthesized in other sites. Thus D/H has decreased from its primordial value; $(D/H)_{observed} \leqslant (D/H)_P$. Boesgaard and Steigman [1] give

$$\left(\frac{D}{H}\right)_P \geqslant 2.0 \times 10^{-5} \tag{5.2}$$

Inside stars D is burned to ^3He and some ^3He survives in stellar evolution. It is therefore possible to obtain an upper bound to the sum of the primordial abundances of D and ^3He

$$\left(\frac{D + {}^3He}{H}\right)_P \leqslant 1.1 \times 10^{-4} \tag{5.3}$$

Nuclei of Li, Be, B are produced in spallation reactions between cosmic ray protons and alphas and interstellar gas nuclei (H, He, C, N, O). Such reactions tend to produce equal amounts of ^6Li and ^7Li but the latter is much more abundant than ^6Li in the solar system. This result indicates that ^7Li is also produced inside stars. The fact that ^7Li has a lower abundance in PopII stars than in PopI stars and meteorites also suggests that ^{7}Li has been produced during galactic chemical evolution. The identification of the ^{7}Li primordial abundance with that of PopII stars gives [7]

$$\left(\frac{^7\mathrm{Li}}{\mathrm{H}}\right)_P \leqslant 2.5 \times 10^{-10}. \qquad (5.4)$$

VI. Conclusions

It is now possible to determine whether or not there are values of η for which the predicted and observed abundances are in agreement. As regards deuterium the combination of eqs. (4.1) and (5.2) gives

$$\eta_{10} \leqslant 6.6.$$

^{7}Li, through eq. (5.4) provides a lower upper bound on η_{10} [5]

$$\eta_{10} \leqslant 5.0$$

Furthermore the upper limit to D+^{3}He leads to a lower bound on the baryon to photon ratio

$$\eta_{10} \geqslant 2.6$$

In conclusion there is a range of values

$$2.6 \leqslant \eta_{10} \leqslant 5.0 \qquad (6.1)$$

for which the nucleon density is consistent with the observational constraints on D, ^{3}He and ^{7}Li. This agreement is very encouraging and strengthens our confidence in the theory of the hot Big-Bang. Moreover we conclude that eq. (4.2) is also compatible with the observed ^{4}He abundance. In fact for $N_\nu=3$ and $\tau_n = 889.8 \pm 4.4$ s eqs. (6.1) and (4.2) give

$$0.235 \leqslant Y_p \leqslant 0.246$$

in good agreement with eq. (5.1).

With the new bounds on η_{10} given by eq. (6.1) we can obtain tighter constraints on Ω_B using eq. (3.2)

$$0.036 < \Omega_B H_{50}^2 < 0.078 \qquad (6.2)$$

For $H_{50} \geqslant 0.8$

$$\Omega_B \leqslant 0.122$$

showing that the Universe cannot be closed by baryons by more than a factor of 8.

The result that baryons alone cannot close the Universe has been very influential in the recent development of cosmology. It can be used as a justification for nonbaryonic dark matter if we favour a flat Universe with $\Omega=1$. An independent estimate of Ω can be reached through the mass to light ratio M/L inferred from the luminous matter in galaxies. Assuming that this is typical of the matter throughout the Universe [11]

$$\Omega_{lum} \leqslant 0.006\ H_{50}^{-1} \tag{6.3}$$

Eqs. (6.2) and (6.3) give

$$\frac{\Omega_B}{\Omega_{lum}} > 6.0\ H_{50}^{-1} \geqslant 3$$

for $H_{50} < 2$. This indicates that the great majority of nucleons in the Universe are dark. Furthermore for small H_{50} the baryonic M/L ratio of the Universe is compatible with the dynamical M/L values inferred from galaxy clusters. This compatibility, if confirmed, would suggest an open Universe without exotic nonbaryonic matter.

The standard model of the Big-Bang assumes that all the constituents of the Universe are uniformly distributed at the time of nucleosynthesis. This would certainly not be true if the quark - hadron phase transition at kT $\sim$ 100 MeV is strongly first order [12]. As the Universe reaches the critical temperature the quark soup supercools until bubbles of the hadronic phase begin to appear. These bubbles grow and leave behind a highly inhomogeneous hadron distribution in a uniform radiation field. After neutrino decoupling the neutrons diffuse towards uniformity while the protons remain locked to the radiation filed. Thus when nucleosynthesis starts the n/p ratio can have strong variations from place to place. These inhomogeneities change the predicted abundances relative to the standard Big-Bang model. In neutron rich regions neutrons must β - decay before synthesis can occur. As a result there is less ^{4}He and more D than in homogeneous models [13]. Furthermore there is considerable more production of ^{7}Li and also of heavier nuclei because of the increased neutron flux.

The most likely reaction sequence in neutron rich zones leading to the production of heavier elements is

$$^{4}\mathrm{He}(^{3}\mathrm{H},\gamma)^{7}\mathrm{Li}\,(\mathrm{n},\gamma)^{8}\mathrm{Li}(\alpha,\mathrm{n})^{11}\mathrm{B}(\mathrm{n},\gamma)^{12}\mathrm{B}(\beta\,\nu)^{12}\mathrm{C}$$

The cross section of most of the reactions in this sequence are fairly well known [14] from experimental studies. However there is still very little information on reactions involving ^{8}Li due to its short half-life of 0.8 s. Present nucleosynthesis calculations within the framework of inhomogeneous Big-Bang models lead to an overproduction of ^{7}Li. This may indicate that such models are unrealistic. To resolve these questions it is essential to improve our knowledge of the relevant nuclear reaction cross sections. It may then become possible to investigate the QCD phase transition in the primitive Universe through a more detailed comparison of predicted and observed abundances.

References

[1] A. M. Boesgaard and G. Steigman, Ann. Rev. Astron. Astrophys. **23**, (1985) 319.

[2] J.C. Mather et al., Ap. J. Lett. (1990), in press.

[3] S. Weinberg, Gravitation and Cosmology, Wiley, New York, 1972

[4] N. Straumann, General Relativity and Relativistic Astrophysics, Springer, Berlin, 1984.

[5] K.A. Olive, D.N. Shramm, G. Steigman and T.P. Walker, Phys. Lett. **236 B** (1990) 454.

[6] G. Steigman, in Primordial Nucleosynthesis (eds., W.J. Thompson, B.W. Carney and H.J. Karwowski; World Scientific) p. 1 (1990).

[7] C.P. Deliyannis, P. Demarque, S.D. Kawaler, L.M. Krauss and P. Romanelli Phys. Rev. Lett **62** (1989) 1583.

[8] F.D. Santos, A. Arriaga, A.M. Eiró and J.A. Tostevin, Phys. Rev. **C 31** (1985) 707.

[9] C.A. Barnes, K. Chang, T.R. Donaghne, C. Rolfs and Kammeraad, Phys. Lett **197 B** (1987) 315.

[10] J. Lin Yun and F.D. Santos, J. Phys. G: Nucl. Part Phys. **15** (1989) 1275.

[11] B.E.J. Pagel, in Baryonic Dark Matter (eds., G. Gilmore and D. Lynden Bell; 1989).

[12] J. Applegate and C. J. Hogan, Phys. Rev **D 31** (1985) 3037.

[13] C.J. Hogan, in Primordial Nucleosynthesis (eds., W.J. Thompson, B.W. Carney and H.J. Karwowski; World Scientific) p. 15 (1990).

[14] G. Caughlan and W. Fowler, At. Dat. Nucl. Dat. Tabl. **40** (1988) 291.

THE COSMOLOGICAL CONSTANT, THIRD QUANTIZATION AND ALL THAT

G. Lavrelashvili
Tbilisi Mathematical Institute, 380093 Tbilisi, USSR

ABSTRACT

The paper consists of two main parts. The first part contains a short review of recent works on the application of the third quantization method to the cosmological constant problem. In the second part critical consideration is given to Dirac's scheme of gravity quantization. We suppose that Dirac's scheme may turn out inadequate for gravity quantization because of anomalies arising in the algebra of constraints.

1. INTRODUCTION

New ways have been proposed recently [1-3] for solving the long-standing cosmological constant problem [4].

It was suggested [1-3] to consider universes with different values of the cosmological constant Λ which are distributed according to some distribution function $P(\Lambda)$*. In this case the most probable value of Λ will be the one that gives a peak of $P(\Lambda)$.

The probability distribution $P(\Lambda)$ was calculated by the euclidean functional integral of quantum gravity. The main contribution to the latter is given by the De Sitter instanton. The instanton exists for $\Lambda>0$ and has the negative action $I_{inst}=-3\pi M_{Pl}^2/\Lambda$. Thus it was found that the distribution function

$$P(\Lambda) \propto \exp(-I_{inst}) = \exp(3\pi M_{Pl}^2/\Lambda), \qquad (1.1)$$

has a peak for $\Lambda=0$.

After summation over multi-instanton contributions [3], an even stronger peak,

$$P(\Lambda) \propto e^{\exp(-S_{inst})} = e^{\exp(3\pi M_{Pl}^2/\Lambda)} \qquad (1.2)$$

was obtained.

---

* As follows from the results of a recent study of baby universes physics [5-7] all coupling constants, including Λ, are really dynamical [8-9]

Another interpretation of eq.(1.1) based not on the euclidean functional integral of quantum gravity but on the third quantization method [10-12] is proposed in refs. [13,14].

Within the framework of this method it is proposed to consider Wheeler-DeWitt equation not as an analog of the Schrödinger equation but as an analog of the Klein-Gordon equation. The notion of creation of universes in superspace [15,13] is naturally introduced in this way and $P(\Lambda)$ is interpreted as the number of large universes created in the superspace from the state containing no small universes.

Subsequently [16-18] the third quantization method was used to investigate more realistic models with matter fields, and to study the case $\Lambda<0$

In Section 2 we consider this method in greater detail on a simple model and give the main results for realistic models.

Another question treated in this paper is the applicability of Dirac's scheme to gravity quantization. As is well-known, gravity is an example of a constrained system. Constraints arise because of the theory being invariant with respect to general-coordinate transformations. At the classical level constraints form a closed algebra. Dirac's scheme is widely used for gravity quantization, in which case Hamiltonian constraints as operators are imposed on allowable state vectors. Then the Wheeler-DeWitt equation is obtained [19]. The use of Dirac's quantization scheme implies that no anomalous terms arise in the algebra of constraints at the quantum level.

This problem is discussed in Section 3 where arguments are given to show that such anomalous terms may appear in the algebra of constraints of quantum gravity. Thus we suppose that Dirac's scheme may be inadequate for gravity quantization.

Our discussion is rather schematic because of the paper's limited space. We consider the main ideas and give results omitting details of calculations.

2. THIRD QUANTIZATION AND THE COSMOLOGICAL CONSTANT PROBLEM

In this section we shall consider the third quantization method and its application to the cosmological constant problem.

The discussion here closely follows [17,18].

2.1 The simplest minisuperspace model

Let us consider the creation of universes in the simplest minisuperspace model with just one dynamical variable, namely, the radius a of the closed Friedmann–Robertson–Walker universe. In this model the Wheeler–DeWitt equation for the wave function of the universe has the form

$$\left(\frac{1}{2} \frac{\partial^2}{\partial a^2} + \frac{1}{2} \mu^2(a) \right) \Psi(a) = 0 \qquad (2.1)$$

where $$\mu^2(a) = - a^2 + \frac{1}{3} \Lambda a^4 + \varepsilon,$$

Λ is the cosmological constant and ε is some positive number introduced for convenience (in the end of calculations we take the limit $\varepsilon \to 0$). Hereafter in this section we shall use the units in which $3\pi M_{Pl}^2/2=1$. In considering model (2.1), we assume $0 < \Lambda \ll 1$.

We tend to use the analogy between eq.(2.1) and the Klein–Gordon equation in (0+1) dimension, a being the analog of time and $\mu^2(a)$ being the time– dependent (mass)^2. μ^2 is positive in two regions (see Figure 1): for small a ($0 < a < a_1$) and for large a ($a \geq a_2$), where $a_1 \cong \varepsilon^{1/2}$, $a_2 \cong \Lambda^{-1/2}$ are two turning points.

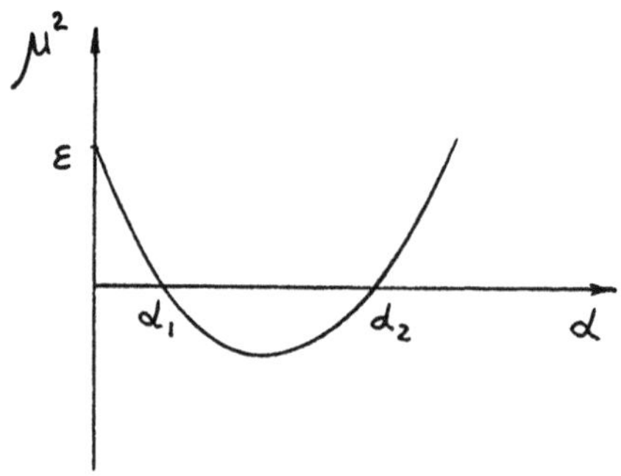

Figure 1

Accordingly, we can define two oscillating semiclassical solutions in each of these regions,

$$\Psi_{in}^{(\pm)}(a) = (2\mu)^{-1/2} e^{\pm S_{in}(a)}, \qquad a < a_1$$

$$\Psi_{out}^{(\pm)}(a) = (2\mu)^{-1/2} e^{\pm S_{out}(a)}, \qquad a > a_2$$

where $$S_{in} = i\int_0^a \mu(a')da' \quad \text{and} \quad S_{out} = i\int_{a_2}^a \mu(a')da' .$$

The standard interpretation [20–24] of Ψ_{in} and Ψ_{out} is that they describe classical Friedmann-like and De Sitter-like universes, respectively.

In the intermediate region $a_1 < a < a_2$ the two linear independent semiclassical solutions to eq.(2.1) exponentially increase and decrease,

$$\Psi_{\text{incr}}(a) = (2\,|\mu|\,)^{-1/2}\cdot e^{I(a)}, \tag{2.2a}$$

$$\Psi_{\text{decr}}(a) = (2\,|\mu|\,)^{-1/2}\cdot e^{-I(a)} \tag{2.2b}$$

where $\quad I(a) = \int_{a_1}^{a} |\mu(a')|\, da'.$

Solutions (2.2) have no interpretation in terms of classical universes; the absence of oscillating solutions corresponds to the fact that no homogeneous and isotropic solutions to the Einstein equations exist in the region $a_1 < a < a_2$.

To obtain the minisuperspace version of the third quantized gravity, one should consider the wave function of the universe $\Psi(a)$ as an operator which later will be denoted by $\hat{\Psi}(a)$. In the regions $a < a_1$ and $a > a_2$, the operator $\hat{\Psi}(a)$ can be represented as follows,

$$\hat{\Psi} = \hat{A}\cdot\Psi^{(-)}_{\text{in}}(a) + \hat{A}^{\dagger}\cdot\Psi^{(+)}_{\text{in}}(a), \qquad a < a_1;$$

$$\hat{\Psi} = \hat{B}\cdot\Psi^{(-)}_{\text{out}}(a) + \hat{B}^{\dagger}\cdot\Psi^{(+)}_{\text{out}}(a), \qquad a > a_2,$$

where the a-independent operators $\hat{A},\hat{A}^{\dagger}$ and $\hat{B},\hat{B}^{\dagger}$ obey the standard commutation relations

$$[\hat{A},\hat{A}^{\dagger}] = [\hat{B},\hat{B}^{\dagger}] = 1.$$

The operator $\hat{A}^{\dagger}$ is interpreted as the creation operator of small Friedmann-like universes, while $\hat{B}^{\dagger}$ creates large De Sitter-like ones.

The state without small universes is the $\hat{A}$-vacuum $\|0\rangle\rangle_F$,

$$\hat{A}\cdot\|0\rangle\rangle_F = 0\ .$$

This state, however, contains a certain number of De Sitter-like universes,

$$N_{\text{DS}} = {}_F\langle\langle 0\|\hat{B}^{\dagger}\hat{B}\|0\rangle\rangle_F$$

N_{DS} can be called the number of De Sitter-like universes created in minisuperspace from the state containing no Friedmann- like ones, and the calculation of N_{DS} coincides technically with the calculation of pair creation in the (0+1)-dimensional scalar field theory in the time-dependent background field $\mu^2(a)$. The operators $\hat{A}$ and $\hat{B}$ are related by the Bogoliubov transformation

$$\hat{B} = u \cdot \hat{A} + v \cdot \hat{A}^{\dagger} ,$$

where

$$v = i[\Psi^{(-)}_{out}]^{*} \overleftrightarrow{\partial}_{a} \Psi^{(+)}_{in} .$$

The number of created universes is

$$N_{DS} = |v|^2 .$$

To perform the actual calculation of v, one has to continue $\Psi^{(-)}_{in}$, via eq.(2.1), to the region $a > a_2$ and then take its projection onto $\Psi^{(+)}_{out}$. In the region $a_1 < a < a_2$, the function $\Psi^{(-)}_{in}$ contains both exponentially increasing and decreasing parts,

$$\Psi^{(-)}_{in} = \frac{1}{2}\, \Psi_{incr} + \frac{1}{2}\, \Psi_{decr} .$$

At small Λ, only the increasing part survives at $a \sim a_2$. At $a > a_2$, the function Ψ_{incr} contains $\Psi^{(+)}_{out}$ and $\Psi^{(-)}_{out}$ with equal weights, while its magnitude is proportional to

$$\Psi_{incr}(a_2) \propto e^{-\frac{1}{2} I_{inst}} ,$$

where $\frac{1}{2}\, I_{inst} = - \int_{a_1}^{a_2} |\mu|\, da$.

Therefore, $v \propto \Psi_{incr}(a_2)$ and, finally,

$$N_{DS} = e^{-I_{inst}} .$$

The notation I_{inst} is not accidental: I_{inst} is equal to the euclidean action of the theory evaluated at the periodic solution to the euclidean Einstein equations (instanton). In the limit $\varepsilon \to 0$, the explicit formulae for $I(a)$ and I_{inst} are

$$I(a) = - \frac{1}{\Lambda}\left[(1- \frac{\Lambda}{3}a^2)^{3/2} - 1\right]$$

and

$$I_{\text{inst}} = -\frac{2}{\Lambda}$$

Note that in this limit the relevant solution is the De Sitter instanton and I_{inst} is its action. Thus the number of created universes is

$$N_{\text{DS}} = e^{2/\Lambda}. \qquad (2.3)$$

If the probability for the universe to have a given value of Λ is assumed to be proportional to the number of universes with a given Λ, then it is easy to see that eq.(2.3) is precisely the Baum-Hawking [1,2] expression (1.1). At the same time we see no way of obtaining the Coleman [3] double-exponential behavior (1.2) in the framework of the third quantized gravity.

2.2 Models with a matter field

The next natural step is introducing matter degrees of freedom in the simplest minisuperspace model.

The Wheeler-DeWitt equation for the minisuperspace model with matter reads

$$\left(\frac{1}{2}\frac{\partial^2}{\partial a^2} + \frac{1}{2}\mu^2(a) + aH_{\text{M}} \right)\Psi = 0$$

where H_{M} is the matter Hamiltonian.

It is convenient to consider as matter the massive scalar field with the Lagrangian

$$\mathcal{L} = (-g)^{1/2}\left(\frac{1}{2}(\partial_\mu\varphi)(\partial^\mu\varphi) - \frac{\mathcal{R}}{12}\varphi^2 - \frac{m^2}{2}\varphi^2 \right) .$$

If we use the conformally rescaled field $\phi = a\varphi$ and decompose ϕ and conjugate momentum π_ϕ over the spherical harmonics on the unit 3-sphere, ϕ_{nlm} , $n=1,2,\ldots$; $l=0,\ldots,(n-1)$; $m=-l,\ldots,l$, then the Wheeler-DeWitt equation will be written in the form

$$\left(\frac{1}{2}\frac{\partial^2}{\partial a^2} - \sum\frac{1}{2}\frac{\partial^2}{\partial\phi^2_{nlm}} + \mu^2(a) + \sum\frac{1}{2}\,\omega_n^2\phi^2_{nlm} \right)\Psi(a,\{\phi_{nlm}\}) = 0 ,$$

where

$$\omega_n^2 = n^2 + m^2 a^2 \quad .$$

The expression

$$M^2(a,\phi_{nlm}) = \mu^2(a) + \sum \omega_n^2(a)\phi_{nlm}^2$$

plays the role of the space-time dependent $(\text{mass})^2$. It is negative in some region of the minisuperspace (see Figure 2)

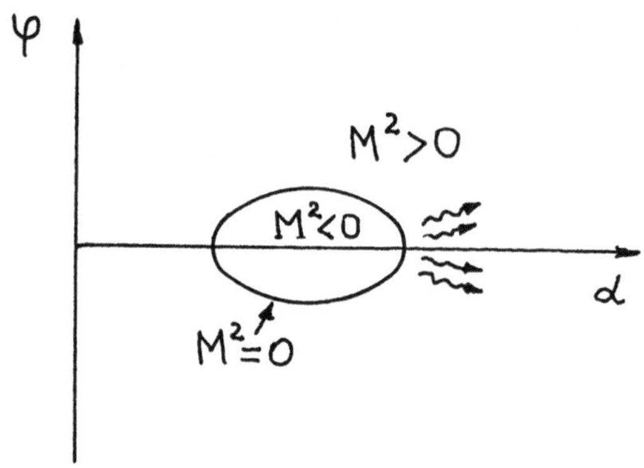

Figure 2

In the region with $M^2<0$ the system is unstable and the creation of universes is exponential. The number of created large universes can be found using the above-discussed Bogoliubov's transformation technique. Then the leading exponential factor in the number of created universes is determined by the value of euclidean action on the instanton (which in this situation always exists [16]) and we again arrive at eq.(1.1).

The matter Hamiltonian can be treated as perturbation and in the approximation where the back reaction of matter on the dynamics of the universe radius is neglected the number of created pairs of matter particles can be found.

The number of created pears is [17]

$$N_{\text{pairs}} \sim \frac{1}{ma_2} \sim \frac{\Lambda^{1/2}}{m} \ll 1$$

This means that large universes created at $\Lambda>0$ in the superspace are mostly empty and do not quite resemble the universe which we live in.

As to the case $\Lambda<0$, the situation is more dramatic [17,18]. In this case the unstable region $M^2<0$ (to which classical universes do not correspond) has a shape shown schematically in Figure 3.

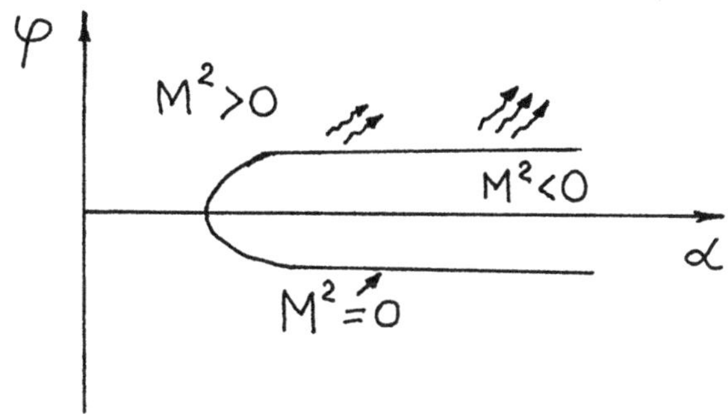

Figure 3

Then the creation and further classical evolution of the universe are possible if the energy of the universe is sufficient for the system to be in a classically allowable region. At $\Lambda<0$ the created universes are filled up with matter [17,18]; however the number of created universes grows with the growing of $|\Lambda|$ and has a peak at $\Lambda=-\infty$.

The interpretation in the Baum-Hawking sense implies that the cosmological constant is most probably negative and infinite.

Thus the cosmological constant problem requires further investigation.

3. ON THE ALGEBRA OF CONSTRAINTS IN QUANTUM GRAVITY

In this section we shall consider whether Dirac's quantization scheme is applicable for gravity quantization.

If instead of the space-time metric $g_{\mu\nu}$ the new variables - lapse and shift functions N and N_i and space metric γ_{ij} - are used according to the equation

$$ds^2=g_{\mu\nu}dx^\mu dx^\nu=(N^2-N_jN^j)dt^2-2N_idx^idt-\gamma_{ij}dx^idx^j ,$$

then from the Einstein action

$$S = -\frac{1}{2}\int\sqrt{-g}\; d^4x\; R$$

one can obtain the canonical Hamiltonian for gravity [19,25]

$$H^c = \int d^3x\;(\,N^iH_i + NH\,) ,$$

where

$$H = \frac{1}{\sqrt{\gamma}}\,(\,\pi_{ij}\pi^{ij} - \frac{1}{2}\,\pi^2 - \gamma\;{}^{(3)}R\,) ,$$

$$H_j = -2\,\pi_i{}^j|_j \;;\; i,j = 1,2,3\ ,$$

π^{ij} is the canonical momentum conjugate to γ_{ij} , ${}^{(3)}R$ denotes a 3-dimensional curvature and $|_j$ a 3-dimensional covariant derivative.

Lapse and shift functions appear as Lagrange multipliers in the Hamiltonian, variation with respect to which gives four Hamiltonian constraints

$$H = 0, \tag{3.1a}$$
$$H_j = 0. \tag{3.1b}$$

Constraints (3.1) form the closed algebra

$$\begin{aligned}
\{\ H(x)\ ,\ H(y)\ \} &= (\ H^j(x) + H^j(y)\ )\ \delta,_j(x-y) \\
\{\ H(x)\ ,\ H_j(y)\ \} &= H(y)\ \delta,_j(x-y) \\
\{\ H_i(x)\ ,\ H_j(y)\ \} &= H_i(y)\ \delta,_j(x-y) + H_j(y)\ \delta,_i(x-y)
\end{aligned} \tag{3.2}$$

with respect to Poisson's brackets

$$\{\ \gamma_{ij},\ \pi^{nm}\ \} = \delta_{ij}{}^{nm}\ \delta(x-y)\ ,\quad \delta_{ij}{}^{nm} \equiv \tfrac{1}{2}\ (\ \delta_i^n\delta_j^m + \delta_i^m\delta_j^n\ ).$$

After quantization, to the dynamical variables γ_{ij} and π^{nm} there correspond the operators $\hat{\gamma}_{ij}$ and $\hat{\pi}^{nm}$ and Poisson's brackets are replaced by commutators

$$[\ \hat{\gamma}_{ij},\ \hat{\pi}^{nm}\ ] = i\ \delta_{ij}{}^{nm}\ \delta(x-y)\ .$$

According to Dirac's quantization scheme we must impose constraints (3.1) as operator conditions on the physical state vectors. In this way one obtains the Wheeler-DeWitt equation [19]

$$\hat{H}\ \Psi = 0 \tag{3.3a}$$

and momentum constraints

$$\hat{H}_j\ \Psi = 0 \tag{3.3b}$$

where Ψ is the wave function of the universe.

This quantization scheme implies that the algebra of constraints (3.2) is supported at the quantum level, i.e. after quantization in this algebra there arise no anomalous terms. In particular, eq.(3.2b) must have the form

$$[\ \hat{H}(x)\ ,\ \hat{H}_j(y)\ ] = \hat{H}(y)\ \delta,_j(x-y) + A_j(x-y) \tag{3.4}$$

with the anomaly $= 0$.
The appearance of such anomalous terms would indicate that Dirac's standard quantization scheme is inapplicable. It is necessary either to somehow modify this scheme or to quantize gravity by another method.

3.1 The (1+1)-dimensional example

The well-known model with such a type of anomaly is the bosonic string [26] (2-dimensional gravity !).
Performing (1+1)-decomposition, from the string action

$$S = -\frac{1}{2}\int d\sigma d\tau\ \sqrt{-g}\ g^{\alpha\beta}\partial_\alpha X^A \partial_\beta X_A\ ;\ \alpha,\beta=0,1$$

one can obtain two constraints [26]

$$H(\sigma) = \frac{1}{2}(\ P_A P^A + X'^A X'_A\ )$$
$$H_1(\sigma) = P_A X'^A \tag{3.5}$$

which is the consequence of the 2-dimensional general-coordinate invariance of the theory.
The Fourier images of these constraints are operators L_n (generators of the Virasoro algebra [26]).
Constraints (3.5) satisfy at the classical level an algebra similar to algebra (3.2). In particular,

$$\{\ H(x)\ ,\ H_1(y)\ \} = (\ H(x)+H(y)\ )\ \delta(x-y). \tag{3.6}$$

The Fourier harmonics L_n satisfy the Virasoro algebra

$$\{\ L_n, L_m\ \} = (\ n-m\ )\ L_{n+m}\ . \tag{3.7}$$

At the quantum level this algebra changes as follows :

$$[\ L_n, L_m\ ] = (\ n-m\ )\ L_{n+m} + \frac{1}{12}\ (n^3-n)\ \delta(n+m)\ . \tag{3.8}$$

The last term in this equation (central extension) corresponds to the appearance of the third derivative of the δ-function in a quantum version of eq.(3.6)

$$[\ \hat{H}(\sigma)\ ,\ \hat{H}_1(\tilde{\sigma})\ ] = i\ (\ \hat{H}(\sigma)+\hat{H}(\tilde{\sigma})\ )\ \delta'(\sigma-\tilde{\sigma}) + \frac{i}{12\pi}\,\delta^{(3)}(\sigma-\tilde{\sigma})\ .$$

On account of (3.8) it is obvious that the conditions

$$L_n\ |\psi\rangle = 0 \text{ for all } n$$

cannot be imposed on physical state vectors, i.e. we cannot require that

$$\hat{H}(\sigma)\ |\psi\rangle = 0 \quad \text{and} \quad \hat{H}_1(\sigma)\ |\psi\rangle = 0,$$

since only the zero vector satisfies these conditions.

3.2 The (3+1)-dimensional example

An analogy with a (3+1)-dimensional models is better suitable for our purposes.

Let us consider a (3+1)-dimensional free massless scalar field as a simple model. Though a scalar field is an unconstrained system, this model is nevertheless interesting because the components of the energy-momentum tensor ,$T_{\mu\nu}$, form the same algebra as (3.2). In this case T_{00} is an analog of H and T_{0j} is that of H_j.

If we define $T_{\mu\nu}$ as normally ordered operators, then for finding anomaly it is enough to calculate the vacuum average of the commutator of two components $T_{\mu\nu}$.

In particular, we are interested in the presence of the anomaly in the analog of eq. (3.4). This anomaly can be calculated by

$$A_j(x-y) = \langle 0|[T_{00}(x),T_{0j}(y)]|0\rangle\ . \tag{3.9}$$

The vacuum average in eq. (3.9) is ill defined and needs redefinition. If we define the commutator in eq. (3.9) as

$$[T_{00}(x),T_{0j}(y)] =(\ [T_{00}(x,x^0),T_{0j}(y,y^0)] + K(x^0-y^0,x-y^0)\ )\big|_{x^0=y^0=0}\ ,$$

where $K(x^0-y^0,x-y^0)$ is some quasilocal operator eliminating divergences, then the calculation of (3.9) reduces to the calculation of the product of the second derivative of two D^-- functions [27] of the scalar field. Direct calculations à la Bogoliubov-Shirkov [27] give

$$<0|[T_{00}(x,x^0),T_{0j}(y,y^0)]|0>\Big|_{x^0=y^0\to 0}=S_F\delta^{(3)}(x-y)+S_D\ \partial_j\delta^{(4)}(x-y)$$

where $S_F=i/(240\pi^2)$ is the finite number and $S_D=(-1/(96\pi))(\delta'_-(t)-\delta'_+(t))$ diverges quadratically as $t=x^0-y^0\to 0$. If we choose a quasilocal operator $K(x^0-y^0,x-y^0)$ so as to cancel this divergence, then we shall finally obtain

$$A_j(x-y) = (i/(240\pi^2))\ \partial_j\delta^{(4)}(x-y)\ .$$

3.3 Concluding remarks

In the same manner as above we can study the algebra of constraints of quantum gravity. One can consider the small perturbation h_{ij} near the background metric η_{ij}

$$\gamma_{ij} = \eta_{ij} + h_{ij}$$

and expand constraints in a power series of small h_{ij}

$$H = H^{(0)}+ H^{(1)}+H^{(2)}+ H^{(3)}+\ldots$$

$$H_j = H_j^{(0)}+ H_j^{(1)}+H_j^{(2)}$$

Then instead of eq. (3.4) we obtain a chain of equations. The case when the operators in the right-hand side of (3.4) are quadratic with respect to h_{ij} is the lowest nontrivial order.

Preliminary calculations indicate that the anomalous term - the fifth derivative of the δ-function actually appears in the algebra of constraints of quantum gravity

Thus we think that the standard operator quantization scheme may be inapplicable because of anomalies arising in the algebra of constraints of quantum gravity.

We hope to discuss this problem in greater detail in a separate publication.

ACKNOWLEDGEMENTS

I am indebted to V.Rubakov and P.Tinyakov for collaboration, to G.Jorjadze, A.Khvedelidze , A.Kuznetsov and B.Magradze for numerous useful discussions, to M.Eliashvili for interest in this work and valuable remarks, and to D.Kodanashvili for checking some formulae.

REFERENCES

1. E.Baum, Phys.Lett. 133B (1983) 185
2. S.W.Hawking, Phys.Lett. 134B (1984) 403
3. S.Coleman, Nucl.Phys. B310 (1988) 643
4. S.Weinberg, Rev.Mod.Phys. 61 (1989) 1
5. S.W.Hawking, Phys.Lett. 195B (1987) 377
6. G.V.Lavrelashvili, V.A.Rubakov and P.G.Tinyakov, JETP Lett. 46 (1987) 167, Nucl.Phys. B299 (1988) 757
7. S.Giddings and A.Strominger, Nucl.Phys. B306 (1988) 890
8. S.Coleman, Nucl.Phys. B307 (1988) 867
9. S.Giddings and A.Strominger, Nucl.Phys. B307 (1988) 854
10.T.Banks, Nucl.Phys B309 (1988) 493
11.S.B.Giddings and A.Strominger, Nucl.Phys. B321 (1988) 481
12.A.Strominger, Baby Universes, UCSB preprint (1988)
13.V.A.Rubakov, Phys.Lett. 214B (1988) 503
14.W.Fischler, I.Klebanov, J.Polchinski and L.Susskind, Nucl.Phys. B327 (1989) 157
15.A.Hosoya and M.Morikawa, Phys.Rev. D39 (1989) 1123
16.G.V.Lavrelashvili, V.A.Rubakov,M.S.Serebryakov and P.G.Tinyakov, Nucl.Phys. B329 (1990) 98
17.V.A.Rubakov and P.G.Tinyakov, Nucl.Phys. B342 (1990) 430
18.G.V.Lavrelashvili, V.A.Rubakov and P.G.Tinyakov, "Wormholes, third quantization and the problem of the negative cosmological constant", To appear in: "Proceedings of the V Seminar on Quantum Gravity", Moscow, 1990, (World Scientific)
19.B.DeWitt Phys.Rev.160 (1967) 1113
20.M.I.Kalinin and V.N.Melnikov, Proc.VNIIFTRI 16 (1972) 43 (State Committe of Standards, Moscow)
21.J.Hartle and S.W.Hawking, Phys.Rev. D28 (1983) 2960
22.A.A.Starobinsky, Talk at P.K.Sternberg State Astronomical Inst. Seminar (1983), unpublished;
A.D.Linde, Lett.Nuovo Cimento 39 (1984) 401
23.V.A.Rubakov, Phys.Lett. 148B (1984) 280
24.A.Vilenkin, Phys.Rev. D30 (1984) 549
25.R.Arnowitt, S.Deser, and C.W.Misner, in Gravitation: An Introduction to Current Research, edited by L.Witten (Willey, New York) 1962
26.M.Heneaux, Principles of string theory,(Plenum Press), 1988
27.N.N.Bogoliubov and D.V.Shirkov, Introduction to the Theory of Quantized Fields, Moscow, Nauka, 1976 (In Russian)

STRUCTURE OF THE INFLATIONARY UNIVERSE

Maxim I. Zelnikov
Department of Theoretisal Physics,
P.N.Lebedev Physical Institute
Leninsky Prospect 53, 117924 Moscow, USSR

In our analysis we distinguish local and global structures of the inflationary Universe. By local structure we mean the space-time geometry in small finite domains of the same comoving size as the observable part of the Universe. We shall call such domains "local regions". Global structure of the Universe is defined as the geometry of the space-time as a whole. As we shall see below there exists a great difference between these aspects of structure. Namely, the inflationary Universe is almost homogeneous locally, being highly inhomogeneous as a whole [1-3]. Since the present structure of the Universe can be derived from the distribution of certain scalar field at final stage of inflation we consider here only this distribution.

The present report is organized as follows. In the first section a brief discussion of scalar field behaviour at the inflationary stage is presented. The second section is devoted to an analysis of the scalar field perturbation spectrum which determines the local structure. In the third section we consider a model in which the inflationary Universe has a nontrivial global dimensional structure.

Scalar field evolution

Let us use for the discussion of scalar field evolution the haotic inflationary model [1-4] with the potential $V(\varphi) = \frac{1}{4}\lambda\varphi^4$. During inflation classical scalar field tends to homogeneity due to cosmological expansion but simultaneously random inhomogeneities are generated with typical wavelength $l \gtrsim H^{-1}$ [5-7]. Here H is the Habble parameter. At sufficiently small scales these perturbations are linear. In the regime of slow rolling in the conformally Newtonian gauge the equation of their evolution is:

$$\delta\dot{\varphi} = - m^2\delta\varphi/(3H) + f , \qquad (1)$$

where $H = \dot{a}/a$, $m^2 = d^2V/d\varphi^2 - (dV/d\varphi)^2/V$, a is the scale factor and f is a random force with zero average and with the following correlation [5]:

$$\langle f(t,\vec{x})f(t',\vec{x}')\rangle = \frac{H^3}{4\pi^2}\,\frac{\sin(Ha|\vec{x}-\vec{x}'|)}{Ha|\vec{x}-\vec{x}'|}\,\delta(t-t')\ .$$

The structure of our local region was forming during the last 60 e-folding of inflation. This period is not sufficiently large for a developement of significant deviations of the scalar field at separate points from its background value. For $\lambda = 10^{-12}$ at the end of inflation in the whole local region we can find on the average only one domain of Habble size where φ exeeds its background value by 1%.

Discussing the global structure of the Universe we have a quite different situation, because we consider *the whole* space which is exponentially large. Unlike local regions, in the whole Universe we can find a lot of domains where scalar field history differs significantly from the classical one [1,3]. In particular, we can find regions where φ permanently *increases* in defiance of the classical law. In such regions there can happen a cardinal restructuring of space-time or of partical physics like a change of space-time dimension [9] or a compactification type change. So the physics of domains with non-classical behaviour of the scalar field is an important aspect of the Universe global structure.

Spectrum of scalar field perturbations

In general an inflationary model can contain several interacting scalar fields. Typically perturbations of one of them make a dominant contribution into the post-inflation distribution of matter. Let us denote this field by χ . The background metric we choose to be:

$$ds^2 = -\,dt^2 + a(t)^2(dx_1^2 + dx_2^2 + dx_3^2)\ .$$

Under certain conditions the equation for $\delta\chi_{\vec{k}} = \int \exp(-i\vec{k}\vec{x})\,\delta\chi(\vec{x},t)\,d^3x$ can be reduced to the following one:

$$\delta\ddot{\chi}_{\vec{k}} + 3H\delta\dot{\chi}_{\vec{k}} + (k^2/a^2 + m^2)\,\delta\chi_{\vec{k}} = 0, \qquad (2)$$

where $m^2(t)$ and $H(t)$ are determined by the evolution of background scalar fields. To take into account the process of perturbations generation we must put certain random force $f_{\vec{k}}$ in the right-hand-side of (2).

It is convenient to characterize the spectrum of $\delta\chi$ by $\delta_k = (4\pi k^3 \langle|\delta\chi_{\vec{k}}|^2\rangle/V)^{1/2}$. (The arrow over k is omitted because in our case δ_k does not depend on the direction of $\vec{k}$.) Here V is the volume of three-dimensional space, $\langle\ldots\rangle$ denotes the statistical average. To calculate δ_k we must make the substitution $\langle|\delta\chi_{\vec{k}}|^2\rangle = |\delta\bar{\chi}_{\vec{k}}|^2$, where $\delta\bar{\chi}_{\vec{k}}$ obeys eq. (2) . Initial conditions for $\delta\bar{\chi}_{\vec{k}}$ are determined by the requirement that at $a \to 0$ $\delta\bar{\chi}_{\vec{k}} = (H_0/(2\pi))(2H_0 a)^{-3/2}\mathcal{H}^{(1)}_{\nu}(k/(Ha))$, where $H_0 = H(a \to 0)$, $\nu = (9/4 - m^2/H^2)^{1/2}$ at $a \to 0$. Further we deal only with $\delta\bar{\chi}_{\vec{k}}$ and omit the bar for brevity.

Now our task is reduced to the investigation of $\delta\chi_{\vec{k}}$ connection with $H(t)$ and $m^2(t)$. Here two different cases must be distinguished. At first we consider the case when

$$|\dot{H}| \ll H^2 , \tag{3}$$

$$m^2 < 9H^2/4 . \tag{4}$$

The connection of $m^2(t)$ and $H(t)$ with δ_k in this case is represented in Figure 1.

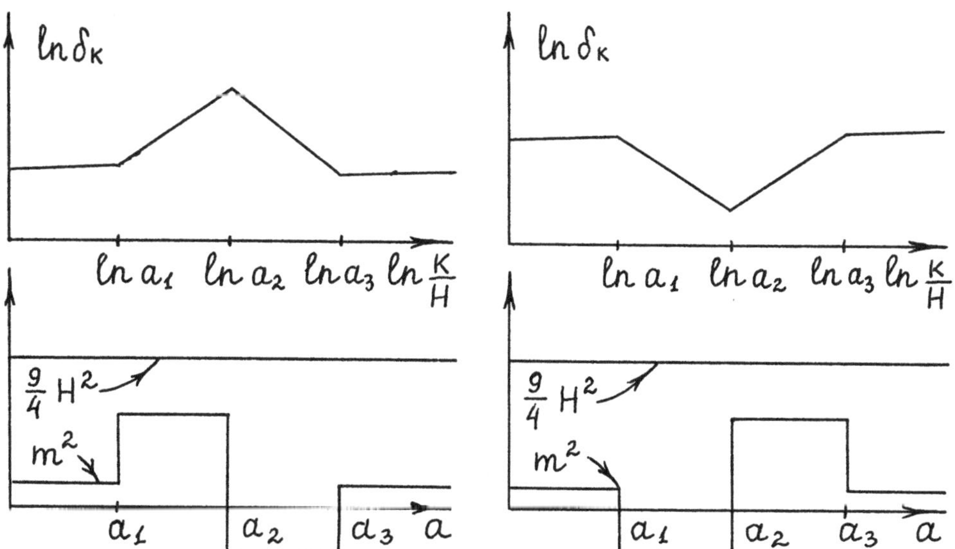

Figure 1

Here $H = const$, $a_1 \ll a_2 \ll a_3$.

Analytic calculations show that in the above case for k/H sufficiently far from a_i the following equation is valid:

$$p \equiv d \ln(\delta_k)/d \ln k = 3/2 - \nu\Big|_{k=Ha}$$

where $\nu = (9/4 - m^2/H^2)^{1/2}$. Analogous formulas has been derived in [10,11]. Note, that under the conditions (3) and (4) the local slope index p is less than 3/2 for all k . Flat Harrison-Zeldovich spectrum appears if $m^2 \ll H^2$.

Now let us analyze cases of (3) and (4) violation separately. When $m^2 > 9H^2/4$ during some time then the spectrum acquires a part with average slope $\langle p \rangle \approx 3/2$. For certain relations between a_i , m_i^2 and H this spectrum also acquires a sharp drop to small k with $p \gg 3/2$ (see Fig.2a).

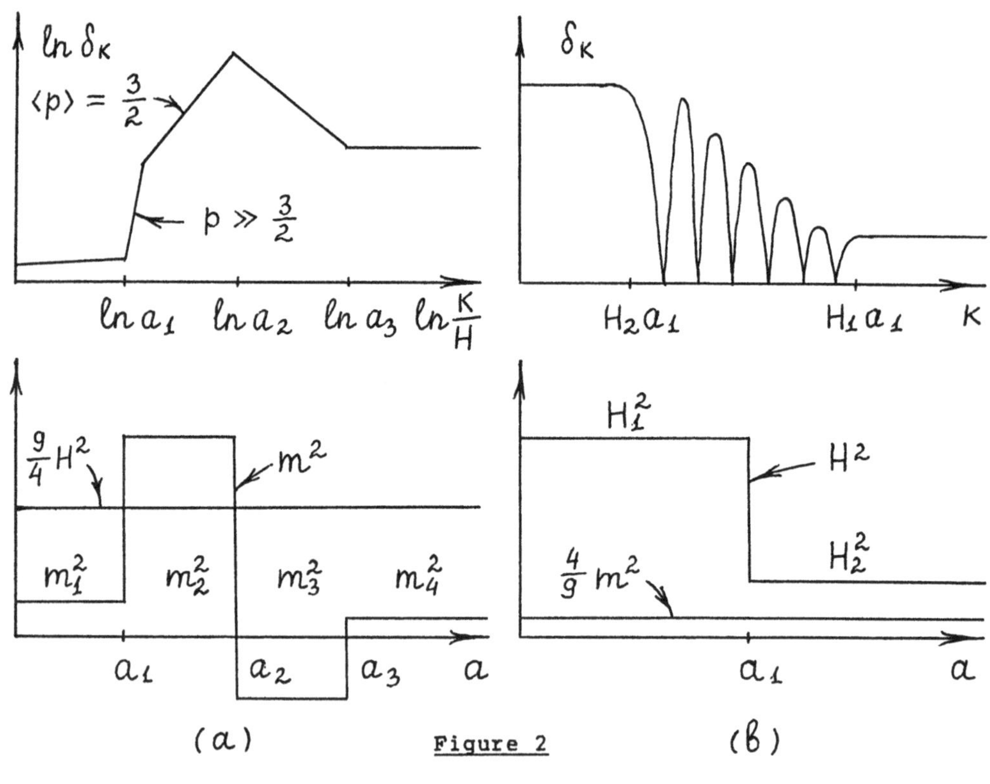

Figure 2

Analytically δ_k for this case is given by the following formulas:

$$\delta_k = (aH)^{\nu_4 - 3/2} \begin{cases} A\, k^{3/2 - \nu_1} |\sin \alpha|, & k \ll Ha_1 \\ B\, k^{3/2} \{ch(\pi|\nu_2|) - \cos \beta\}^{1/2}, & Ha_1 \ll k \ll Ha_2 \\ C\, k^{3/2 - \nu_3}, & Ha_2 \ll k \ll Ha_3 \\ D, & Ha_3 \ll k \end{cases}$$

where $a_1 \ll a_2 \ll a_3$, $\beta = 2|\nu_2|\ln(k/(Ha_2)) + \psi$, $\alpha = |\nu_2|\ln(a_2/a_1) - \mathrm{arctg}\{|\nu_2|(\nu_1 + \nu_3)/(|\nu_2|^2 - \nu_1\nu_3)\}$, $\nu_i = (9/4 - m_i^2/H^2)^{1/2}$, $|\nu_2| = (m_2^2/H^2 - 9/4)^{1/2}$, A,B,C,D,ψ are some constants.

When $m^2 < 9H^2/4$ for all t and H drops rapidly the spectrum acquires a comb-like part (see Fig.2b). Asymptotics of δ_k at $a \to +\infty$ are:

$$\delta_k = (aH_2)^{\nu_2 - 3/2} \begin{cases} A'\, k^{3/2 - \nu_1}, & k \ll H_2 a_1 \\ B'\, k^{1 - \nu_1 - \nu_2} |\cos \gamma|, & H_2 a_1 \ll k \ll H_1 a_1 \\ C'\, k^{3/2 - \nu_2}, & H_1 a_1 \ll k \end{cases}$$

where $H_1 \gg H_2$, $\gamma = k/(H_2 a_1) - \pi\nu_2/2 - \pi/4$, A',B',C' are some constants, $\nu_i = (9/4 - m^2/H_i^2)^{1/2}$, $m^2 \neq 0$.

We see that for a number of k δ_k is exactly zero. Certainly, such an abrupt drop of H is an idealization, but in more realistic double inflation models the spectrum is qualitatively the same [12].

Thus our brief analysis of scalar field perturbations shows how non-flat spectra can be obtained from inflation. In particular, we see that in general inflationary models there can naturally arise a mountain in the spectrum and a suppression of modes with large wavelengths which seem to be required by observations [11,12].

Change of space-time dimension

As was clamed in the first section of this report, in the regions of the inflationary Universe where the scalar field evolves non-classically the space-time dimension can change. Here we present a model [9] in which this process really takes place. We consider a theory with the following action:

$$S = N\int d^6x\,(-g_6)^{1/2}\{\,R\cdot M_p^2/(16\pi) - \Lambda - \tfrac{1}{2}\,\varphi_{;m}\varphi^{;m}$$

$$- V(\varphi) - \tfrac{1}{4}\,F_{mn}F^{mn}\,\}$$

where $F_{mn} = \nabla_m A_n - \nabla_n A_m$, $V(\varphi) = \frac{1}{4}\lambda\varphi^4$, N , Λ = const , m,n = 0,1,2,3,4,5. For the metric we use the ansatz:

$$ds^2 = -\,dt^2 + a(x_j,t)^2(dx_1^2 + dx_2^2 + dx_3^2)$$

$$b(x_j,t)^2(d\theta^2 + \sin^2\theta\,d\alpha^2)$$

where $j = 1,2,3$. This theory admits a stable compactification of the type (four-dimensional Minkowski space-time)×(two-sphere of a constant radius) if $\varphi = 0$,if all components of A_m are zero exept $A_\alpha = h(1 - \cos\theta)$ and if the following relations are valid [14]: $h = b_0M_p/(8\pi)^{1/2}$ and $\Lambda = M_p^2/(16\pi b_0^2)$. Here b_0 is the compactification radius in the ground state. We choose $b_0 \approx O(10)M_p^{-1}$. Consequently, the two extra dimensions have no influence on low energy physics and the effective space-time dimension in this state is 4.

Taking into account that at the stage of inflation spatial gradients of a, b and φ quickly become negligible we obtain from the Einstein equations :

$$\ddot{\varphi} + \dot{\varphi}\,(3\dot{a}/a + 2\dot{b}/b) + \partial V/\partial\varphi = 0 \quad ,$$

$$\ddot{a}/a + 2\dot{a}^2/a^2 + 2(\dot{a}/a)\dot{b}/b = (16\pi/M_p^2)[(\Lambda + V)/4 - h^2b^{-4}/8] \quad ,$$

$$\ddot{b} + \dot{b}(3\dot{a}/a + \dot{b}/b) + \partial W(b,\varphi)/\partial b = 0 \quad ,$$

where $W(b,\varphi) = 3b_0^2/(8b^2) + \ln(b/b_0) - (1 + V(\varphi)/\Lambda)b^2/(8b_0^2)$.
Here we have supposed that the slow rolling conditions are fulfiled.

One can easyly check that a stable compactification exists if only $V(\varphi) < \Lambda/3$ (see Fig.3). Classically homogeneous scalar field energy density must decrease. So if in the local domain $V(\varphi)$ is initially less than $\Lambda/3$ then the effective space-time dimension will remain 4 there during the whole subsequent evolution. However, as we have stated previously this is not true for the Universe as a whole. Somwhere there exist regions of Habble size where the potential energy of almost homogeneous scalar field evolves from $V(\varphi) < \Lambda/3$ to $V(\varphi) > \Lambda/3$. As soon as $V(\varphi)$ exeeds $\Lambda/3$ the local minimum of W disappears and the extra dimensions decompactify (see Fig.4). It is easy to show that the geometry in such regions quickly becomes locally indistinguishable from the geometry of six-dimensional de Sitter space-time. Thus we find that in our model some domains of the inflationary Universe have space-time dimension d = 6 while in other regions d = 4.

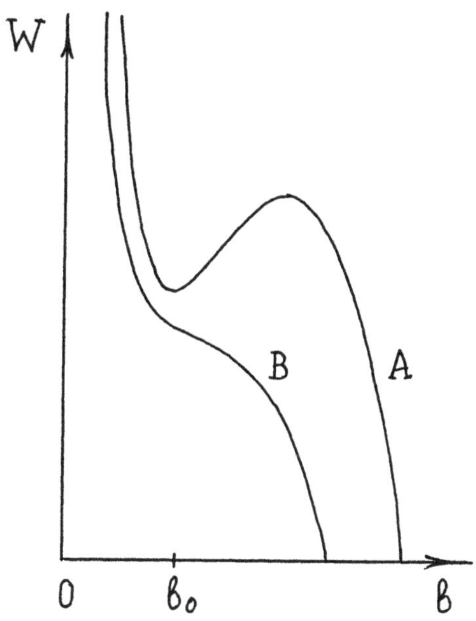

Figure 3

The shape of the effective potential W
A: $V(\varphi) = 0$,
B: $V(\varphi) > \Lambda/3$.

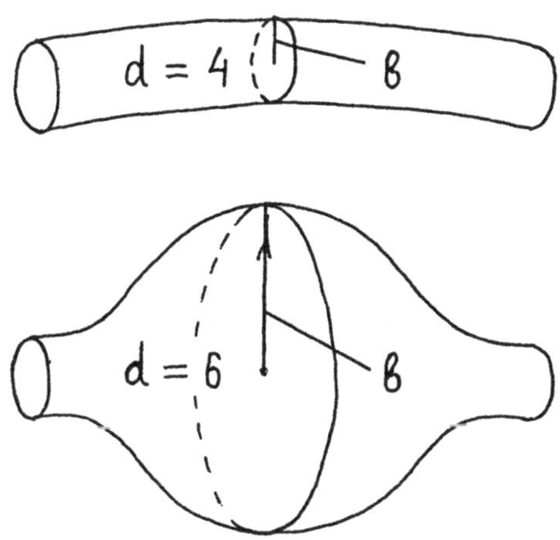

Figure 4

A sketch of the local decompactification process.

Conclusions

We briefly analyze the structure of the Universe at the stage of inflation. It turns out that fluctuations of the scalar field lead to small perturbations at the scale of the present Habble horizon while at large scales the same process causes very strong inhomogeneity. Discussing the local structure of space-time we investigate the spectrum of generated perturbations . Analyzing the global structure of the inflationary Universe we consider the dynamics of space-time dimension on the example of a particular model.

Acknowledgement

The author is grateful to A.D.Linde and V.F.Mukhanov for collaboration, and to Centro de Fisica da Materia Condensada for kind hospitality.

References

[1] Linde,A.D. 1986, *Phys.Lett.*, **B175**, 395.
[2] Linde,A.D. 1986, *Mod.Phys.Lett.*, **A1**, 81.
[3] Goncharov,A.S.,Linde,A.D.,and Mukhanov,V.F. 1987, *Int.J.Mod.Phys.*, **A2**, 561.
[4] Linde,A.D. 1983, *Phys.Lett.*, **B129**, 177.
[5] Starobinsky, A.A. 1986, in *Lecture Notes in Physics* (Springer-Verlag), **246**, p. 107.
[6] Vilenkin,A.,and Ford,D. 1982, *Phys.Rev.*, **D26**, 1231.
[7] Linde,A.D. 1982, *Phys.Lett.*, **B116**, 335.
[8] Starobinsky,A.A. 1982, *Phys.Lett.*, **B117**, 175.
[9] Linde,A.D.,and Zelnikov,M.I. 1988, *Phys.Lett.*, **B215**, 59.
[10] Kofman,L.A.,and Pogosyan,D.Yu. 1988, *Phys.Lett.*, **B214**, 508.
[11] Salopek,D.S.,Bond,J.R.,and Bardeen,J.M. 1989, *Phys.Rev.*, **D40**, 1753.
[12] Mukhanov,V.F.,and Zelnikov,M.I.(in preparation).
[13] Bardeen,J.M.,Bond,J.R.,and Efstathiou,G. 1987, *A.J.*, **321**, 28.
[14] Randjbar-Daemi,S.,Salam,A.,and Strathdee,J. 1983, *Nucl.Phys.*, **B214**, 491.

GPSR Compliance

The European Union's (EU) General Product Safety Regulation (GPSR) is a set of rules that requires consumer products to be safe and our obligations to ensure this.

If you have any concerns about our products, you can contact us on ProductSafety@springernature.com

In case Publisher is established outside the EU, the EU authorized representative is:

Springer Nature Customer Service Center GmbH
Europaplatz 3
69115 Heidelberg, Germany

Batch number: 10640288

Printed by Printforce, the Netherlands